BIOINDICATORS AND ENVIRONMENTAL MANAGEMENT

edited by
D. W. Jeffrey
and
B. Madden

*Department of Botany,
Trinity College, Dublin, Ireland*

ACADEMIC PRESS
Harcourt Brace & Company, Publishers
LONDON SAN DIEGO NEW YORK BOSTON
SYDNEY TOKYO TORONTO

This book is printed on acid-free paper

ACADEMIC PRESS LIMITED
24-28 Oval Road
London NW1 7DX

United States Edition published by
ACADEMIC PRESS INC.
San Diego, CA 92101

A catalogue record for this book is available from the British Library
ISBN 0-12-382590-3

Printed in Great Britain by
T. J. Press (Padstow) Ltd., Padstow, Cornwall

Bioindicators and Environmental Management

CONTENTS

Introduction

I.U.B.S. Bioindicators Commission Members

List of Participants

I. Bioindicators, Industry and Administration

II. Environmental Radioactivity and Biomonitoring of the Chernobyl Accident

III. Monitoring Long Term and Large Scale Environmental Trends

IV. Basic Research in Biomonitoring

Introduction

This volume contains material presented at the 6th International Bioindicators Symposium held in Trinity College, Dublin, 23-28th September 1990. The Symposium was organized by the Commission for Bioindicators, International Union for Biological Sciences, in association with the National Committee for Biology, Royal Irish Academy.

A key role of the Commission for Bioindicators, IUBS, is to promote the use of bioindicators in environmental management. This means encouraging the transfer of ideas regarding potential bioindicators, and originating in laboratories, into the harsher realities of field environmental monitoring. Although the concept of biomonitoring is ancient, its application to current monitoring problems is relatively slow to develop. Hence it is valuable to convene international scientific meetings which enable approaches developed in all parts of the world to be discussed.

It is hoped that this collection of nearly forty papers will provide both environmental administrators and research scientists with a valuable sense of proportion of the state of the art in their particular field.

The Symposium was grant-aided by the Commission of the European Community, and sponsored by the Department of the Environment, Republic of Ireland; Aer Lingus and the Pfizer Chemical Corporation.

I.U.B.S. Bioindicators Commission Members

J. Salánki, Hungary - Chairman
D.W. Jeffrey, Ireland - Secretary
R. Baudo, Italy
A. Bevan, U.K.
M. Bianchi, France
J. Bohác, Czechoslovakia
K. Chang, Taiwan
R.R. Colwell, U.S.A.
G.D. Floodgate, U.K.
E.D. Goldberg, U.S.A.
O. Hänninen, Finland
G.M. Hughes, U.K.
K. Kogure, Japan
S. Kozuharov, Bulgaria
K.J.M. Kramer, The Netherlands
D.A. Krivolutzky, U.S.S.R.
J.G. Nemcsók, Hungary
R. Norris, Australia
P. Oftedal, Norway
D. Pascoe, U.K.
A.C. Posthumus, The Netherlands
K. S.-Rózsa, Hungary
W. Schmitz, Germany
A.K. Sharma, India
E. Weinert, Germany
B.A. Whitton, U.K.
M. Yasuno, Japan
S. Yu, China

List of Participants

Mr. D. Atkin, Department of Biosciences, Polytechnic of East London, Romford Road, Stratford, London E15 4LZ, England.

Dr. I. Benedeczky, Department of Zoology, József Attila University, Szeged, P.O. Box 659, Hungary.

Dr. M. Bianchi, CNRS Microbiologie Marine, Facult des Sciences de Luminy, Case 907, Avenue de Luminy, 13288 Marseille Cedex 9, France.

Dr. P. Birch, Department of Biosciences, Polytechnic of East London, Romford Road, Stratford, London E15 4LZ, England.

Dr. J. Boháč, Institute of Landscape Ecology, Czechoslovak Academy of Sciences, Na sadkach 7, 370 05 Ceské Budejovice, Czeckoslovakia.

Dr. C.E. Booth, PES(E) School of Agriculture, University of Nottingham, Sutton Bonington, Loughborough, Leicestershire LE1Z 5RD, U.K.

Ms. J. Botterweg, DBW/RIZA, PO Box 17, NL-0200 AA Lelystad, The Netherlands.

Mr. O.C. Boyle, Department of the Environment, Custom House, Dublin 1, Ireland.

Dr. A.A. Callaghan, Department of Applied Sciences, Staffordshire Polytechnic, Stoke-on-Trent, ST4 2DE, U.K.

Dr. A. Carballeira, Area de Ecología, Universidad de Santiago de Compostela, 15071 Santiago de Compostela, Spain.

Mr. D. Casey, Donegal County Council, County House, Lifford, Co. Donegal.

Dr. P.A. Colgan, Nuclear Energy Board, 3 Clonskeagh Square, Dublin 14, Ireland.

Dr. A. Cooper, Department of Environmental Studies, University of Ulster, Coleraine, Co. Londonderry BT52 1SA, Northern Ireland.

Dr. J. Coosen, RWS Tidal Waters Division, PO Box 8039, 4330 EA Middelburg, The Netherlands.

Dr. M.J. Costello, Environmental Science Unit, Trinity College, Dublin 2, Ireland.

Dr. P. Dowding, Department of Botany, Trinity College, Dublin 2, Ireland.

Ms. R. Dwyer, Department of Botany, Trinity College, Dublin 2, Ireland.

Dr. B. Elkaim, Laboratoire d'Hydrobiologie, Universit P. et M. Curie Paris VI, 12 rue Cuvier, 75005 Paris, France.

Mr. P. Fay, Department of Botany, Trinity College, Dublin 2, Ireland.

Dr. G.D. Floodgate, School of Ocean Sciences, University College of Wales, Bangor, Menai Bridge, Gwynedd, LL59 5EY, U.K.

Mr. I. Galan, Department of Zoology, Trinity College, Dublin 2, Ireland.

Dr. M.J. Gormally, Writtle Agricultural College, Chelmsford CM1 3RR, U.K.

Prof. O. Hänninen, Department of Physiology, University of Kuopio, PO.Box. 6, SF-70211 Kuopio, Finland.

Mr. A.C. Hoekstra, NV. Dune water works Zuid-Holland, Builinom 18, 2512 XA The Hague, The Netherlands.

Prof. G.M. Hughes, Research Unit for Comparative Animal Respiration,

Biological Sciences Building, University of Bristol, Woodland Road, Bristol BS8 1UG, U.K.

Dr. Rong-Quen Jan, Institute of Zoology, Academia Sinica, Nankang, Taipei 11529, Taiwan, Republic of China.

Dr. D.W. Jeffrey, Department of Botany, Trinity College, Dublin 2, Ireland.

Dr. M.B. Jones, Department of Botany, Trinity College, Dublin 2, Ireland.

Dr. J.M. Kees-Kramer, Laboratory for Applied Marine Research, P.O. Box 57, 1780 AB Den Helder, The Netherlands.

Ms. J.N. Khan, Environmental Sciences Unit, Trinity College, Dublin 2, Ireland.

Dr. D.A. Krivolutzkii, National Committee of Soviet Biologists, Moscow W-71, 117071, USSR.

Dr. R.M. Lunnon, Department of Zoology, Trinity College, Dublin 2, Ireland.

Dr. B. Madden, Department of Botany, Trinity College, Dublin 2, Ireland.

Dr. B.F. Masterson, Department of Biochemistry, University College, Belfield, Dublin 4, Ireland.

Dr. S. Maund, School of Pure and Applied Biology, University of Wales College of Cardiff, PO Box 915, Cardiff, CF1 3TL, U.K.

Dr. E.J. McGee, Nuclear Energy Board, 3 Clonskeagh Square, Dublin 14, Ireland.

Ms E.A. Moorkens, Environmental Sciences Unit, Trinity College, Dublin 2, Ireland.

Dr. J. Nemcsók, Department of Biochemistry, József Attila University, Szeged, P.O.Box 659, Hungary.

Dr. B. O'Connor, Aqua-fact Ltd, 16 Newcastle Park, Galway, Ireland.

Prof. P. Oftedal, Department of Biology, Division of General Genetics, University of Oslo, PO Box 1031 Blindern, 0315 Oslo 3, Norway.

Dr. D. Pascoe, School of Pure and Applied Biology, University of Wales College of Cardiff, P.O. Box 915, Cardiff CF1 3TL, U.K.

Ms. J. Peacock, Department of Botany, Trinity College, Dublin 2, Ireland.

Dr. R.C. Peters, University of Utrecht, Laboratory of Comparative Physiology, Padualaan 8 NL-3584 CH Utrecht, The Netherlands.

Dr. D.K. Phelps, US Environmental Research Laboratory, US EPA, 27 Tarzwell Drive, Narragansett RI 02882, USA.

Dr. B. Rafferty, Department of Botany, Trinity College, Dublin 2, Ireland.

Dr. M. Vikram Reddy, Environmental Biology Laboratory, Department of Zoology, Kakatiya University, Warangal - 506 009, Andhra Pradesh, India.

Dr. R. Retuerto, Department of Botany, University of Cambridge, Downing St., Cambridge CB2 3EA, U.K.

Dr. J.D. Reynolds, Department of Zoology, Trinity College, Dublin 2, Ireland.

Prof. D.H.S. Richardson, Department of Botany, Trinity College, Dublin 2, Ireland.

Ms. A. Rower, Environmental Sciences Unit, Trinity College, Dublin 2, Ireland.

Prof. Katalin S.-Rózsa, Balaton Limnological Research Institute of the Hungarian Academy of Sciences, Tihany, Hungary H-8H-8237.

Prof. János Salánki, Chairman I.U.B.S. IBC, Balaton Limnological Research Institute of the Hungarian Academy of Sciences, Tihany, Hungary H-8H-8237.

Dr. Y. Shaaltiel, Galilee Technological Centre, South Industrial Zone, Kiryat Shmona 10200, Israel.

Dr. D. Sheenan, Department of Biochemistry, University of Cork, Lee Maltings, Prospect Row, Cork, Ireland.

Mr. A. Sides, Riofinex, 22 Tattyreagh Road, Omagh BT 781TZ, Northern Ireland.

Dr. G.C. Smith, School of Ocean Sciences, University College of Wales, Bangor, Menai Bridge, Gwynedd, LL59 5EY, U.K.

Mr. L. Stapleton, Environmental Research Unit, St. Martin's House, Dublin 4, Ireland.

Ms. M.R. Sullivan, Environmental Sciences Unit, Trinity College, Dublin 2, Ireland.

Dr. E.J. Taylor, School of Pure and Applied Biology, University of Wales College of Cardiff. PO Box 915, Cardiff, CF1 3TL, U.K.

Dr. P.C. Thomas, School of Pure and Applied Biology, University of Wales College of Cardiff, PO Box 915, Cardiff, CF1 3TL, U.K.

Dr. H.M. Thompson, Central Science Laboratory, MAFF, Hook Rise South, Tolworth, Surbiton, Surrey, KT6 7NF, U.K.

Dr. C. Turner, School of Pure and Applied Biology, University of Wales College of Cardiff. PO Box 915, Cardiff, CF1 3TL, U.K.

Mr. K.P. Twomey, Department of Pharmacology, Trinity College, Dublin 2, Ireland.

Prof. E. Weinert, Martin-Luther-Universitat Halle-Wittenberg, Sektion Biowissen Schaften, DDR-4020 Halle, Neuwerk 21, Germany.

Dr. B.A. Whitton, Department of Biological Sciences, University of Durham, Durham DH1 3LE, U.K.

Dr. B.R.H. Williams, Brixham Laboratory, ICI PLE, Freshwater Quarry, Brixam, Devon. U.K.

Dr. J.G. Wilson, Environmental Sciences Unit, Trinity College, Dublin 2, Ireland.

Dr. M. Yasuno, National Institute for Environmental Studies, Onogawa, Tsukuba 305, Japan.

Prof. S. Yu, 300 Fonglin Road, Shanghai Institute of Plant Physiology, Academia Sinica, Shanghai 200032, China.

Monitoring of Benthic Flora and Fauna in Channels Draining a Sewage Plant

M. Yasuno[1], S. Fukushima[2] and Y. Sugaya[1]

[1]National Institute for Environmental Studies, Onogawa, Tsukuba, 305 Japan.
[2]Yokohama City Institute for Environmental Research, Takigashira, Isogo-ku, Yokohama, 235 Japan.

Key words: attached algae, zoobenthos, chironomids, artificial substrates, sewage treatment plant, water quality, Japan.

Abstract

A sewage plant started to discharge treated water into open channels which had not been used for some years. In one channel monitoring started in December 1984 soon after the start of discharge and continued until December 1989. The discharged water was secondary treated and the quality was quite close to that of clean water. In 1987, nitrogen content, particularly ammonium nitrogen was reduced by improving the treatment. The algal flora on the gravel bed changed drastically after the change in water quality. The flora was compared with those in similar channels draining a sewage plant in Osaka and those in a polluted stream in Yokohama.

Zoobenthos in these channels were limited in the number of species. Since the channels were simple in structure, we placed three kinds of artificial substrates at several locations in an attempt to increase species diversity; the artificial substrates were concrete blocks, leaf litter bags and water plants. Those were very effective, respectively, in increasing the densities of zoobenthos but did not increase species diversity. There were some differences in the composition of benthos among these substrates. The chironomids collected from two channels receiving the same treated water were species typical of polluted streams, except in the lower region of one channel where we found some other species, indicating slightly better water quality. This zone was extended in 1989.

Introduction

Sewage treatment in Japan is still mainly to secondary level, hence the quality of water returned to rivers is below modern expectations. However, such plants are

Bioindicators and Environmental Management
ISBN 0-12-382590-3

usually able to reduce concentrations of organic matter, particularly organic nitrogen, and some changes could still be expected in the flora and fauna. Many studies have been carried out on the flora and fauna of polluted water, particularly tolerance of species of diatoms to pollution (Lange-Bertalot 1979; Kobayashi and Mayama 1982; Savater *et al*. 1987). Archibald (1972) and Schoemann (1973) indicated that organic nitrogen is essential for the appearance of some diatom species. William *et al*. (1973) and Keithan (1988) showed experimentally the importance of nitrogen in determining the nature of diatom communities.

We have had an opportunity to monitor benthic flora and fauna in channels draining a sewage plant, comparing these with a stream directing receiving domestic waste water. This comparative study will be useful in clarifying the relationship between water quality and flora and fauna.

Study site and methods

A sewage treatment plant in the north-western area of Tokyo has a treatment capacity of 150 000 tons per day. The Tokyo metropolitan authority started to drain 20 000 tons of treated water per day, after sand filtration, into one channel (Nobidome) in 1984, 23 000 tons per day to another channel (Tamagawa) in 1986 and 10 000 tons per day was diverted from the second channel to a third (Senkawa) in 1988 (Fig. 1.). No domestic sewage effluent enters these channels. Sampling stations were selected at the point of influx and 7.8 km downstream in the Nobidome channel, at the influx point and 6.6 km downstream in the Tamagawa channel and at the influx point and 2.5 km downstream in the Senkawa channel. The average widths of the channels are 1.6 m, 2.0 m and 1.5 m, respectively. The depths range from 10-40 cm and the current is 10-60 cm s^{-1}. The bed of the Nobidome channel is of clay or gravel, that of the Tamagawa is totally of clay, while the Senkawa has a concrete bed. The first two channels are mostly in the shade of deciduous trees, excepting the downstream sampling point in the Nobidome, while the Senkawa channel is mostly in full sunlight.

The water quality of these channels has been monitored by Kawahara *et al*. (1987) and Fukushima *et al*. (unpublished). Various forms of nitrogen and phosphorus have been measured, as well as Biochemical Oxygen Demand (BOD). The attached algal flora and fauna have been partly reported by Kawahara *et al*. (1988) and the recent changes in the flora will be published elsewhere. Artificial substrates were used to sample algae and zoobenthos, with additional zoobenthos samples being taken using a Surber sampler. Because of the simplicity of the channels, it was expected that biodiversity would be increased by the artificial substrates.

The artificial substrates were concrete blocks (20cm x 10cm x 6cm), leaf litter bags (100 g dry weight of leaves of deciduous trees in a 3mm mesh bag measuring 50cm x 30cm) and bundles of the water plant *Egeria densa* (24 g wet weight each; approximately 17 plants of 1m each). Two hundred concrete blocks were placed as a group within a 10 m length near the sampling points of the first two channels. Thirty-six leaf litter bags were anchored similarly within a 10 m long region at each of four places. Four of each type of artificial substrate were collected twice from each site at intervals of 40 and 87 days. Forty weighted bundles of water plant were placed in the third channel (Senkawa) and collected after 30 days. Algal samples were taken from 5 cm x 5 cm areas on the surface of the bricks and the enumerated cell numbers for each species were expressed in mm^2. All zoobenthos on the artificial substrates were enumerated in the laboratory.

As references, data from the Tsurumi River of Yokohama, the Imagwawa channel of Osaka and the Nikaryou channel of Kawasaki were used. The first one directly receives domestic sewage and is known as a seriously polluted river. The second channel receives the effluent from the sewage treatment plant in Osaka. The last one has been used as an irrigation channel but at present receives domestic sewage (Fig. 1).

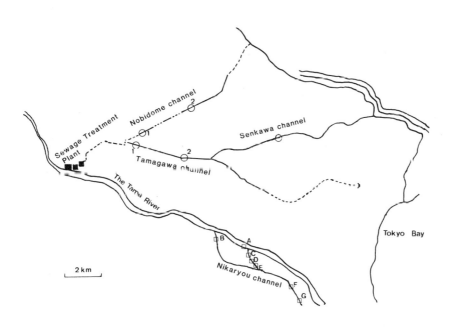

Figure 1. Channels used for biological monitoring of water quality. Circles indicate fixed sampling sites. Open squares indicate the sites selected for adult chironomid collections.

Results

Similarity in the flora between a stream in the urban area and the treated water channels

We selected a typical urban stream in Yokohama City for a comparison. The average values of characters indicated water quality of the Tsurumi River were: BOD - 19 mg l^{-1}; NH_4-N - 9.4 mg l^{-1}; NO_3-N - 2.28 mg l^{-1}; and PO_4-P 0.83 mg l^{-1}. Morisita's (1959) similarity index of algal communities between the Nobidome channel and the Tsurumi River was relatively high in 1984, 1985 and 1986 but decreased drastically in December 1987 and afterwards (Table 1). The commonest species in the urban streams were *Nitzschia palea, Gomphonema parvulum, Chlamydomonas* spp., *Navicula seminulum, Navicula veneta* and *Achnanthes minutissima*. Most of these were abundant at the beginning of the Nobidome channel indicating that this channel, which receives treated water, did not differ from the polluted streams in the urban area. The decrease in the similarity value in 1987 was due to the improvement of quality in the treated water. The value of BOD was reduced from 13 ± 8 to 4 ± 3 mg l^{-1} and that of NH_4-N from 5.1 ± 3.7 to 0.4 ± 0.6 mg l^{-1}. Other water quality characteristics did not change (Table 2). The result is reflected in the decrease in the number of *Navicula seminulum*, the most dominant species throughout the season. Other species thus became predominant and furthermore new species appeared in the channels. This result suggests that *Navicula seminulum* favours the water containing high concentrations of ammonium.

Table 1. Similarity index of attached algal communities between the Tsurumi River in an urban area, and the Nobidome channel (N1) or Tamagwa channel (T2). Data of the Tsurumi River were taken in January 1988).

	Dec 1984	Feb 1985	Dec 1986	Dec 1987	Jan 1988	Mar 1989
N-1 vs Tsurumi	0.623	0.697	0.630	0.039	0.092	0.012
T-2 vs Tsurumi	-	-	-	0.742	0.748	0.169

Succession in the algal flora in the channels

The similarity indices between the communities observed from time to time in successive years in the Nobidome channel are shown in Table 3. This shows that there are two distinct groups of records with high similarity, those before December 1986 and those after December 1987. Each group is distinctly different, however. As noted already, this is due to the changes in treated water quality. The changes in the abundance of the main species are shown in Figure 2. *Navicula seminulum* Grun. and *Nitzschia palea* (Kuetz.) tended to decrease, and

Table 2. Water quality of the Tsurumi River, Nobidome channel and Imagawa channel (mean $\pm$ SD at mg l^{-1}).

	Tsurumi Jan 88	Nobidome Dec 84-Dec 86 (n=13)	Nobidome Jun 87-Feb 88 (n=5)	Imagawa Jan-Dec 87 (n=12)
BOD	19.0	13.0 ± 8.0	4.0 ± 3.0	3.2
NH_4-N	9.4	5.1 ± 3.7	0.4 ± 0.6	4.3
NO_2-N	-	-	0.8 ± 0.6	0.29
NO_3-N	2.3	-	6.9 ± 0.7	11.0
NO_x-N	-	7.4 ± 1.6	-	11.3
TIN	-	12.5 ± 3.2	8.0 ± 2.0	12.0
T-N	-	12.3 ± 4.3	8.9 ± 2.0	17.0
PO_4-P	0.83	1.4 ± 0.7	1.7 ± 0.3	-
T-P	-	1.9 ± 0.8	1.8 ± 0.4	1.4
TIN/PO_4-P	-	12.0 ± 7.0	5.0 ± 1.0	-

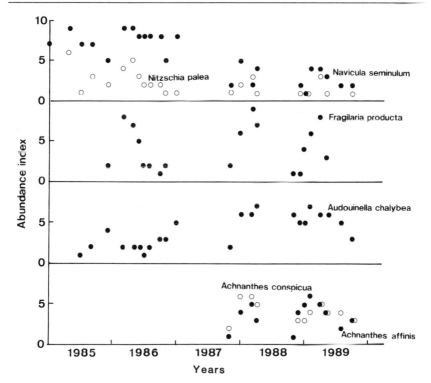

Figure 2. Changes in the abundance of selected attached algal species in the Nobidome Channel. Abundance index is based on the number of cells per mm^2.

Table 3. Similarity index of attached algal communities between different years in the Nobidome channel.

	Dec 1984	Aug 1985	Feb 1986	Sep 1986	Dec 1986	Dec 1987	Jan 1989
Aug 1985	0.992						
Feb 1986	0.863	0.891					
Sep 1986	0.997	0.998	0.887				
Dec 1986	1.000	0.995	0.867	0.999			
Dec 1987	0.011	0.081	0.080	0.079	0.077		
Jan 1989	0.031	0.039	0.050	0.032	0.032	0.717	
Sep 1989	0.000	0.000	0.000	0.000	0.000	0.819	0.547

Fragilaria producta Grun. showed a marked seasonal fluctuation, but did not change in its trends. *Audouinella chalybea* Bory increased after the change in the water quality and became a predominating species. Newly encountered species were *Achnanthes affinis* Kuetz. and *Achnanthes conspicua*.

The results of similar analysis for the Tamagawa channel are not so clear, because the water quality had been improved at the time of the start of draining (Table 4.). In this channel, *Navicula seminulum* and *Nitzschia palea*, species indicating highly polluted water, were dominant for the first year, although the flowing water contained less NH_4-N. This phenomenon suggests that a period of time is required to attain a stable state. It is not known whether *Navicula seminulum* or *Nitzschia palea* are important as precursors in the succession, or the absence of a source of other species explains the phenomenon. However, the results rather suggest the competitive replacement of the species.

Table 4. Similarity index of attached algal communities between different years in the Tamagawa channel.

	Dec 1987	Mar 1988	Oct 1988	Jan 1989	Mar 1989
Mar 1988	0.871				
Oct 1988	0.313	0.286			
Jan 1989	0.864	0.825	0.299		
Mar 1989	0.148	0.259	0.061	0.210	
Sep 1989	0.004	0.005	0.811	0.087	0.016

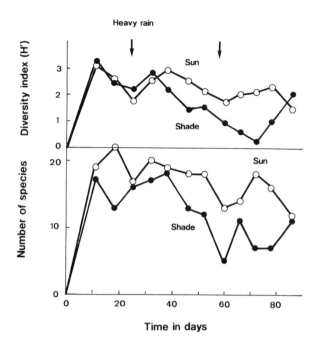

Figure 3. Changes in the number of species and diversity of algal community on bricks placed in the Nobidome channel.

The importance of competitiveness in determining the algal flora was suggested from the observation that the species diversity of the algal community on bricks developed rapidly after placement and gradually decreased. It rose temporarily again when the community was perturbed by heavy rains (Fig. 3).

In the Senkawa channel, *Naviclula seminulum* and *Nitzschia palea* were absent or extremely scarce. *Cocconeis pediculus*, however, an indicator of moderate water quality, became abundant in December 1989, indicating self-purification of the water there. Unfortunately, we have no records of water quality.

Characteristic zoobenthos in the channels
The fauna of zoobenthos in these channels draining treated water from the sewage treatment plant was very simple. The most common species were *Chironomus* spp. and *Rheotanytarsus kyotoensis*, followed by about another 10 species of Chironomudae (Table 5). The species were essentially similar before and after the change in the water quality. Other organisms occurring were *Cheumatopsyche brevilineata*, *Asellus hirgendorfi* and tubicifids. The fauna was similar to that in urban streams receiving raw domestic waste water (Sasa *et al.* 1980; Sasa 1983). Some of the species are, however, not necessarily characteristic of polluted water but are inhabitants of the lower reaches of rivers.

Table 5. Zoobenthos communities of three channels receiving effluent from a sewage treatment plant (no. m^{-2}).

Species	Senkawa			Nobidome		Tamagawa	
	Sep	Oct	*Egeria*	1	2	1	2
Cheumatopsyche brevilineata	25.0	12.5	1.3	3300.0	150.0	19300.0	2566.7
Oligochaeta	12.5	12.5	9.6	216.7	616.7	300.0	416.7
Baetis sahoensis	431.3	68.8	2.8	-	-	-	150.0
Asellus hilgendorfii	-	-	-	-	16.7	-	-
Theinemanniella majuscula	75.0	75.0	156.7	16.7	-	-	266.7
Cricotopus bicinctus	93.8	50.0	50.3	16.7	25.0	183.3	-
Paratrichocladius rufiventris	-	6.3	2.8	16.7	-	33.3	-
Rheocricotopus chalybeatus	118.8	112.5	48.5	33.3	150.0	50.0	2083.3
Nanocladius tamabicolor	37.5	6.3	2.4	1433.3	733.3	33.3	33.3
Chironomus circumdatus	-	-	-	-	33.3	-	-
Glyptotendipes tokunagai	18.8	-	1.1	1233.3	100.0	50.0	-
Plypedilum ureshinoense	-	-	-	116.7	-	-	-
Cryptochironomus spp.	37.5	18.8	-	-	-	-	-
Reotanytarsus kyotoensis	-	12.5	26.3	-	33.3	33.3	466.7
Tanytarsus tamanonus	-	-	0.2	-	-	-	33.3

The chironomid survey in the Nikaryou channel, running into the opposite site of the Tama River, demonstrated some interesting results (Table 6, and see the sampling sites in Fig. 1). The eight or nine species of chironomids were collected from the channel bank close to the water intake point, and then decreased abruptly to one (*Chironomus yoshimatsui* only) at the site 100 m downstream where the BOD value increased to 15 mg l^{-1}. Therefore, BOD seems to be important in determining the chironomid fauna. *Chironomus yoshimatsui* occupies a niche similar to that of *Chironomus thummi* in Europe and is a good indicator of water polluted with organic matter in Japan. In the Nobidome, Tamagawa and Senkawa channels, this species was absent. However, four of the ten species found in these channels were common to the species collected from the Nikaryou channel.

Table 6. Chironomidae (adults) collected from the bank of the Nikaryou channel.

Species				Sampling site			
	A	B	C	D	E	F	G
Corynoneura spp.	-	2	14	-	-	-	-
Cricotopus bicinctus	36	22	-	2	-	5	3
C. sylvestris	1	-	-	-	-	-	1
C. tamapullus	15	-	-	-	-	-	-
C. triannulatus	6	2	12	1	-	-	2
C. trifasciatus	3	-	-	-	-	-	-
Limnopyes tamakitanaides	-	2	-	-	-	1	-
Nanocladius tamabicolor	4	15	6	2	-	-	-
Paratrichocladius rufiventris	10	1	6	1	-	-	19
Rheocricotopus chalybeatus	20	5	1	-	-	-	-
Smittia spp.	-	13	4	3	-	-	-
Chironomus yoshimatsui	-	-	-	3	122	62	6
C. nippodorsalis	-	-	-	-	-	1	3
Rheotanytarsus kyotoensis	-	88	209	68	-	-	-
No. of species	8	9	7	7	1	4	6
No. of individuals	95	150	252	80	122	69	34
BOD ($mg\ l^{-1}$)	5.7	6.7	5.3	-	15	21	15

Zoobenthos on artificial substrates

As the channel bed is mostly clay, inhabitants would be limited, and it was though that artificial substrates would increase habitats. Figure 4 shows the numbers of principal species collected from leaf litter-bags after 32 days, concrete blocks after 55 days and Surber net samples. No new inhabitants appeared, but the artificial substrates certainly provided good habitats as a marked increase in numbers was observed. *Asellus hirgendorfi* could live in leaf litter-bags, although only a few individuals were found there before. It is difficult to detect the differences in the efficacy of the artificial substrates, between leaf litter-bags and concrete blocks, for other organisms. Some were found more on concrete blocks at one site and more on litter-bags at another.

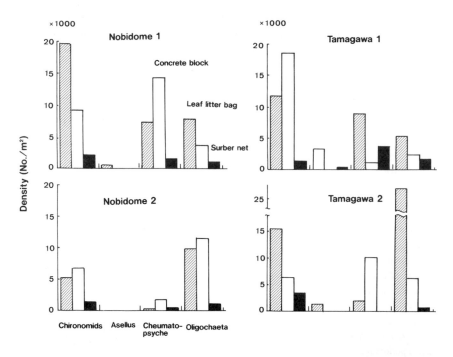

Figure 4. Density of zoobenthos found in or on artificial substrates in the Nobidome and Tamagawa Channel. Solids bars indicate the densities of respective species collected by Surber net at respective sites as reference.

The bundles of water plant placed in the third channel (Senkawa), which had a concrete bed, showed similar results (see Table 5). Here the numbers of benthos on a bundle of *Egeria densa* are shown, and they should be multiplied by a conversion factor of 5.5 (estimated from the leaf areas), if actual density per square metre is required). No new organisms were found on the bundles when collected after one month, but some species such as *Thienemanniella majuscula*, utilized them efficiently. The appearance of a species of mayfly, *Baetis sahoensis*, in the lower reaches of the second and third channel possibly indicates a slight change in water quality.

Discussion

The water quality of the effluent from a sewage treatment plant was not initially satisfactory with respect to BOD. However, it contained little organic nitrogen. Thus, in terms of nitrogen sources, it was different from raw sewage water, and the dominance of *Nitzschia palea*, which is well known to demand a high concentration of organic nitrogen (Archibald 1972), could not be explained. The reduction of BOD was significant in the change of the flora in the Nobidome channel. However, the reduction of NH_4-N seems to be more important. In the Imagawa channel, where BOD was as low as 3.2 mg l^{-1} (ranging from 1.0 to 4.7 mg l^{-1}), but with a higher concentration of NH_4-N (4.3 mg l^{-1}), *Nitzschia palea* was dominant. Kobayashi and Mayama (1982) studied the diatoms in severely polluted rivers in the vicinity of Tokyo and found *Nitzschia palea, Navicula seminulum* and *Navicula atomus* were the dominant types in every river. They could not find any correlation between the occurrence of these three species and environmental factors expressed as pH, O_2-saturation, conductivity and BOD. This may be because the rivers surveyed were all seriously polluted and there were no differences between them.

Audouinella chalybea, belongs to the Rhodophyceae, and has never been found in urban streams polluted with raw sewage. Thus, this channel had a unique algal flora after the change in the water quality, although the influence of domestic sewage remained. This can be ascribed partly to the relatively high water temperatures, 9.2°C in winter to 24.0°C in summer. However, temperature is not an absolute condition for the occurrence of this species, since *Nitzschia palea* is also found in warm waters (Kobayashi and Mayama 1982).

The decrease in BOD from 10-12 to 4 mg l^{-1} in the Nobidome channel did not cause any significant change in the number of species, or the species composition, of benthic invertebrates. In the case of invertebrates, NH_4-N would be a significant factor determining the benthic fauna. In fact, we found a more simple chironomid fauna dominated by *Chironomus yoshimatsui, Cricotopus sylvestris, Cricotopus bicinctus* and *Paratanytarsus* spp. in the Imagawa channel draining treated water from a sewage plant in Osaka, where NH_4-N was high but BOD was low, as mentioned above. On the contrary, in the three channels, the Nobidome, Tamagawa, and Senkawa, there was also a unique fauna as *Chironomus yoshimatsui,* one of the pollution indicators of chironomids (Sasa et al. 1980), was lacking. *Cheumatopsyche brevilineata* inhabits rivers with slow currents, which usually run through urban areas, but are not necessarily polluted (Yasuno et al. 1985). However, it is certain that this species is tolerant of highly polluted waters, as it was observed both before and after the change in the water quality of the Nobidome channel.

We did not intend to develop artificial substrates as sampling units and, therefore, the samples on them contained more zoobenthos than on the natural substrate. The species compositions were different between the artificial substrates according to their characteristics. However, in the channels with clay or concrete beds, some kinds of artificial substrates are useful and convenient, particularly for periphyton and chironomids.

Acknowledgements

We wish to express our thanks to Dr T. Sato of Kawasaki City Health Institute for his assistantance in carrying out the survey in the Nikaryou channel.

References

Archibald, R.E. (1972) Diversity in some South African diatom associations and its relation to water quality. *Water Research*, 6, 1229-1238.

Kawahara, H., Okada, M., Fukushima, S. and Mutoh, A. (1987) Introduction of secondary treated domestic wastewater into a small and dry stream - Chemical analysis of Nobidome channel. *Japan Journal of Water Pollution Research*, 11, 642-630.

Kawahara, H., Fukushima, S., Mutoh, A. and Okada, M. (1988) Introduction of secondary treated domestic wastewater into a small and dry stream - Periphytic algae and small benthic invertebrates in the Nobidome channel. *Japan Journal of Water Pollution Research*, 11, 231-239.

Keithan, E.D. (1988) Benthic diatom distribution in a Pennsylvania stream : Role of pH and nutrients. *Journal of Phycology*, 24, 581-585.

Kobayasi, H. and Mayama, S. (1982) Most pollution-tolerant diatoms of severely polluted rivers in the vicinity of Tokyo. *Japanese Journal of Phycology*, 30, 188-196.

Lange-Bertalot, H. (1979) Pollution tolerance of diatoms as a criterion for water quality estimation. *Nove Hedwigia, Beih*, 64, 285-304.

Morisita, M. (1959) Measuring of interspecific association and similarity between communities. *Memoir of Faculty of Science, Kyushu University, Series E (Biology)*, 3, 65-80.

Sasa, M., Yasuno, M., Ito, M. and Kikuchi, T. (1980) Studies on chironomid midges of the Tama River. Part 1. The distribution of chironomid species in a tributary in relation to the degree of pollution with sewage water. *Research Report from the National Institute for Environmental Studies, Japan*, No.13, 1-8.

Sasa, M. (1983) Studies on chironomid midges of the Tama River. Part 5. An observation on the distribution of Chironominae along the main stream in June, with description of 15 new species. *Research Report from the National Institute for Environmental Studies, Japan*, No.43, 1-67.

Savater, S. Savater, F. and Tomas, X. (1987) Water quality and diatom communities in two Catalan Rivers (N.E. Spain). *Water Research*, 21, 901-911.

Schoemann, F.R. (1973) *A systematical and ecological study of the diatom flora of Lesotho with special reference to the water quality*. National Institute for Water Research. Pretoria, South Africa.

Williams, S.L., Colon, L.E.M., Kohlberger, R. and Clesceri, N.L. (1973) Response of plankton and periphyton diatoms in Lake Gregore (N.Y.) to the input of nitrogen and phosphorus. In G.E. Glass (ed.) *Biassay Techniques and Environmental Chemistry*, 441-466. Ann Arbor Science Publishers Inc., Ann Arbor, Michigan.

Yasuno, M., Yoshio, S. and Iwakuma, T. (1985) Effects of insecticides on the benthic community in a model stream. *Environmental Pollution*, 38, 31-43.

Use of Biotransformation Activity in Fish and Fish Hepatocytes in the Monitoring of Aquatic Pollution Caused by Pulp Industry

O. Hänninen[1], P. Lindström-Seppä[1], M. Pesonen[1], S. Huuskonen[1, 2] and P. Muona[3]

[1]Department of Physiology and [2]Department of Applied Zoology, University of Kuopio, P.O.B. 6, SF-70211 Kuopio, Finland.
[3]Water Protection Association of Savo-Karjala, Yrittäjäntie 24, SF-70150 Kuopio, Finland.

Key words: biomonitoring, cytochrome P-450 monooxygenases, fish, hepatocytes, pulpmill effluents, Finland.

Abstract

The polysubstrate monooxygenase system appears to be highly responsive to chemical pollution. The present study summarizes findings on biomonitoring of the waste waters released by the pulp and paper industry with the aid of fish and their hepatic biotransformation enzymes. The 7-ethoxyresorufin O-deethylase activity is clearly induced in fish hepatocytes exposed to unbleached and bleached pulp and paper mill effluent fractions in vitro. Furthermore, the increase of polysubstrate monooxygenase activities is observed both in feral fish and in the cultivated rainbow trout caged in waters downstream of these mills.

Introduction

The wood processing (especially pulp and paper) industry is one of the most significant sources of aquatic pollution in many countries in the boreal region, like in Finland, through the release of copious volumes of waste water. Chemicals released by this energy intensive industry, even into the air and to the soil end up in the waters of the region. The effluents from pulp and paper mills contain a wide variety of harmful substances in different concentrations (deSousa et al. 1988). Resin and fatty acids, as well as chlorophenols, are the major compounds in bleached pulpmill effluents (BKME) responsible for fish mortality. Biotreatment of effluents normally reduces the levels of resin and fatty acids in mill effluents to sub-lethal levels (Hutchins 1979). Chlorophenols and other chlorinated compounds, however, remain as toxicants because of their high resistance to biodegradation. Even when diluted in receiving waters the effluents can cause sub-lethal disturbances, e.g. to fish living downstream of the sewer outlet (Oikari *et al.* 1985), which could lead, for example, to population changes.

Fish are extensively exposed to chemicals in their environment. The lipid nature of their membranes (e.g. in gills and elsewhere) attract the lipophilic compounds. The food of fish (plankton etc.) also take up chemicals, and they enter the fish via the gastrointestinal tract. These chemicals can be excreted only after conversion into water soluble forms. Of the many biotransformation enzymes, cytochrome P-450 and polysubstrate monooxygenase activities (e.g. 7-ethoxyresorufin O-deethylase = EROD) appear to be most sensitive in fish to the chemical loading of their environment (Payne 1984; Lindström-Seppä 1990). In experimental laboratory and field studies the biotransformation of xenobiotics in fish species has been shown to be affected, for example by petroleum or bleached kraft pulp mill effluents (Payne *et al.* 1987; Andersson *et al.* 1988; Lindström-Seppä and Oikari 1989; 1990a; 1990b).

In fish, as in mammals, the liver is the main organ involved in biotransformation of xenobiotics to water-soluble metabolites. The biotransformation can be oxidative, mediated by cytochrome P-450 monooxygenases, and the product may be conjugated by, for example glucuronic acid. The conjugated products are hydrophilic and can be excreted in urine and bile.

Fish respond to environmental pollutants by increasing (induction) their metabolism. Induction of cytochrome P-450 monooxygenase activities can be used as a sensitive indicator of exposure to environmental chemicals such as polyaromatic hydrocarbons (PAH, TCDD, PCB).

The aims of the present study were to investigate the use of cytochrome P-450 monooxygenase activities of fish in monitoring the early arising effects of pulp and paper mill effluents. The use of feral and caged fish as well as cultures of fish hepatocytes was tested. Two different processes of pulp production and effluent treatment were compared as for studying the effect of different effluent types. Furthermore, a preliminary comparison of waste water processing in a bleached pulp mill at Baikalsk, Lake Baikal, USSR is made with the treatments above.

Pulp production and bleaching

In the kraft process, which is the main process used in chemical pulping, wood chips are digested or cooked under pressure with a mixture of hot caustic soda and sodium sulfide. Lignin and wood extractives are then solubilized, leaving the insoluble cellulose fibers, i.e. pulp. The pulp then enters the bleaching process which contains several steps. These steps and their order vary considerably between mills. The different steps of conventional chlorine bleaching are: chlorination (C), treatment with alkali (E), chlorination with Na-hypochlorite (H), and treatment with chlordioxid (D).

The two mills studied in this report, one producing unbleached pulp/cellular board and the other bleached pulp/paper, are situated in the lake district of Finland (Lake Kallavesi and Lake Saimaa, respectively).

The bleaching sequence at the mill in Lake Saimaa is D/C - E - D - E - D. The effluents pass through a mechanical - microbiological (aerobic process with 2-3 day retention time) waste-water treatment system. The unbleached pulp (Lake Kallavesi) is made using an ammonium-based sulphite process. The effluents are mechanically and biologically treated (anaerobic process).

Primary cultures of fish hepatocytes

Primary cultures of rainbow trout hepatocytes were used as a screening method prior to a larger-scale field monitoring study.

Hepatocytes were isolated from immature rainbow trout with collagenase perfusion described previously by Pesonen and Andersson (1990). Before experiments the cells were counted and the cell viability was always determined by trypan blue dye exclusion test. The cells with mean viability over 90% were seeded on plastic petri dishes (10^6 cell ml^{-1}) and incubated at 10°C.

Two paper mill effluents and control water from a clean area were extracted with organic solvent. One litre of each studied water was extracted with diethylether, evaporated and the residues were dissolved in 1 ml dimethylsulfoxide (DMSO). The prepared extracts were then added to cultures in 10 µl fractions per 10 ml of culture medium.

Both unbleached and bleached effluent fractions increased significantly cytochrome P-450 dependent EROD-activity by low concentrations of the fractions (0.1 - 0.5 µl ml^{-1}) (Fig. 1). In higher concentrations the unbleached effluent fraction abolished EROD-activity entirely. This may be a direct toxic effect. Bleached effluents, however, inhibited EROD-activity in a dose-dependent manner when concentration of the fraction was raised in the culture medium.

Comparison of feral and caged fish in monitoring the effect of bleached and unbleached effluents

Based on the findings from the cell culture studies, the monooxygenase enzyme activities in feral perch *Perca fluviatilis* living in waters polluted by the unbleached pulp/paperboard mill effluents were investigated. The effects seen were compared to the effects of bleached pulp and paper mill effluents (BKME) studied earlier (Lindström-Seppä and Oikari 1990b). Population samples from perch were caught from several locations downstream from the pulpmill sewers and from control sites upstream. In addition rainbow trout *Oncorhynchus mykiss* were caged in Lake Kallavesi and southern Lake Saimaa (Lindström-Seppä and Oikari 1990a) at various locations downstream from the effluent source and at control sites.

The one year old rainbow trout were caught from a fish farm, transported to the lakes in PVC-bags and put into cages. The exposure time was 3 weeks.

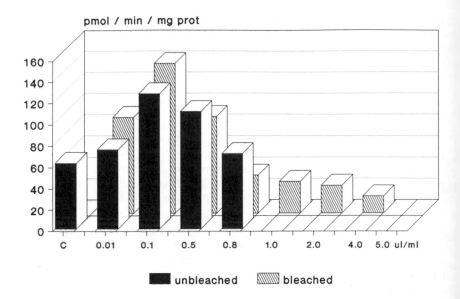

Figure 1. 7-Ethoxyresorufin O-deethylase (EROD) activity in primary culture of rainbow trout hepatocytes. The cells were exposed for 48 hours to various concentrations of liposoluble fractions extracted from unbleached and bleached pulp and paper mill effluents. Each bar represents the mean of 3-4 petri dishes.

The fish were captured and killed by a blow on the head. The gall bladder was removed intact and the liver was taken and put into liquid nitrogen. In the laboratory the livers were thawed, the microsomal fraction prepared and the samples analysed for monooxygenase enzyme activities as reported earlier (Lindström-Seppä and Oikari 1990a; 1990b).

Biotreated unbleached effluent increased EROD activities in feral perch, compared to controls, down to the last study site, 33%, 52% and 44% at 1 km, 3 km and 5.5 km, respectively (Fig. 2).

Perch captured downstream from the BKME source in southern Lake Saimaa showed greater increases of liver EROD activities than perch in Lake Kallavesi. The effect was observed at least up to 13 km from the sewer compared to controls; increases of EROD activity were 179% at 1 km, 18% at 7 km and 77% at 13 km (Fig. 2). The general pattern of the induced state of cytochrome P-450 system in fish living in this watercourse is further supported by observations on other biotransformation activities, for example benzo(a)pyrene hydroxylase, which was showing similar or sometimes even stronger responses (results not shown). Rainbow trout caged on the same lake area showed elevated enzyme activities with a distance related response (Fig. 3).

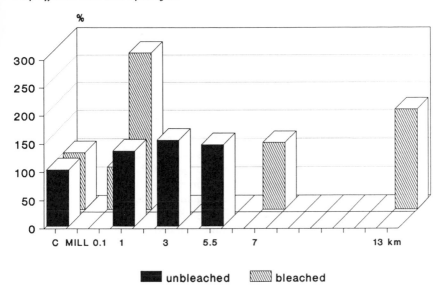

Figure 2. Relative 7-Ethoxyresorufin O-deethylase activities (control = 100) in the liver of feral perch *Perca fluviatilis* caught at waters contaminated with active sludge treated unbleached kraft pulp/paperboard mill effluents at Lake Kallavesi and with bleached kraft pulp/paper mill effluents after mechanical - aerobic bond treatment at Lake Saimaa, Finland.

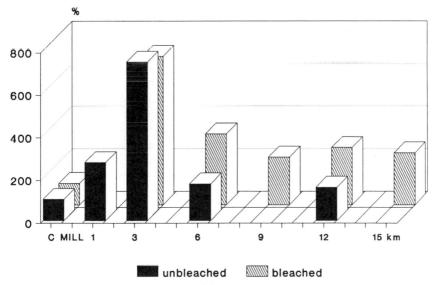

Figure 3. Relative activities of hepatic 7-ethoxyresorufin O-deethylase in rainbow trout caged at various distances from the sewers of unbleached and bleached kraft pulp and paper mills at Lake Kallavesi and Lake Saimaa, Finland.

Preliminary caging experiments at Lake Kallavesi showed a similar inducing effect of unbleached effluents on caged rainbow trout as seen in feral perch.
However, the induction coefficients observed in rainbow trout were higher and
comparable to the results with caged fish at the southern Lake Saimaa.

Pulpmill effluents and fish in Lake Baikal

The pulpmill of Baikalsk produces super-strong dissolving pulp for the
automobile, aeroplane and rayon textile industry. The bleaching process applied
includes the use of chlorine (C), sodium hydroxide (E), hypochlorite (H) and
chlordioxide (D). Waste waters from the pulpmill as well as from Baikalsk City
(to provide nutrients for microbes) undergo biological, chemical and mechanical
treatments (anaerobic biological - mechanical - chemical - mechanical - aerobic
biological) with final stabilization basins and dilution before discharge into Lake
Baikal. The released water was clean with a slight taste of chemicals (personal
experience of O. Hänninen). Endemic Baikal fish are living in waters in the
discharge area and no visible effect of waste waters was observed at the release
site or elsewhere.

Some endemic Baikal fish, *Cottus kessleri* and *Paracottus kneri*, were collected
in summer 1987 (Lindström-Seppä and Oikari 1990b). There were no statistical
differences between control (central Baikal), reference (just upstream to the pipe)
and exposed fish (not illustrated), which was in accordance with earlier
observations (Beym 1986).

Discussion

Fish liver monooxygenase activities can serve as a monitoring tool for the effects
of pulpmill effluents. Both the fish liver cell culture as well as the liver
microsomal fractions of caged rainbow trout showed clear induction of EROD
activities with the Finnish effluent. No significant difference was seen between
the bleached and unbleached processes. Both effluents contain compounds
which are able to affect the monooxygenase enzyme system. The chlorinated
compounds of the bleached process are not the only ones responsible for the
inducing capacity. The waste water treatment systems in the Finnish pulp mills
appear to be less efficient than the system used in Baikalsk, if compared by
measuring the metabolic responses.

However, the influence of seasonal and other basic factors affecting these
activities cannot be excluded (Koivusaari *et al.* 1981; 1983; 1984).

Because the collecting of feral fish and even the caging of cultivated fish
have their problems and are rather time consuming activities, the cell cultures of
fish hepatocytes offer a number of advantages. The fish liver cells in primary
cultures maintain xenobiotic metabolizing enzyme activities at stable levels in
culture conditions better than mammalian hepatocytes. Fish liver cells retain also
their ability to respond to PAH-type enzyme inducers, such as PCBs and TCDD.

In addition to the inducibility, the checking of the direct toxicity of the compounds is possible in the cell cultures. The method complements the direct studies in fish and makes it possible to screen rather fast the effects of the different components of complex pollutant mixtures. This method could be used in preliminary screening of effluents before more complex field experiments are conducted.

Conclusions

Caged rainbow trout were shown to respond with induced hepatic EROD-activity to unbleached and bleached pulpmill effluents in a more pronounced manner than feral perch. It is suggested that caged fish should be used in biomonitoring programmes whenever possible. The primary culture of fish hepatocytes can be used as a convenient *in vitro* method for a screening of potentially hazardous xenobiotics in effluents.

Acknowledgements

These studies has been supported by the Academy of Finland/Research Council for the Environmental Sciences, Finnish Ministry of Agriculture and Forestry, Kymenlaakso Fund of the Finnish Cultural Foundation and Water and Environment District of Kuopio.

References

Andersson, T., Förlin, L., Härdig, J. and Larsson, Å. (1988) Physiological disturbances in fish living in coastal water polluted with bleached kraft pulp mill effluents. *Canadian Journal of Fisheries and Aquatic Sciences*, **45**, 1525-1536.

Beym, A. M. (1986) Biological testing of industrial effluent. In *Problems of aquatic toxicology, biotesting and water quality management*. Proceedings of USA-USSR symposium, 122-135. Borok, Jaroslavl Oblas, 1984 EPA Environmental Research.

DeSousa, F., Strömberg, L. and Kringstad, P. (1988) The fate of spent bleach liquor material in receiving waters. *Water Science and Technology*, **20**, 35-42.

Hutchins, F. E. (1979) *Toxicity of pulp and paper mill effluent: A literature review*. Report of Corvallis Environmental Research Laboratory, Office of Research and Development, U.S. Environmental Protection Agency EPA-600/3-79-013, 1-43.

Koivusaari, U., Harri, M. and Hänninen, O. (1981) Seasonal variation of hepatic biotransformation in female and male rainbow trout (*Salmo gairdneri*). *Comparative Biochemistry and Physiology*, **70C**, 149-157.

Koivusaari, U. (1983) Thermal acclimatization of hepatic polysubstrate monooxygenase and UDPglucuronosyltransferase of mature rainbow trout (*Salmo gairdneri*). *Journal of Experimental Zoology*, **227**, 35-42.

Koivusaari, U., Pesonen, M. and Hänninen, O. (1984) Polysubstrate monooxygenase activity and sex hormones in pre- and postspawning rainbow trout (*Salmo gairdneri*). *Aquatic Toxicology*, **5**, 67-76.

Lindström-Seppä, P. (1990) *Biotransformation in fish: monitoring inland water pollution caused by pulp and paper mill effluents.* Ph.D. Thesis, University of Kuopio, Kuopio, Finland.

Lindström-Seppä, P. and Oikari, A. (1989) Biotransformation and other physiological responses in whitefish caged in a lake receiving pulp and paper mill effluents. *Ecotoxicology and Environmental Safety*, **18**, 191-203.

Lindström-Seppä, P. and Oikari, A. (1990a) Biotransformation and other toxicological and physiological responses in rainbow trout (*Salmo gairdneri* Richardson) caged in a lake receiving effluents of pulp and paper industry. *Aquatic Toxicology*, **16**, 187-204.

Lindström-Seppä, P. and Oikari, A. (1990b) Biotransformation activities of feral fish in waters receiving bleached pulp mill effluents. *Environmental Toxicology and Chemistry*. (in press).

Oikari, A., Holmbom, B., Ånäs, E., Miilunpalo, M., Kruzynski, G. and Castren, M. (1985) Ecotoxicological aspects of pulp and paper mill effluents discharged to an inland water system: Distribution in water, and toxicant residues and physiological effects in caged fish (*Salmo gairdneri*). *Aquatic Toxicology*, **6**, 219-239.

Payne, J. F. (1984) Mixed function oxygenase in biological monitoring programs: Review of potential usage in different phyla of aquatic animals. In G Persoone, E. Jaspers and C. Claus (eds.) *Ecotoxicological testing for the marine environment*, Vol 1, 625-650. State University of Ghent and Institute of Marine Scientific Research, Bredene, Belgium, 625-650.

Payne, J.F., Fancey, L.L., Rahimtula, A.D. and Porter, E.L. (1987) Review and perspective on the use of mixed-function oxygenase enzymes in biological monitoring. *Comparative Biochemistry and Physiology*, **86C**, 233-245.

Pesonen, M. and Andersson, T. (1990) Characterization and induction of xenobiotic metabolizing enzyme activities in a primary culture of rainbow trout hepatocytes. *Xenobiotica*. (in press).

A Monitoring Study of the Succession of Marine Sessile Macro-organisms Five Years Before and After the Operation of a Nuclear Power Plant

Rong-Quen Jan and Kun-Hsiung Chang

Institute of Zoology, Academia Sinica, Nankang, Taipei 11529, Taiwan, Republic of China.

Key words: bioindicator, sessile organism, thermal pollution, soft coral, stony coral, succession, nuclear power plant, Taiwan.

Abstract

In assessing the ecological impact of the operation of the third Nuclear Power Plant on the marine environment, a collaborative project, organized by the National Scientific Committee on Problems of the Environment (SCOPE/ROC), has been in progress since 1979, five years before the commercial operation of the plant was launched. In the present study, which is among thirteen monitoring studies in this project, sessile macro-organisms along three fixed transects (at three stations, namely, Stations A, B and D respectively) were surveyed four to five times a year by scuba diving. For data collection, the macro-organisms were divided into four categories, i.e. algae, soft corals, hard corals, and "other-invertebrates". Data collected between October 1979 and June 1989 are used in the present paper. There was no marked change in the coverage by each category of organism when the data collected before and after the operation of the plant were compared. Results from the analysis of correlation of the time series data between transects showed that both algae and "other-invertebrates" were highly correlated between Stations A and B. However, a similar phenomenon was not found in either soft corals or hard corals between stations. Overall, with the inconsistency between trends of macro-organism coverage on transects, it is suggested that information provided thus far by results from this part of the study on the influence of the operation of the nuclear power plant on the local marine environment is inconclusive.

Introduction

The Taiwan Power Company constructed its third nuclear power plant at the

Bioindicators and Environmental Management
ISBN 0-12-382590-3

southern tip of Taiwan in 1979. This nuclear power plant is furnished with two generating units, and seawater in the Nanwan Bay is used as a cooling medium. Commercial operation of this nuclear power plant started in 1984, with the operation of one of its two generating units. In the next year, the operation of the second generating unit was launched.

With the large quantity of cooling water routinely used by the power plant, it is possible that the operation of the plant could modify the marine environment. This may happen because of predictable change between intake and discharge including increase in temperature, changes to trace element concentrations and low level radioactive discharge. Variations of these abiotic factors may influence biological activities in the marine ecosystem, and hence cause changes to the biotic community (Johannes 1975). Moreover, drastic environmental changes may also damage specific marine biological resources.

A collaborative project has therefore been organized by the National Scientific Committee on Problems of the Environment, Academia Sinica (SCOPE/ROC), with financial support from the Atomic Energy Council of the Republic of China, to assess the ecological impact of the operation of this nuclear power plant on the marine environment. This project includes surveys on both non-biological and biological factors, and has been in progress since 1979, five years before the launch of the commercial operation of the power plant.

The present study, which is one of thirteen monitoring studies involved in this project, is an attempt to use sessile macro-organisms, which are very abundant in local waters (Jones *et al.* 1972), as a bioindicator. On the one hand, the results may provide information on the succession of the macro-organisms of the local waters. On the other hand, we hope the information obtained from this part of the study may help to assess the environmental changes associated with the power plant operation.

Materials and methods

This long-term study was initiated in July 1979. Field trips were undertaken bimonthly to the southern tip of Taiwan (120°45'E, 21°57'N) (Fig. 1). However, owing to weather conditions, data from only four to five collections were available each year. Sessile macro-organisms at three sub-tidal stations, namely, Stations A, B and D, were surveyed using the transect-line method by scuba diving. (Station C was located on a reef-flat at the outlet of the water discharge canal and was abandoned in the second year of this study for the construction of the canal.) Among the three stations, Stations A and B were located close to the water discharge canal of the power plant, whereas Station D was located 500 meters away from the water intake of the plant. These study sites are described in detail below.

Station A was assigned to an angular block, 3 m in height and 4 m in width, which was among the many blocks located outside the surge zone. The water depth to the top of the block was between 5 and 7 m.

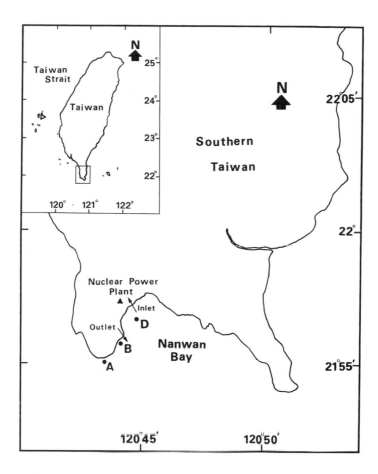

Figure 1. Map showing Nanwan Bay and the study sites. A, B, D are sub-tidal stations where the survey was undertaken; Inlet, inlet of the water intake constructed by the nuclear power plant; Outlet, outlet of the water discharge canal.

Station B was located on a wide non-limestone terrace at the reef slope. This terrace slanted southeasterly and was 60 m away from the rocky coast. The depth to the highest part of the rock was 6 m and it was 13 m at the seaward end.

Station D was composed of six adjacent rocks on the sea bed. These rocks were of different sizes. The largest was 2 m in height. The depth was between 10 and 12 m.

At each station, a yellow plastic mark of 3 cm x 3 cm was fixed at intervals of one metre on the reef surface to provide a permanent track for a transect. Only one transect was made in each station. Fouling organisms on the marks were removed; missing marks were replaced in the consecutive surveys.

Plate 1. Underwater view of the survey. During the survey, a tapeline was connected to marks previously fixed to the reef surface to make up a temporary transect. Divers used the tapeline to measure the coverage of macro-organism of different categories.

During the survey, a tapeline was connected to the marks to make up a temporary transect (Plate 1). The length of the transect was determined mainly by the underwater topography. However, some adjustments were made in the second year to ensure that the long-term study was feasible. The transect at Station A was approximately 10 m; at Station B 20 m; and at Station D 13 m (Table 1). Macro-organisms occurring along the transect were divided into four categories, i.e., algae, soft corals (alcyonaceans), hard corals (scleractinians) and other-invertebrates. The coverage in length of each category on the transect was read from the tapeline, and the percentages of coverage by different organisms along the transect were calculated.

Table 1. Length (in cm) of transect lines.

		Stations	
	A	B	D
No. of measurements	46	41	44
Maximum length	1200	2200	1440
Minimum length	810	1860	1100
Average	1008	2002	1335
Standard deviation	92	89	89

The data collected before the operation of the power plant were tentatively used as the control to test against the data collected after the launch of the commercial operation. Data were analysed in two ways. First, coverages from the sequential observations on various macro-organisms were tested for randomness using the runs up and down test based on the assumption that the sequence followed a random walk (Sokal and Rohlf 1981). The ten-year succession of one organism at one station was also used to compare with that of its counterpart of other stations, and to study whether the successions between stations were correlated. Secondly, to reduce the risks of Type II errors, the data were pooled against months in a year cycle to show the yearly trend of the biota succession. Also, the monthly variations before and after the power plant operation were compared to study whether differences existed between temporal patterns.

Results

The composition of macro-organism communities varied between stations, as shown from data obtained in the survey (Fig. 2). The major faunistic and floristic components occupying space among the benthic communities of the study area are hard (stony) corals, soft corals and algae. Moreover, there is a certain proportion of substrates not occupied by any observable organisms (i.e. blanks). The coverages at the studied stations were all dominated by corals; Station A by hard corals such as *Platygyra* and *Pocillopora*; Station B by soft corals such as *Sarcophyton, Lobophytum* and *Sinularia*; and, Station D by hard corals such as *Favia, Favites* and *Platygyra*. This pattern of dominance persisted throughout the study. The algae found in the study area are characterized by a predominance of calcified species and filamentous algal turfs.

Sea urchins, giant clams, sea stars, crinoids (some are not sessile though) and polychaetes were the major type in the category of "other invertebrates". Apart from a few erratic mass occurrences of some invertebrates, in most of the surveys these marine animals occurred sporadically (Fig. 2). Parts of the substrate were free of sessile organisms. This was most conspicuous at the lower side of the block of Station A and on the top of the rocks of Station D.

Results from tests for randomness of successions of the macro-organisms in terms of percentage coverage show that except for the category of "other invertebrate", all the successions followed a random walk (Table 2). This characteristic persisted both before and after the launch of the operation.

On the other hand, the ten-year succession of most macro-organisms did not significantly correlate to each other between stations, as indicated in Table 3. Eleven of the fifteen correlation coefficients obtained from this treatment, including those for both hard corals and soft corals, are not significantly different from zero. Significant correlation was found in successions of both algae and "other invertebrates" between Stations A and B. The other two correlated successions occurred in the category of blanks, between Stations A and B, and A and D respectively (Table 3).

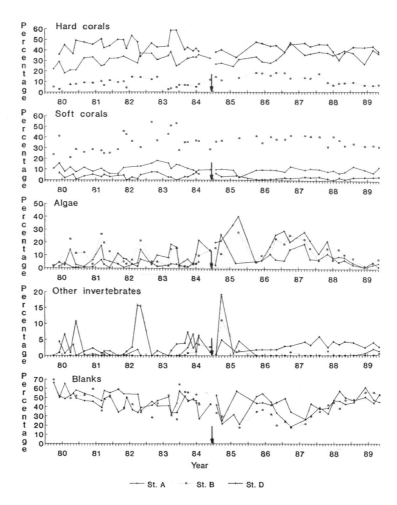

Figure 2. Successions of percentage coverage of various kinds of macro-organism surveyed at three stations. Arrow indicates the time the operation of the power plant was launched. Note scales on the Y-axes are not unified.

Monthly variations, obtained by arranging data points in accordance with an annual cycle, are shown in Figure 3. The coverage by algae varied seasonally. The seasonal patterns were similar between Stations A and B, where the algal coverage peaked in April-May, declined in summer, and was lowest during winter. Nevertheless, at Station D an additional algal bloom was found during October-November. Except for the seasonality in algal growth mentioned above, no consistent patterns on the monthly variation were found among macro-organism categories before and after the power plant operation.

Table 2. Results of tests for randomness of sequential observations based on runs up and down, where * denotes that the sequence of observations do not occur in a random order. Data were divided into two parts in this treatment. One includes those collected during July 1979 - June 1984 (before operation of the nuclear power plant). The other includes those collected between July 1984 and June 1989 (when the plant was in operation).

Station	Category	Before operation No.of Observ.	Prob.	In operation No.of Observ.	Prob.
	Hard corals	25	0.933	21	0.652
	Soft corals	25	0.566	21	0.928
A	Algae	25	0.935	21	0.928
	Other invertebrates	25	0.761	21	0.017 *
	Blanks	25	0.934	21	0.928
	Hard corals	25	0.566	20	0.164
	Soft corals	25	0.935	20	1.000
B	Algae	25	0.681	20	0.164
	Other invertebrates	25	0.001 *	20	0.696
	Blanks	25	0.163	20	0.781
	Hard corals	25	0.566	20	0.404
	Soft corals	25	0.935	20	0.164
D	Algae	25	0.681	20	0.404
	Other invertebrates	25	0.933	20	0.404
	Blanks	25	0.681	20	1.000

Since the majority of the successions followed a random walk (Table 2), and displayed no clear seasonality (Fig. 3), the data collected before and after June 1984, when operation of the plant was launched, were grouped separately to test for differences of the means. The results show that at Station A the mean percentage coverages of hard corals before and after the launch were significantly different (Table 4). The mean value was higher after the launch. Similar situations were found at Station B. By contrast, soft corals at these two stations showed no significant differences. Comparatively, at Station D the means for both hard corals and soft corals decreased. Other invertebrates also showed no significant difference between stations (Table 4). In view of the seasonality of algal growth (Fig. 3), comparisons based on the means were not made. However, it is clear that the percentage coverages increased at all the three stations after the launch (Fig. 3).

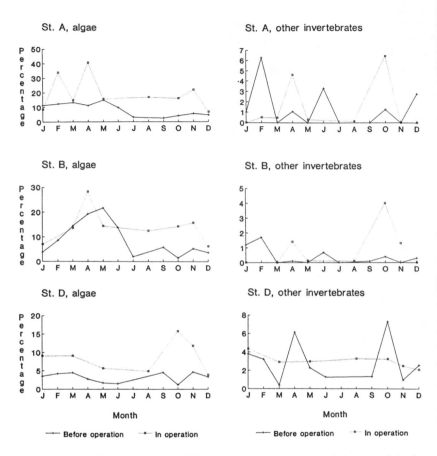

Figure 3. Monthly variations of the percentage coverage of algae and "other invertebrates". Note scales on the Y-axes are not unified.

Table 3. Correlation coefficients based on percentage coverages of different categories of macro-organism between stations (data pairs which include missing values are ignored). The original correlation matrices contain 15 coefficients. Only the significant non-zero coefficients are listed.

Algae between Stations A and B	r = 0.7238	p < 0.0001
Other invertebrates between Stations A and B	r = 0.8536	p < 0.0001
Blanks between Stations A and B	r = 0.7277	p < 0.0001
Blanks between Stations A and D	r = 0.3408	p = 0.0337

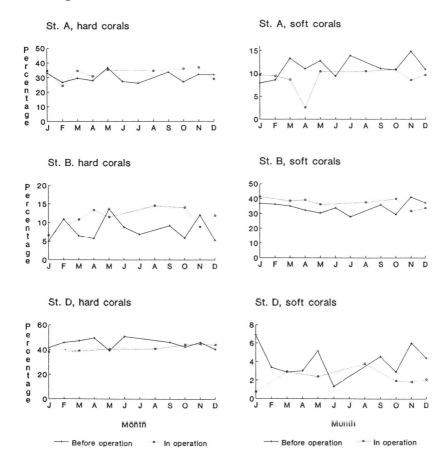

Figure 3 (continued). Monthly variations of the percentage coverage of hard corals and soft corals. Note scales on the Y-axes are not unified.

Discussion

In this study we found that except for the seasonality of the algal blooms, successions of macro-organisms in the past ten years are temporally variable. In addition, successions on different transects did not follow a common trend. In discussion of the factors underlying the variability of these successions, some aspects are considered to be important. For example, natural fluctuations of environmental factors have a profound influence on the biotic successions on the coral reefs (Huston 1985); interspecific interactions also play an important role in the variations of biotic succession (Williams 1981; Anderson 1986, p. 280); and, the successions could have been affected by impacts from the power plant operation.

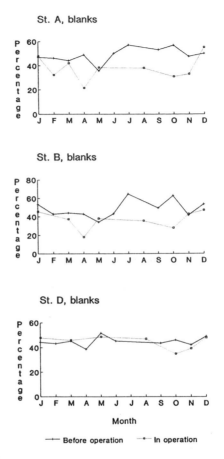

Figure 3 (continued). Monthly variations of the percentage coverage of blanks. Note scales on the Y-axes are not unified.

Southern Taiwan is located on a typhoon path. There have been 309 typhoons in the Taiwan area between 1897 and 1984; during the same period, the study area suffered direct hits by 1.17 typhoons per year (CWB 1985). Most of these typhoons occurred in summer and autumn. During 1986-1987, southern Taiwan experienced six typhoon hits. The physical disturbances caused by these typhoons have contributed much to the damage of the biota in the study area (Su *et al.* 1980 p. 16; Su *et al.* 1988, p. 50), and consequently increased the percentage coverage of the category "blank" found in some surveys in the present study. Such disturbance created open substrates for the colonization by macro-organisms (Connell 1978; Connell and Keough 1985; Huston 1985; Harmelin-Vivien and Laboute 1986), and hence might account for some of the variability of the percentage coverage of the various categories.

Table 4. Results of t-tests on the difference between means of the data collected before and after the operation of the power plant, where * denotes significant at 5 %; ** at 1%.

Station Category	Mean percentages Before	After	Difference	t-value	Prob.
No. of observations	25	21			
Hard corals	29.77	33.60	-3.83	-2.318	0.025 *
Soft corals	11.08	9.54	1.54	1.823	0.075
A Algae	not available				
Other invertebrates	1.67	1.36	0.31	0.300	0.765
Blanks	49.36	38.95	10.41	3.284	0.002 **
No. of observations	25	20			
Hard corals	8.01	12.07	-4.06	-3.418	0.001 **
Soft corals	34.64	37.41	-2.77	-1.243	0.221
B Algae	not available				
Other invertebrates	0.39	0.78	-0.39	-0.746	0.460
Blanks	47.65	36.23	-11.42	-3.281	0.002 **
No. of observations	25	20			
Hard corals	44.94	41.16	3.78	2.207	0.033 *
Soft corals	3.87	2.59	1.28	2.227	0.028 *
D Algae	not available				
Other invertebrates	3.21	3.01	0.20	0.203	0.840
Blanks	44.45	44.98	-0.53	-0.229	0.820

Moreover, the substratum is often a limiting resource for sessile marine invertebrates, and competition for substrate has been considered one of the major factors underlying community structure (Jackson and Buss 1975; Buss 1981; 1986; Goodall 1986). The competitive interaction between macro-organisms over substrate is most likely to occur in the non-typhoon seasons. In the present study, several types of interactions between alcyonacean (soft corals) and scleractinian corals (hard corals) occurred during their colonization. The interactions might include direct interference, overgrowth, and colony movement (Dai 1988). In the case of direct interference, alcyonacean corals are subordinate and frequently suffer from the attacks of scleractinian corals. The most abundant species of alcyonaceans at Station B, i.e. *Sarcophyton trocheliophorum, Sarcophyton crassocaule, Lobophytum sarcophytoides, Sinularia exilis* and *Sinularia facile*, often suffered severe tissue damage while contacting scleractinian corals. The damages included partial mortality of the polyps, tissue disintegration, and tissue necrosis (Dai 1988). By contrast, alcyonaceans, through their specific morphology and their faster growth rates, frequently could overgrow scleractinians, particularly the less aggressive species (Dai 1988). In the present study the coverage by hard corals at Stations A and B was higher after June 1984 than it was before. However, by putting this result and the occurrence of interspecific interactions together, it seems insufficient to relate the difference of the percentage coverage to the potential impact from the power plant operation (Dollar 1982).

Similar to findings of Gaines and Lubchenco (1982), the predominant algal fauna in the study area were calcified species and filamentous algal tufts. In addition, seasonality of algal growth was evident. Standing crops of algae may influence the succession of other macro-organisms (Dean and Connell 1987). For example, they may cover or overtop hard coral and consequently cause their death (Potts 1977). Algae may also entrap sediment and render substrate unsuitable for coral larvae to settle, or cause deterioration of the coral colony (Johannes 1975). Furthermore, filamentous algae can rapidly colonize newly denuded surfaces or vacant substrate and prevent the settlement of coral larvae (Morrissey 1980). Conversely, grazing activity by sea urchins and herbivorous fish might suppress algal monopolization of space (Ogden and Lobel 1978; Lewis 1986). Therefore, the coverage variation of macro-organisms other than algae is likely to be further affected by the seasonality of algal growth.

Seawater temperature is one of the physical factors most likely to be altered by the discharge of thermal effluent from the power plant, and hence may affect the colonization by macro-organisms. The majority of reef-building corals live between 23-28°C (Wells 1957). Extreme water temperature can be a limiting factor for the growth of corals (Sheppard 1982). In the study area, corals collected from the sea would bleach in water at 30-31°C in the laboratory, and die consequently. They would die quickly when water temperature reached 33°C (Yang *et al.* 1980).

Normally the temperature of the discharged water is 4-5°C higher than that of the ambient seawater near the outlet. However, this high temperature is subject to dilution from the local water movement. The tidal regime of southern Taiwan is semi-diurnal, with a cycle of 12 hours and 25 minutes. Spring tides alternate regularly with neap tides. In the study area, the mean spring tide range is 1.35 m and the mean neap tide range is 0.63 m. The tides have profound influences on current patterns of inshore waters of the study area. The current flows from east to west during flood tide and reverses during ebb tide. Partly due to the strong Kuroshio current, the current speed in the Nanwan Bay is relatively fast (Fan and Yu 1981). Hence, the dilution effect might be crucial in the alleviation of the acute impact from the thermal effluent. Even so, seawater temperature increased dramatically around the outlet of the discharge canal. In summer, the temperature measured in the water 30 m away from the outlet of the canal could reach as high as 36.5°C (Su *et al.* 1988). Meanwhile, a temperature gradient was formed in the water column, with the upper layer warmer than the lower layer. In the summers of 1987 and 1988, bleached corals were widely found around the outlet of the discharge canal (Hung 1989). Most corals distributed in the depth zone of 0-3 m were found dead after the bleaching event. The unusually high water temperatures have been thought to be responsible for the high mortality of corals (Su *et al*. 1988). By contrast, very few corals bleached in the summer of 1989, despite the continual operation of the plant (personal observation).

In the present study, the transects were located at a depth of more than 5 m. At such depths influence of thermal discharge on the water temperature might not be significant. For example, seawater temperatures recorded during 1979-1988 at a depth of 3 m adjacent to Station B were in the range between 22°C and 30°C (Su *et al*. 1988, p. 337). Surprisingly, the peak temperature, which almost reached the upper tolerance limit for coral survival, occurred in the period before the operation of the power plant rather than after it. Accordingly, it is not evident that water temperatures might be responsible for the change of some of the succession patterns of the macro-organisms.

As conditions are ever changing, communities are never stationary (Williamson 1987). It should be therefore emphasised that if the plant operation has an impact on the environment, the impact is unlikely to occur independently. It is likely to have been broadly integrated with the natural variations. Thus, the environmental pressure related to succession of the macro-organisms may be complex. Overall, with the inconsistency between trends of macro-organism coverage on transects, it is suggested that information provided so far, by results from this part of the study, is inconclusive; the same applies to findings from studies on hydrographical (except seawater temperature) and chemical properties, phytoplankton, zooplankton, and fish community (Su *et al*. 1988). Currently, this study is still in progress, as information obtained from such a study might be essential in the elucidation of the future long-term effect of the power plant operation.

Acknowledgements

Many thanks are due to Dr. C. P. Chen for his help in the initiation of this study; Y. S. Chen, G. C. Chen, C. C. Chen and M. S. Jeng for help with fieldwork. Dr. C.-F. Dai kindly loaned us references; we are also grateful for his comments on the manuscript. This project was organized by the National Scientific Committee on Problems of the Environment (SCOPE/ROC), and financially supported by the Atomic Energy Council of the Republic of China.

References

Anderson, D.J. (1986) Ecological succession. In J. Kikkawa and D. J. Anderson (eds.) *Community Ecology*, 269-285. Blackwell Scientific Publications.
Buss, L.W. (1981) Group living, competition, and the evolution of cooperation in sessile invertebrates. *Science*, **213**, 1012-1014.
Buss, L.W. (1986) Competition and community organization on hard surfaces in

the sea. In J. Diamond (ed.) *Community Ecology*, 517-546. Harper and Row.

Connell, J.H. (1978) Diversity in tropical rainforests and coral reefs. *Science, N.Y.*, **199**, 1302-1310.

Connell, J.H. and Keough, M.J. (1985) Disturbance and patch dynamics of sub-tidal marine animals on hard substrata. In S.T.A. Pichett and P.S. White (eds.) *The Ecology of Natural Disturbance and Patch Dynamics*, 125-151. Academic Press, New York..

CWB (1985) *Questions on Typhoons*. Central Weather Bureau Publication, Taipei. 28-31 (in Chinese).

Dai, C.-F. (1988) *Community ecology of corals on the fringing reefs of southern Taiwan*. Ph.D dissertation, Yale University.

Dean, R.L. and Connell, J.H. (1987) Marine invertebrates in an algal succession. I. Variation in abundance and diversity with succession. *Journal of Experimental Marine Biology and Ecology*, **109**, 249-273.

Dollar, S.J. (1982) Wave stress and coral community structure in Hawaii. *Coral Reefs*, **1**, 71-81.

Fan, K.L. and Yu, C.Y. (1981) A study of water masses in the seas of south-ernmost Taiwan. *Acta Oceanographica Taiwanica*, **12**, 94-111.

Gaines, S.D. and Lubchenco, J. (1982) A unified approach to marine plant-herbivore interactions. II. Biogeography. *Annual Review of Ecology and Systematics*, **13**, 111-138.

Goodall, D.W. (1986) Biotope structure and patterning. In J. Kikkawa and D.J. Anderson (eds.) *Community Ecology*, 30-40. Blackwell Scientific Publications.

Harmelin-Vivien, M.L. and Laboute, P. (1986) Catastrophic impact of hurricanes on atoll outer reef slopes in the Utmost (French Polynesia). *Coral Reefs*, **5**, 55-62.

Hung, T.-C (1989) The preliminary report for the assessment of ecological impact of the operation of the third nuclear power plant on the Nanwan Bay. National Scientific Committee on the Problems of Environment, Academia Sinica. 34pp. (in Chinese).

Huston, M.A. (1985) Patterns of species diversity on coral reefs. *Annual Review of Ecology and Systematics*, **16**, 149-177.

Jackson, J.B.C. and Buss, L.W. (1985) Adaptive strategies of coral reef invertebrates. *American Scientist*, **73**, 265-274.

Johannes, R.E. (1975) Pollution and Degradation of Coral Reef Communities. In E.J. Ferguson-Wood and R.E. Johannes (eds.) *Tropical Marine Pollution*, 13-51. Elsevier, New York.

Jones, O.A., Randall, R.H., Cheng, Y.M., Kami, H.T. and Mak, S.M. (1972) *A marine biological survey of southern Taiwan with emphasis on corals and fishes*. Institution of Oceanography, National Taiwan University. Special Publication No. 1, 93pp.

Lewis, S.M. (1986) The role of herbivorous fishes in the organization of a Caribbean reefy community. *Ecological Monograph*, **56**, 183-200.

Morrissey, J. (1980) Community structure and zonation of macroalgae and corals on a fringing reef of magnetic Island (Queensland, Australia). *Aquatic Botany*, **8**, 91-139.

Ogden, J.C. and Lobel, P.S. (1978) The role of herbivorous fishes and urchins in coral reef communities. *Environmental Biology of Fishes*, **3**, 49-63.

Potts, D.C. (1977) Suppression of coral populations by filamentous algae within damselfish territories. *Journal of Experimental Marine Biology and Ecology*, **28**, 207-216.

Sheppard, C.R.C. (1982) Coral populations on reef slopes and their major controls. *Marine Ecology Progress Series*, **7**, 83-115.

Sokal, R.R. and Rohlf, F.J. (1981) *Biometry*. 2nd ed. W.H. Freeman and Company, New York.

Su, J.-C., Hung, T.-C., Chiang, Y.-M., Tan, T.-H., Chang, K.-H., Yang, R.-T., Cheng, Y.-M., Fan, K.-L. and Chang, S.-D. (1980) An ecological survey on the waters adjacent to the nuclear power plant in southern Taiwan. I. The progress report of the first year study (1979-1980). (Before operation of the plant). National Scientific Committee on the Problems of Environment, Academia Sinica, Special Publication No. 7, 115pp. (in Chinese).

Su, J.-C., Hung, T.-C., Chiang, Y.-M., Tan, T.-H., Chang, K.-H., Shao, K.-T., Huang, P.-P., Lee, K.-T., Huang, C.-Y., Fan, K.-L. and Yeh, S.-Y. (1988) An ecological survey on the waters adjacent to the nuclear power plant in southern Taiwan. IX. The progress report of the ninth year study (1987-1988) and the preliminary report for the assessment of ecological impact of the operation of the power plant (July 1979 - June 1988). National Scientific Committee on the Problems of Environment, Academia Sinica, Special Publication No. 59, 394pp. (in Chinese).

Wells, J.W. (1957) Coral reefs. *Geological Society of America*. Memoir **67**, 609-631.

Williams, A.H. (1981) An analysis of competitive interactions in a patchy back-reef environment. *Ecology*, **61**, 1107-1120.

Williamson, M. (1987) Are communities ever stable? In A.J. Gray, M.J. Crawley and P.J. Edwards (eds.) *Colonization, Succession and Stability*, 353-371 Blackwell Scientific Publications, Oxford, London.

Yang, R.-T., Yen, S.-Z. and Sun, C.-L. (1980) Effects of temperature on reef corals in the Nay-wan Bay, Taiwan. Institution of Oceanography, National Taiwan University. Special Publication No. 23, 27pp. (in Chinese).

The Use of Indicator Organisms for the Protection of Recreational Users of Estuarine and Coastal Waters from Risks to Health

Boyle, O.C.[1], Masterson, B.F.[2] and Stapleton, L.[3]

[1]Department of the Environment, Custom House, Dublin 1, Ireland.
[2]Environmental Institute, University College, Belfield, Dublin 4, Ireland.
[3]Environmental Research Unit, St. Martin's House, Dublin 4, Ireland.

Key words: indicator organism, health risk, swimmer, marine, bathing-water quality, epidemiology.

Abstract

The international primary literature concerning the risk to health for recreational users of marine waters is reviewed in the context of a water quality management plan being developed for Dublin Bay. Clinical reports of illness associated with the use of marine waters are few, and the illnesses are mostly of a minor nature. There is some small risk of contracting such minor illness from bathing in marine waters, even where levels of sewage contamination are within strict regulatory limits. The risk, although not alarming, is greater with higher levels of sewage contamination. Epidemiological studies which test association between indicator organism levels in marine waters and hazard to the health of recreational users are reviewed, and it is argued that the evidence for association is weak. Uncertainties which have compromised the interpretation of the results of the epidemiological studies are considered; these relate to water sampling, organism isolation and enumeration, and measurement of illness incidences.

Novel emerging techniques based on the new biotechnologies are referred to, and these may enable in the future, the direct quantitation of pathogenic microorganisms and viruses in marine waters. The growing public imperative for improvement of all aspects of water quality is adverted to.

Introduction

Bays, estuaries and other sea inlets have traditionally been used for the disposal of sewage and other effluents from onshore settlements and developments.

Bioindicators and Environmental Management
ISBN 0-12-382590-3

In the present century, the dramatic increase in the quantities of sewage and industrial effluents has led to significant pollution problems in various parts of the developed world. Parallel trends have included growing recreational demands for bathing, boating and other water sports, a growing appreciation of the wildlife conservation value of many such inlets, and a growing environmental awareness generally. The demands being made on the marine environment are many, and equally there are few if any uses that do not at least potentially conflict with other uses. Management strategies are needed that will anticipate and minimize such conflict. In Ireland the Water Pollution Act 1977 empowers the local authority for an area to make a Water Quality Management Plan (WQMP) for any waters situated in its functional area, or which adjoin that area. Currently the Environmental Research Unit (ERU) of the Department of the Environment is developing a WQMP for Dublin Bay on behalf of the relevant local authorities.

Sewage enters Dublin Bay at a number of discharge points; details are given in Table 1. Engineering works to divert those of the southern bay to the central treatment facility at the mouth of the Liffey are in progress or are planned.

Microbiological monitoring surveys have made it clear that discharged sewage causes some contamination of recreational sites in Dublin Bay, and recently developed hydrological models support this. Recreational users of Dublin Bay have expressed concern about published microbiological data

Table 1. Details of sewage discharges impacting on Dublin Bay.

Location	Estimated present contributing population	Sewage treatment; outfall construction; discharge regime
Nose of Howth	310 000	Screening; seabed without diffuser; continuous
Ringsend	613 000	Preliminary & primary sedimentation; estuarine; continuous to power-station cooling water outfall
West Pier Dun Laoghaire*	54 500	Tidal tank; seabed without diffuser; discontinuous
Bullock Harbour*	16 500	Tidal tank; seabed without diffuser; discontinuous
Shanganagh	47 000	Comminution & screening; sea outfall with diffuser (2 km; 1200 mm diameter); continuous

* To be diverted to the Ringsend treatment works.

(Clarke *et al.* 1985), and about the significance of patent aesthetic problems such as those derived from algal growths, visibly ailing wildlife, and beach litter. There is an evident public perception that the use of the bay is gravely compromised by sewage pollution, which has given rise to uncertainty about the risk of infection to bathers and other contact users. Also, sightings of sewage-derived materials from time to time in the water and on the beaches have had a substantial off-putting effect on members of the public.

As a response to public concern about health hazards an up-to-date review of the international primary literature concerning the risks to health for recreational users of marine waters, including swimming and other contact activities has been completed as part of a programme of investigations and studies being undertaken for the WQMP. The retrieved literature items have been assembled as a data base allowing detailed literature analysis.

Clinical reports of swimming-related illness

Survey of the literature reveals only a very small number of outbreaks or cases of illness associated with use of recreational waters which have been reported in a clinical context and brought under medical surveillance. It should be noted, however, that a proportion of outbreaks of a minor nature may not be reported or their origin may not be identified. Infections such as **leptospirosis**, arising from animal faecal contamination, and which are primarily encountered by freshwater users, are not considered here. There is increasing concern about **Campylobacters**, as agents of infective diarrhoea (Franco 1989; Skirrow 1987); although widely found in recreational waters, there is no evidence to date to implicate these sources with human infection (Jones *et al.* 1990b; Mawer 1988a; 1988b). The illnesses reviewed may be categorized as gastrointestinal (enteric) infections, respiratory infections, eye, ear, and skin infections, and wound infections (Table 2).

(a) Enteric outbreaks may be associated with contamination of the water by human faeces either arising from sewage pollution, or from "interbather" contamination, that is by the shedding, as Robinton and Mood (1966) have demonstrated, of faecal organisms by the bathers themselves.

There have been a small number of minor **typhoid** outbreaks as listed by Shuval (1988), all of which involved fresh or marine waters at sites which were grossly polluted with sewage to an extent unlikely to be encountered in recreational waters (close to sewage outfalls for instance).

There are only a few putative cases of the transmission of **poliovirus** infection to bathers, which are considered inconclusive (Mosley 1975). Cabelli (1989) states that there are no cases, in accord with the findings of Moore (1959).

The sole reported outbreak of **hepatitis-A** (Bryan *et al.* 1974) occurred among a scout troop camped at a fresh water lake, but it is likely that the lake-water was used by mistake for drinking purposes.

Table 2. Infections considered with regard to recreational water use.

(a) Enteric Infections	
Typhoid	Shuval (1988) *
Polio Virus	Mosley (1975) *
Hepatitis-A	Bryan *et al.* (1974)
Shigellosis	Makintubee *et al.* (1987)
	Rosenberg *et al.* (1976)
Coxsackievirus	Shuval (1988)
	Denis *et al.* (1974)
	Hawley *et al.* (1973)
Norwalk Virus	Baron *et al.* (1982)
Enterovirus-Like	Lenway *et al.* (1989)
	Holmes *et al.* (1989)
(b) Respiratory, Eye, Ear, and Skin Infections	
Adenovirus	D'Alessio *et al.* (1981)
	D'Angelo *et al.* (1979)
	Foy *et al.* (1968)
	Kaji *et al.* (1961)
Otitis Externa	Sullivan and Barron (1989) *
	Springer and Shapiro (1985) *
	Seyfried and Cook (1984)
	Calderon and Mood (1982)
	Reid and Porter (1981)
	Alcock (1977) *
	Weingarten (1977)
	Hoadley and Knight (1975)
Swimmers'-Itch	Eklu-Natey *et al.* (1985)
(c) Wound Infections by Autochthonous Pathogenic Bacteria	
	Auerbach (1987)
	Delbeke *et al.* (1985)
	Larsen *et al.* (1981)
	Smith (1980)
	Joseph *et al.* (1979) *
	Fulghum *et al.* (1978) *
	Hanson *et al.* (1977)

* Marine waters

Only two small swimming-associated outbreaks of **shigellosis**, a bacterial cause of gastroenteritis, have been reported, one at a lake (Makintubee *et al.* 1987) for which no source of infective pollution was identified, and one at a stretch of river (Rosenberg *et al.* 1976) for which very high levels of sewage pollution were evident.

Three incidences in the 1960's of **coxsackievirus** gastroenteritis outbreaks on a small scale in wading and swimming pools are listed by Shuval (1988). A small outbreak for five children who bathed in a lake was reported by Denis *et al.* (1974), and a larger one by Hawley *et al.* (1973) in which 33 campers at a boys' summer lakeside camp became infected; however, the evidence that these outbreaks were swimming related has been criticised (Mosley 1975).

Baron *et al.* (1982) reported an outbreak involving over 300 persons who suffered gastroenteritis most likely induced by **Norwalk** virus infection. They claimed that infection was significantly associated with swimming in an artificial lake in the recreational park visited by the patients, although the source of pollution was not discovered.

An **enterovirus-like** outbreak at a community wading pool was reported recently by Lenway *et al.* (1989), which was attributed to ineffective pool-water chlorination, and a similar outbreak was reported by Holmes *et al.* (1989) at a swimming pool and spa.

(b) The infective organisms causing respiratory, eye, ear, and skin infections are unlikely to be introduced to sewage to any significant degree by infected members of the community. Outbreaks are unlikely to be sewage related; rather, the infective sources are likely, therefore, to be found among the users themselves. Though there are some reports of outbreaks suggesting a relationship with sewage contamination, these are for waters which were probably heavily polluted.

There have been frequent reports of outbreaks of throat and eye infections caused by **adenovirus** (for example, Kaji *et al.* 1961; Foy *et al.* 1968; D'Angelo *et al.* 1979). Such incidents, mainly in swimming pools, are usually found to be associated with the use of inadequately chlorinated water, which facilitated interbather infection.

With regard to acute viral infections, D'Alessio *et al.* (1981) demonstrated a significant association in children under 16 years of age between swimming at freshwater beaches and incidence of respiratory illness (and gastroenteritis). The degree of beach pollution may have been understated, so that the question of whether illness arose through sewage contamination or through interbather infection remains unsettled.

Otitis externa, an outer-ear infection, has been long associated with recreational water use (Hoadley and Knight 1975; Reid and Porter 1981), and the most prominent causative organism is *Pseudomonas aeruginosa* (Calderon and Mood 1982). Its sustained presence in swimming pools has been attributed to poor chlorination practice (Weingarten 1977). The infection is readily transmitted between individuals via the water, and also in the commercial diving environment via the person-to-person contact route (Alcock 1977). For frequent swimmers, the condition is likely to be encountered more in fresh than in swimming-pool and sea water (Springer and Shapiro 1985). Studies by Seyfried and Cook (1984) have indicated that *P. aeruginosa* can reside in lake-water sediments and cause infections; possibly the organisms may be excreted into sewage in small amounts and multiply there, and subsequently become autochthonous in the sewage-contaminated lake sediment. Among illnesses arising from recreational wateruse otitis externa is the one most frequently observed by clinicians. "Honest case" evidence of this has been reported by Sullivan and Barron (1989) with lifeguards receiving compensation awards for work-related illness claims.

Although a sizeable outbreak of **"swimmers'-itch"** in a large lake has been reported (Eklu-Natey *et al.* 1985), the occurrence of skin ailments is rare, and in illness surveys low incidences are usually recorded.

(c) Traumatic injuries incurred in the aquatic environment may become infected by autochthonous pathogenic bacteria present as part of the natural microbial fresh or marine water ecology (Larsen *et al.* 1981; Auerbach 1987). There are only a few reported cases (Delbeke *et al.* 1985; Smith 1980; Joseph *et al.* 1979; Fulghum *et al.* 1978; Hanson *et al.* 1977), and such wound infections are regarded as rare (Cabelli 1989).

The above reports represent clearcut clinical incidences of illnesses associated with the use of recreational water, which are mostly of a minor nature. The majority of reports evidence the transmission of infection between bathers, especially in poorly managed swimming pools (Galbraith 1980). The firm implication of sewage as a source of infection is rare, only arising where a major degree of sewage pollution existed; as an example, DeWailly *et al.* (1986) showed an association between gastroenteritis, ear, eye and skin infections among competitive windsurfers using waters judged unsafe for swimming. The burden of the evidence is that reported outbreaks at marine water sites are very rare indeed. It is clear though, that some risk attaches to the use of marine recreational waters, however small that risk may be. The risks may be somewhat different for different user categories, such as windsurfers and lifeguards considered in reports above, and for scuba divers who are protected by diving suits and hoods (Coolbaugh *et al.* 1981), and for snorklers (Philipp *et al.* 1985). It should be noted that diseases which are not present in the community, such as typhoid or poliomyelitis, cannot be transmitted by sewage contamination, and where there are low levels in the community this is reflected in correspondingly low levels of risk. The question is can risk to health, even though small, be quantified?

Quantifying the risk to health

In the case of normal incidences of human disease encountered in developed countries, the levels of infectious organisms found in sewage receiving waters are extremely low, and for those particular pathogens for which methods of enumeration are available, the levels are generally below those measurable. The practical approach is to rely on "Indicator Organisms" (Elliot and Colwell 1985; Olivieri 1982) which are shed with human faeces into sewage, and survive in sufficient numbers to be measurable by standard methods. The rationale is that recorded levels of these organisms represent or index the levels of the accompanying pathogens. Many organisms have been proposed as suitable for this role, and many studies have been undertaken to assess their capabilities as indicators of infective sewage contamination of marine waters.

The challenge of finding an ideal indicator organism for marine conditions is a considerable one; relevant published criteria (Table 3) have been reviewed by Dufour (1984).

While all of the criteria are important, the most important and cogent criterion which supersedes others, is that measured levels of the chosen indicator organism should correlate with the observed levels of illness as determined by epidemiological survey. Referring to epidemiological data from studies conducted in the United States of America, Dufour (1984) concluded that enterococci measurements best met this criterion for marine waters, and this view has been maintained by Cabelli (1989). However, the results from studies elsewhere (Table 4) are not always consistent with this; for this reason the case for widespread application of any of the present models which associate indicator organism levels to illness incidences must be regarded as weak. The studies fail to meet in varying degrees the classical nine viewpoints of Hill (1965), which he considered supportive of a cause-effect relationship between two associated variables. The differences which have arisen can be attributed at least in part to shortcomings in technical aspects of the studies; some of the relevant aspects are now considered.

Table 3. Ideal indicator organism characteristics.*

Be present where pathogens are.

Occur in greater numbers than pathogens.

Be unable to replicate in aquatic environment.

Survive as well in aquatic environment as pathogens.

Be more resistant to disinfection than pathogens.

Be easy to isolate and enumerate.

Be applicable to all types of water.

Be exclusively associated with sewage only.

Indicator density should correlate with
  - degree of faecal contamination,
  - degree of health risk.

---

* After Dufour (1984).

Table 4. International reports which associate marine illness incidences with indicator organism levels.

Illness/Indicator association	Report	Country
Enteric		
Enterococci	Cabelli *et al.* (1982)	USA
	El-Sharkawi and Hassan (1982)	Egypt
	Fattal *et al.* (1986)	Israel
Escherichia coli	El-Sharkawi and Hassan (1982)	Egypt
	Fattal *et al.* (1986)	Israel
	Holmes (1989)	Hong Kong
Staphylococci	Fattal *et al.* (1986)	Israel
	Holmes (1989)	Hong Kong
Respiratory, Ear, Eye, Skin		
Escherichia coli	Holmes (1989)	Hong Kong
Staphylococci	Holmes (1989)	Hong Kong

1. Uncertainties about sampling in marine conditions

The sampling protocol must be adequate to counter variability (El-Shaarawi and Pipes 1982) from causes such as are outlined below. The sewage discharge characteristics can be variable, for instance because of compositional variations or tidal influences, and the dispersion may be heterogeneous on account of hydrological variations driven by tides and weather. Mallmann (1962) found that in swimming pools, pollution from bathers occurred in a non-homogeneous way as discrete pockets of contamination, with consequent difficulties for the design of proper sampling procedures. The die-off rates of the indicator organisms, upon which relationships with pathogen numbers and water infectivity depend, are greatly influenced by variations in sunlight levels, either seasonal or brought about by changes in the water turbidity; likewise variations in water temperatures are influential (Evison and Tosti 1980b; Vasl *et al.* 1981). Weather conditions or activity by bathers may cause resuspension of sediment, producing organism levels which are untypical and less amenable to control (Seyfried *et al.* 1985; Foster *et al.* 1971). The bathers themselves can be a significant source of infective water contamination, so that the sampling should be done on and during days when recreational use is prevalent, and numbers using the water (bather load) at the time of sampling should be considered (Sherry 1986; Brenniman *et al.* 1981; Geldreich 1974; Foster *et al.* 1971). In deep-water sites, as used by scuba divers, multi-depth sampling may be required because of stratification effects on the water column (Garber 1956). Organisms may be introduced sporadically from small local sources such as septic tanks and passing pleasure craft, or from storm drains carrying land run-off after heavy rain (Geldreich 1970).

The sampling protocol should be sufficiently randomized and of sufficient frequency to permit proper application of statistical techniques (O'Kane 1983). Additionally, data processing techniques such as the use of the mean, mode, median, or the geometric mean of replicate measurements, and of logarithmic transformations should have a rational basis which best represents the level of pathogens present in the water, or its infectivity.

2. Uncertainties about organism isolation and enumeration

Standard methods have been adopted for the isolation and enumeration of many of the indicator organisms used for water quality monitoring, giving fully explicit experimental instructions (Clesceri *et al.* 1989). For some organisms however, the methods are less well described, and as interlaboratory comparisons are apparently rare for such analyses, some doubt must attach to comparisons of the results of studies conducted in laboratories at a distant national or international remove from each other. Numerical relationships between indicator organism and infection levels are influenced considerably by treatment, such as chlorination, which the sewage may have received before discharge (Jones *et al.* 1990a). Likewise, variable environmental stress-effects may cause some organisms though viable, to become nonculturable (Roszak and Colwell 1987), and consequently a distortion of the infectivity relationship may be generated. Antibiotic-resistant bacteria are widely found in the aquatic environment (Baya *et al.* 1986; Jones *et al.* 1986). The proportion of resistant bacteria in sewage effluent may be increased by treatment such as chlorination (Murray *et al.* 1984), and they have demonstrated higher survival rates. Their presence in marine waters is long known (Smith 1971), and the danger of spread of antibiotic resistance in the human community by bathers has been adverted to (Grabow *et al.* 1974). These observations would suggest that morbidity associated with the use of polluted marine waters might relate to the degree of antibiotic resistance of infective organisms present, but this point does not seem to have been investigated and is not pursued further here.

3. Uncertainties about the measurement of illness incidences

The epidemiological study seeks to correlate measured indicator organism densities with degree of water infectivity determined by measurement of the level of illness incidences. Threshold or "acceptable" levels of illness can be set, allowing the corresponding indicator organism densities to be declared as water quality standards. Shortcomings in the approach are increasingly being identified. The most widely accepted studies have been prospective, that is, designed as follow-up studies in which a cohort including recreational water users and non-swimmers were interviewed on the beach and then queried some days later concerning any subsequent illness experience. Jones *et al.* (1990a) have referred to these as "perception" studies, to advert to the subjectivity of personal illness assessment by the interviewees. Although efforts have been made to establish illness credibility levels in some studies, the more rigorous approach would be to have medical confirmation of illnesses, along the lines of D'Alessio *et al.* (1981). The exposure of users has generally been quantified unsatisfactorily , usually just in terms of the incidence or non-incidence of "head immersion" (Brown *et al.* 1987; Cabelli *et al.* 1979). The water "dose" would depend on the degree and duration of body contact and extent of water ingestion, and eludes easy determination especially in the case of young children. Other influential factors may be listed, such as the exacerbating effects on infection levels of pathogen transmission via resuspended sediments and surf aerosols (Baylor *et al.* 1977), and the differing immunological competencies of young and old and of local and visiting users.

In the light of the many difficulties at all levels which afflict efforts to quantify health risks, variously reviewed by other authors (Godfree *et al.* 1990; Jones *et al.* 1990a; Lacey and Pike 1989; Moore 1975) it is hardly surprising that there is as yet no ubiquitously adopted indicator organism or group of organisms for recreational marine waters, nor agreement on the quantitative relationship between organism densities and risk of illness. This is evident from the results of a number of elaborate epidemiological studies of marine waters which have been undertaken in various parts of the World, and which are summarized in Table 4. Other studies have been reported which are less competent by design in quantifying the health risks. Studies in the USA (Coye and Goldoft 1989; Stevenson 1953) and in France (Foulon *et al.* 1983) did not discover any firm association between indicators and illnesses, perhaps reflecting the greater difficulty in detecting such an association where the indicator densities encountered complied with adopted water quality standards. A study in the United Kingdom found that incidences of enteric fever (typhoid and paratyphoid) were attributable to swimming at marine beaches only where gross sewage pollution was present (Moore 1959); the study was not concerned with relatively minor, non-notifiable illnesses of current concern. In a smallscale study Philipp *et al.* (1985) found a high incidence of gastroenteritis symptoms among snorkel swimmers competing in dockland waters. Brown *et al.* (1987) reported increased levels of gastroenteritis among swimmers at "highly polluted" beaches in the United Kingdom; indicator organism levels were not given. Comprehensive illness studies were conducted at Spanish beaches (Mujeriego *et al.* 1982), but water quality data were not associated adequately with the results. Studies improved in design are underway (Jones *et al.* 1990a), which are aimed at eliminating at least some of the previous uncertainties.

It is probable that in the not too distant future, new biotechnological approaches to microbiological monitoring, including molecular biological and novel cytometric techniques (Table 5) will enable direct identification and enumeration of pathogens in marine waters, and so reduce reliance on the use of indicator organisms. New devices based on one of these techniques or a number in combination, are under development. It is probable that some of the approaches at least will be successful, and will meet the design objective of permitting onsite, direct pathogen detection and enumeration. However, the need to develop proper sampling-protocol design and epidemiological survey techniques will remain, but at least procedures for monitoring pathogens directly should make studies in the future less difficult to undertake.

Conclusion

Unless serious illness is endemic in the population, there is no appreciable risk of contracting such serious illness from bathing in marine waters, but the risk is increased if a sewage outfall is nearby or the water is otherwise grossly contaminated with sewage. There is some small risk of contracting minor illness from bathing in marine waters even where levels of sewage contamination are within strict regulatory limits. The risk, although not alarming, is greater with higher levels of sewage contamination. The case that indicator organism densities can index this low risk is weak, because the results of epidemiological studies are inconsistent and there is no agreement on which organism best serves this function, and because various shortcomings in technical aspects of the studies frustrate clearcut interpretations. New techniques and improved study protocols are expected to improve matters in the future, but for the present, monitoring of indicator organisms is useful, at least as a management tool (Wolfe 1980), to perceive sewage pollution at recreational water sites. It should be sufficient and cost-effective, to monitor for one selected indicator organism where no additional information is yielded by monitoring for two or more (Pedersen *et al.* 1980;

Table 5. New emerging techniques for marine pathogen monitoring.

Technique	Reference
Epiflourimetry	Pettipher (1989); Xu *et al.*(1984)
Image Analysis	Singh *et al.* (1989)
Gene Probes/Polymerase Chain Reaction	Gerba *et al.* (1989); Hurst *et al.* (1989); Gerba *et al.* (1988); Ogram & Sayler (1988)
Monoclonal Antibodies	Joret *et al.* (1989)
Biosensors	Hall (1990)
Laser Cytometry	Jensen and Horan (1989)

Evison and Tosti 1980a). The sanitary status of shellfish should be borne in mind in any event; the presence of pathogens in shellfish would signal persistence water contamination, even where the organisms were not detectable in the water body (Hugues *et al.* 1988).

It may indeed be argued that the risks of illness associated with the use of sewage contaminated recreational waters are trivial in comparison to general community health problems, and that the concern about monitoring is unwarranted and the costs involved unjustified. Nevertheless, public expectations of environmental quality are increasingly expressed in eclectic terms, so improvements in the sanitary and aesthetic standards for amenity waters are strongly demanded and associated with enhanced quality of life. In turn, it can be expected that public perception of these improvements will increase the use and enjoyment of recreational marine waters (Moore *et al.* 1979).

Acknowledgements

We wish to thank David Kay and Frank Jones (Cardiff), and Paul Toner (Dublin) for assistance and useful discussion, and Catriona Scaife for help with the preparation of the SPIRES data base.

References

Alcock, S.R. (1977) Acute otitis externa in divers working in the North Sea: a microbiological survey of seven saturation dives. *Journal of Hygiene*, **78**, 395-409.

Auerbach, P.S. (1987) Natural microbiologic hazards of the aquatic environment. *Clinics in Dermatology*, **5**, 52-61.

Baron, R.C., Murphy, F.D., Greenberg, H.B., Davis, C.E., Bregman, D.J., Gary, G.W., Hughes, J.M. and Schonberger, L.B. (1982) Norwalk gastrointestinal illness: an outbreak associated with swimming in a recreational lake and secondary person-to-person transmission. *American Journal of Epidemiology*, **115**, 163-172.

Baya, A.M., Brayton, P.R., Brown, V.L., Grimes, D.J., Russek-Cohen, E. and Colwell, R.R. (1986) Coincident plasmids and antimicrobial resistance in marine bacteria isolated from polluted and unpolluted Atlantic ocean samples. *Applied and Environmental Microbiology*, **51**, 1285-1292.

Baylor, E.R., Baylor, M.B., Blanchard, D.C., Syzdek, L.D. and Appel, C. (1977) Virus transfer from surf to wind. *Science*, **196**, 575-580.

Brenniman, G.R., Rosenberg, S.H. and Northrop, R.L. (1981) Microbial sampling variables and recreational water quality standards. *American Journal of Public Health*, **71**, 283-289.

Brown, J.M., Campbell, E.A., Rickards, A.D. and Wheeler, D. (1987) Sewage pollution of bathing waters. *Lancet II*, **(8569)**, 1208-1209.

Bryan, J.A., Lehmann, J.D., Setiady, I.F. and Hatch, M.H. (1974) An outbreak of hepatitis-A associated with recreational lake water. *American Journal of Epidemiology*, **99**, 145-154.

Cabelli, V.J. (1989) Swimming-associated illness and recreational water quality criteria. *Water Science and Technology*, **21**, 13-21.

Cabelli, V.J., Dufour, A.P., Levin, M.A., McCabe, L.J. and Haberman, P.W. (1979) Relationship of microbial indicators to health effects at marine bathing beaches. *American Journal of Public Health*, **69**, 690-696.

Cabelli, V.J., Dufour, A.P., McCabe, L.J. and Levin, M.A. (1982) Swimming-associated gastroenteritis and water quality. *American Journal of Epidemiology*, **115**, 606-616.

Calderon, R. and Mood, E.W. (1982) An epidemiological assessment of water quality and swimmer's ear. *Archives of Environmental Health*, **37**, 300-305.

Clarke, J., Masterson, B., Max, M. and O'Connor, P. (1985) *A Study of the Impact of Sewage Discharges on the Recreational Quality of South Dublin Bay.* Comhairle fo Thuinn, Dublin.

Clesceri, L.S., Greenberg, A.E. and Trussell, R.R. (eds.) (1989) *Standard Methods for the Examination of Water and Wastewater.* 17th Edition, American Public Health Association, Washington DC.

Coolbaugh, J.C., Daily, O.P., Joseph, S.W. and Colwell, R.R. (1981) Bacterial contamination of divers during training exercises in coastal waters. *Marine Technology Society Journal*, **15**, 15-21.

Coye, M.J. and Goldoft, M. (1989) Microbiological contamination of the ocean, and human health. *New Jersey Medicine*, **86**, 533-538.

D'Alessio, D.J., Minor, T.E., Allen, C.I., Tsiatis, A.A. and Nelson, D. B. (1981) A study of the properties of swimmers among well controls and children with enterovirus-like illness shedding and not shedding an enterovirus. *American Journal of Epidemiology*, **113**, 533-541.

D'Angelo, L.J., Hierholzer, J.C., Keenlyside, R.A., Anderson, L.J. and Martone, W.J. (1979) Pharyngoconjunctival fever caused by adenovirus type 4: report of a swimming pool-related outbreak with recovery of virus form pool water. *Journal of Infectious Diseases*, **140**, 42-47.

Delbeke, E., DeMarcq, M.J., Roubin, C. and Baleux, B. (1985) Contamination aquatique de plaies par Aeromonas sobria apres bain de riviere. *La Presse Medicale*, **14**, 1292.

Denis, F.A., Blanchouin, E., DeLignieres. A. and Flamen, P. (1974) Coxsackie A16 infection from lake water. *Journal of the American Medical Association*, **228**, 1370-1371.

DeWailly, E., Poirier, C. and Meyer, F.M. (1986) Health hazards associated with windsurfing on polluted waters. *American Journal of Public Health*, **76**, 690-691.

Dufour, A.P. (1984) Bacterial indicators of recreational water quality. *Canadian Journal of Public Health*, **75**, 49-56.

Eklu-Natey, D.T., Al-Khudri, M., Gauthey, D., Dubois, J.P., Wuest, J., Vaucher, C. and Huggel, H. (1985) Epidemiologie de la dermatite des baigneurs et morphologie de Trichobilharzia cf. ocellata dans le lac Leman. *Revue Suisse de Zoologie*, **92,** 939-953.

El-Shaarawi, A.H. and Pipes, W.O. (1982) Enumeration and statistical inferences. In W.O. Pipes (ed.) *Bacterial Indicators of Pollution*, 43-66. CRC Press Inc., Boca Raton.

El-Sharkawi, F. and Hassan, M.N.E.R. (1982) The relation between the state of pollution in Alexandria swimming beaches and the occurrence of typhoid among bathers. *Bulletin of the Higher Institute for Public Health Alexandria*, **12,** 337-351.

Elliot, E.L. and Colwell, R.R. (1985) Indicator organisms for estuarine and marine waters. *FEMS Microbiology Reviews*, **32,** 61-79.

Evison, L. and Tosti, E. (1980a) Bathing water quality in the North Sea, and the Mediterranean. *Marine Pollution Bulletin*, **11,** 72-75.

Evison, L.M. and Tosti, E. (1980b) An appraisal of bacterial indicators of pollution in sea water. *Progress in Water Technology*, **12,** 591-599.

Fattal, B., Peleg-Olevsky, E., Yoshpe-Purer, Y. and Shuval, H.I. (1986) The association between morbidity among bathers and microbial quality of seawater. *Water Science and Technology*, **18,** 59-69.

Foster, D.H., Hanes, N.B. and Sabin, M.L. (1971) A critical examination of bathing water quality standards. *Water Pollution Control Federation Journal*, **43,** 2229-2241.

Foulon, G., Maurin, J., Quoi, N.N. and Martin-Bouyer, G. (1983) Relationship between the microbiological quality of bathing water and health effects: a preliminary survey. *Revue Francaise des Sciences de l'Eau*, **2,** 127-143.

Foy, H.M., Cooney, M.K. and Hatlen, J.B. (1968) Adenovirus type 3 epidemic associated with intermittent chlorination of a swimming pool. *Archives of Environmental Health*, **17,** 795-802.

Franco, D.A. (1989) Campylobacteriosis: the complexity of control and prevention. *Journal of Environmental Health*, **52,** 88-92.

Fulghum, D.D., Linton, W.R. and Taplin, D. (1978) Fatal Aeromonas hydrophilia infection of the skin. *Southern Medical Journal*, **71,** 739-741.

Galbraith, N.S. (1980) Infections associated with swimming pools. *Environmental Health*, **81,** 31-33.

Garber, W.F. (1956) Bacteriological standards for bathing waters. *Sewage and Industrial Wastes*, **28,** 795-808.

Geldreich, E.E. (1974) Microbiological criteria concepts for coastal bathing waters. *Ocean Management*, **3,** 225-248.

Geldreich, E.E. (1970) Applying bacteriological parameters to recreational water quality. *Journal of the American Water Works Association*, **62,** 113-120.

Gerba, C.P., Margolin, A.B. and Hewlett, M.J. (1989) Application of gene probes to virus detection in water. *Water Science and Technology*, **21,** 147-154.

Gerba, P., Margolin, B. and Trumper, E. (1988) Enterovirus detection in water with gene probes. *Zeitschrift fuer die gesampte Hygiene und Ihre Granzgebiete*, **34,** 518-519.

Godfree, A., Jones, F. and Kay, D. (1990) Recreational water quality: the management of environmental health risks associated with sewage discharges. *Marine Pollution Bulletin*, **21**, 414-422.

Grabow, W.O.K., Prozesky, O.W. and Smith, L.S. (1974) Drug resistant coliforms call for review of water quality standards. *Water Research*, **8**, 1-9.

Hall, E. (1990) *Biosensors*. Open University Press, Milton Keynes.

Hanson, P.G., Standridge, J., Jarrett, F. and Maki, D.G. (1977) Freshwater wound infection due to *Aeromonas hydrophila*. *Journal of the American Medical Association*, **236**, 1053-1054.

Hawley, B.H., Morin, D.P., Geraghty, M.E., Tomkow, J. and Phillips, C.A. (1973) Coxsackievirus B epidemic at a boys' summer camp. *Journal of the American Medical Association*, **226**, 33-36.

Hill, A.B. (1965) The environment and disease: association or causation? *Proceedings of the Royal Society of Medicine*, **58**, 295-300.

Hoadley, A.W. and Knight, D.E. (1975) External otitis among swimmers and nonswimmers. *Archives of Environmental Health*, **30**, 445-448.

Holmes, P.R. (1989) Research into health risks at bathing beaches in Hong Kong. *Journal of the Institution of Water and Environmental Management*, **3**, 488-492.

Holmes, S.E., Pearson, J.L., Kinde, M.R. and Hennes, R.F. (1989) Gastroenteritis outbreak: disease linked to swimming pool and spa use. *Journal of Environmental Health*, **51**, 286-288.

Hugues, B., Pietri, Ch., Puel, D., Crance, J.M., Cini, C. and DeLoince, R. (1988) Research of enterovirus, hepatitis A virus in a bathing area over a six month period and their salubrity impact. *Zentralblatt für Bakteriologie Mikrobiologie und Hygiene Serie B*, **185**, 560-568.

Hurst, C.J., Benton, W.H. and Stetler, R.E. (1989) Detecting viruses in water. *Journal of the American Water Works Association*, **81**, 71-80.

Jensen, B.O. and Horan, P.K. (1989) Flow cytometry: rapid isolation and analysis of single cells. *Methods in Enzymology*, **171**, 549-581.

Jones, F., Kay, D., Stanwell-Smith, R. and Wyer, M. (1990a) An appraisal of the potential health impacts of sewage disposal to UK coastal waters. *Journal of the Institution of Water and Environmental Management*, **4**, 295-303.

Jones, J.G., Gardener, S., Simon, B.M. and Pickup, R.W. (1986) Antibiotic resistant bacteria in Windermere and two remote upland tarns in the English Lake District. *Journal of Applied Bacteriology*, **60**, 443-453.

Jones, K., Betaieb, M. and Telford, D.R. (1990b) Thermophylic campylobacters in surface waters around Lancaster, UK: negative correlation with campylobacter infections in the community. *Journal of Applied Bacteriology*, **69**, 758-764.

Joret, J.C., Cervantes, P., Levi, Y., Dumoutier, N., Cognet, L., Hasley, C., Husson, M.O. and Leclerc, H. (1989) Rapid detection of *E. coli* in water using monoclonal antibodies. *Water Science and Technology*, **21**, 161-167.

Joseph, S.W., Daily, O.P., Hunt, W.S., Seidler, R.J., Allen, D.A. and Colwell, R.R. (1979) Aeromonas primary wound infection of a diver in polluted waters. *Journal of Clinical Microbiology*, **10**, 46-49.

Kaji, M., Kimura, M., Kamiya, S., Tatewaki, E., Takahashi, T., Nakajima, O., Koga, T., Ishida, S. and Majima, Y. (1961) An epidemic of pharyngoconjunctival fever among school children in an elementary school in Fukuoka prefecture. *Kyushu Journal of Medical Science*, **12**, 1-8.

Lacey, R.F. and Pike, E.B. (1989) Water recreation and risk. *Journal of the Institution of Water and Environmental Management*, **3**, 12-18.

Larsen, J.L., Farid, A.F. and Dalsgaard, I. (1981) Occurrence of *Vibrio parahaemolyticus* and *Vibrio alginolyticus* in marine and estuarine bathing areas in Danish coast. *Zentralblatt für Bakteriologie Mikrobiologie und Hygiene I. Abt. Orig. B*, **173**, 338-345.

Lenway, D.D., Brockman, R., Dolan, G.J. and Cruz-Uribe, F. (1989) An outbreak of an enterovirus-like illness at a community wading pool: implications for public health inspection programmes. *American Journal of Public Health*, **79**, 889-890.

Makintubee, S., Mallonee, J. and Istre, G.R. (1987) Shigellosis outbreak associated with swimming. *American Journal of Public Health*, **77**, 166-168.

Mallmann, W. L. (1962) Cocci tests for detecting mouth and nose pollution of swimming pool water. *American Journal of Public Health*, **52**, 2001-2008.

Mawer, S.L. (1988a) Campylobacters in man and the environment in Hull and East Yorkshire. *Epidemiology and Infection*, **101**, 287-294.

Mawer, S.L. (1988b) The pathogenicity of environmental campylobacters - a human volunteer experiment. *Epidemiology and Infection*, **101**, 295-300.

Moore, B. (1975) The case against microbial standards for bathing beaches. In A.L.H. Gameson (ed.) *Discharge of Sewage from Sea Outfalls*, 103-109. Pergamon Press, Oxford.

Moore, B. (1959) Sewage contamination of coastal bathing waters in England and Wales. *Journal of Hygiene*, **57**, 435-472.

Moore, J.L., Perin, D.E. and Maiden, B.G. (1979) Estimating the effect of water quality improvement on public swimming. *Water Resources Research*, **15**, 1323-1328.

Mosley, J.W. (1975) Epidemiological aspects of microbial standards for bathing beaches. In A.L.H. Gameson (ed.) *Discharge of Sewage from Sea Outfalls*, 85-93. Pergamon Press, Oxford.

Mujeriego, R., Bravo, J.M. and Feliu, M.T. (1982) Recreation in coastal waters: public health implications. In *Workshop on Pollution of the Mediterranean: VIes Journee Etudes Pollutions, Cannes. Commission Internationale pour l'Exploration Scientifique de la Mer Mediterranean*, Monaco. 585-594.

Murray, G.E., Tobin, R.S., Junkins, B. and Kushner, D.J. (1984) Effect of chlorination on antibiotic resistance profiles of sewage-related bacteria. *Applied and Environmental Microbiology*, **48**, 73-77.

O'Kane, J.P. (1983) An examination of the EEC bathing water directive. In J. Blackwell and F. Convery (eds.) *Promise and Performance: Irish Environmental Policies Analysed*, 13-23. University College, Dublin.

Ogram, A.V. and Sayler, G.S. (1988) The use of gene probes in the rapid analysis of natural microbial communities. *Journal of Industrial Microbiology*, **3**, 281-292.

Olivieri, V.P. (1982) Bacterial indicators of pollution. In W.O. Pipes (ed.) *Bacterial Indicators of Pollution*, 21-41. CRC Press Inc., Boca Raton.

Pedersen, H., Stadil, U. and Dietz, H.H. (1980) Bathing water control according to the EEC directive with a critical evaluation of coliforms as pollution indicators in marine environments. *Zentralblatt für Bakteriologie Mikrobiologie und Hygiene I. Abt. Orig. B*, **171**, 195-200.

Pettipher, G.L. (1989) The direct epifluorescent technique. *Progress in Industrial Microbiology*, **26**, 19-30.

Philipp, R., Evans, E.J., Hughes, A.O., Grisdale, S.K., Enticott, R.G. and Jephcott, A.E. (1985) Health risks of snorkle swimming in untreated water. *International Journal of Epidemiology*, **14**, 624-627.

Reid, T.M.S. and Porter, I.A. (1981) An outbreak of otitis externa in competitive swimmers due to *Pseudomonas aeruginosa*. *Journal of Hygiene*, **86**, 357-362.

Robinton, E.D. and Mood, E.W. (1966) A quantitative and qualitative appraisal of microbial pollution of water by swimmers: a preliminary report. *Journal of Hygiene*, **64**, 489-499.

Rosenberg, M.L., Hazlet, K.K., Schaefer, J., Wells, J.G. and Pruneda, R.C. (1976) Shigellosis from swimming. *Journal of the American Medical Association*, **236**, 1849-1852.

Roszak, D.B. and Colwell, R.R. (1987) Survival strategies of bacteria in the natural environment. *Microbiological Reviews*, **51**, 365-397.

Seyfried, P.L., Tobin, R.S., Brown, N.E. and Ness, P.F. (1985) A prospective study of swimming-related illness. 2. Morbidity and the microbiological quality of water. *American Journal of Public Health*, **75**, 1071-1075.

Seyfried, P.L. and Cook, R.J. (1984) Otitis externa infections related to *Pseudomonas aeruginosa* levels in five Ontario lakes. *Canadian Journal of Public Health*, **75**, 83-91.

Sherry, J.P. (1986) Temporal distribution of faecal pollution indicators and opportunistic pathogens at a Lake Ontario bathing beach. *Journal of Great Lakes Research*, **12**, 154-160.

Shuval, H.I. (1988) The transmission of virus disease by the marine environment. *Schriftenreihe Verein fuer Wasser-, Boden und Lufthygiene*, **78**, 7-23.

Singh, A., Pyle, B.H. and McFeters, G.A. (1989) Rapid enumeration of viable bacteria by image analysis. *Journal of Microbiological Methods*, **10**, 91-101.

Skirrow, M.B. (1987) A demographic survey of campylobacter, salmonella and shigella infections in England. *Epidemiology and Infection*, **99**, 647-657.

Smith, H.S. (1971) Incidence of R+ *Escherichia coli* in coastal bathing waters of Britain. *Nature*, **234**, 155-156.

Smith, J. A. (1980) Aeromonas hydrophila: analysis of 11 cases. *Canadian Medical Association Journal*, **122**, 1270-1272.

Springer, G.L., and Shapiro, E.D. (1985) Freshwater swimming as a risk factor for otitis externa: a case-control study. *Archives of Environmental Health*, **40**, 202-206.

Stevenson, A.H. (1953) Studies of bathing water quality and health. *American Journal of Public Health*, **43**, 529-538.

Sullivan, C.S.B. and Barron, M.E. (1989) Acute illnesses among Los Angeles County lifeguards according to worksite exposures. *American Journal of*

Public Health, **79**, 1561-1563.

Vasl, R., Fattal, B., Katzenelson, E. and Shuval, H. (1981) Survival of enteroviruses and bacterial indicator organisms in the sea. In M. Goddard and M. Butler (eds.) *Viruses and Wastewater Treatment*, 113-116. Pergamon Press, Oxford.

Weingarten, M.A. (1977) Otitis externa due to *Pseudomonas* in swimming pool bathers. *Journal of the Royal College of General Practitioners*, **27**, 359-360.

Wolfe, D.A. (1980) Discussion on microbial indicators of environmental quality. In D. Schlessinger (ed.) *Microbiology-1980*, 382-383. American Society for Microbiology, Washington DC.

Xu, H.-S., Roberts, N.C., Adams, L.B., West, P.A., Siebeling, R.J., Huq, A., Huq, M.I., Rahman, R. and Colwell, R.R. (1984) An indirect fluorescent antibody staining procedure for detection of *Vibrio cholerae* serovar 01 cells in aquatic environmental samples. *Journal of Microbiological Methods*, **2**, 221-231.

Indicator Organisms as a Guide to Estuarine Management

[1]D.W. Jeffrey, [1]B. Madden, [1]B. Rafferty, [1]R. Dwyer and [2]J.G. Wilson

[1]Department of Botany and [2]Environmental Sciences Unit, Trinity College, Dublin 2, Ireland.

Key words: estuarine management, indicator organisms, eutrophication, Dublin Bay.

Abstract

The Water Quality Management Plan (W.Q.M.P.) is the main focus for organizing the management of estuaries and other extensive areas of water in the Republic of Ireland. The plan entails maximizing the utilization of a series of 'beneficial uses' which are listed for each area. In Dublin the beneficial uses include water contact sports, nature conservation, major port activities, electricity generation and sewage discharge.

From intensive studies of intertidal and benthic ecosystems, it is now possible to suggest a series of biomonitoring systems, for management under a W.Q.M.P. These provide information of general environmental quality; quality of small areas; specific information on eutrophication; and the environmental quality of nature conservation and amenity areas. Furthermore it is possible to suggest bioindicator systems of potential value should an environmental accident occur.

The bioindicator systems include: indices based on community type; population density; presence or absence of key species; bioaccumulating species; measurement of microbial process rates; use of biological assay.

Introduction

The statutory basis for water pollution control is the Local Government (Water Pollution) Act (1977). Amongst its provisions it enables a local government authority to draw up a Water Quality Management Plan. In the case of Dublin Bay, three local authorities - Dublin Corporation, Dublin County Council, and the County Borough of Dun Laoghaire, have pooled their resources to draw up a joint W.Q.M.P. This paper is based on studies commissioned as part of the plan.

Under a W.Q.M.P. listed 'beneficial uses' must be facilitated, and an

Bioindicators and Environmental Management
ISBN 0-12-382590-3

optimum balance struck between any conflicts. In this case, it entails the resolution of conflicts between uses and users, which are listed below. See also the map, Figure 1.

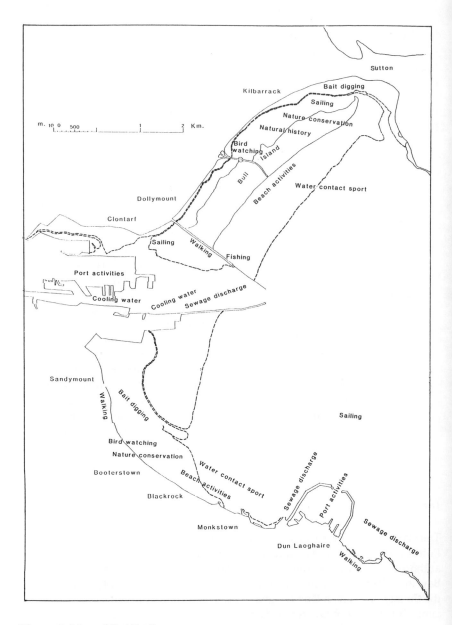

Figure 1. Map of Dublin Bay.

A. Discharge of sewage. The major discharge point is shown, serving a population of about 600 000. Another outfall is located just north of the area mapped, serving an additional 300 000.

B. Cooling water is withdrawn for two power stations, with a total generating capacity of 788 MW.

C. Port activities are mainly conducted in a limited area to the north of the main channel of the River Liffey. The principal bulk cargo is oil, with most other cargoes being containerized. A notable export cargo is lead and zinc ore-concentrate.

D. Recreational uses. These range from the intense use of the shorelines for walking and birdwatching; a range of beach-based activities; water-contact activities, including bathing, board-sailing, SCUBA-diving and dinghy sailing; yachting and motor boating; fishing from the shoreline and boats; bait digging.

E. Nature conservation. The whole Bay is of strategic importance for the conservation of migratory wildfowl and waders. However, two sites are of particular importance: a) the North Bull Island which is a large dune, saltmarsh and mudflat complex. Bull Island is recognized in several ways. It was the first statutory bird sanctuary in the State; it is a nature reserve; a UNESCO World Biosphere Reserve; and a Ramsar Convention registered site of international importance for 4 species of bird. The history and ecology of the island has been described in detail in Jeffrey (1977); and b) Booterstown marsh in the south bay which is a small (4.3 ha) brackish marsh with a diverse vegetation that includes a protected grass *Puccinellia fasiculata*. Its development and flora has been reviewed by Reynolds and Reynolds (1990).

The task of an at-present-hypothetical management authority is to optimize on the above uses, at the same time complying with relevant E.C. directives.

The main preoccupations of the manager, in order of priority, are:-

a) Microbial quality - to comply with the E.C. Bathing Waters Directive. Microbial Quality has been dealt with by Masterson *et al.* (in this volume), and will not be considered further in this paper.

b) Overall Estuarine and Coastal Environmental Quality.

c) Eutrophication, as manifested by nuisance algal growths.

d) Nature Conservation Management - in view of the special importance of Dublin Bay.

e) Effects of spillage.

Overall estuarine quality

In the 1970's, the Irish Estuarine Research Programme (IERP) came to the conclusion that a relatively simple and robust indexing system was the best way to integrate biological information relating to estuarine quality. An estuarine Biological Quality Index (B.Q.I.) was devised (Jeffrey and Wilson 1985). The background and practical applications of the index are reviewed in Wilson and Jeffrey (1987). In this system, the distribution of three community types is mapped. They are:-

A) 'Abiotic' areas, in which no macro-organisms are detectable. Clearly, micro-organisms and meiofauna may be present.

B) Opportunistic communities, dominated by short-lived and rapidly repro-ducing species, e.g. most green macroalgae and many Annelida.

C) Stable communities, dominated by long living and rather slowly reproduc-ing organisms - for example, fucoid algae and bivalves.

The proportional areas of the estuary occupied by each type is determined, such that A + B + C = 1.0. The Biological Quality Index is then calculated using the formula:

$$B.Q.I. = antilog_{10} (C - A)$$

The index scale has a maximum value of 10 - 100% stable, down to 0.1 - 100% abiotic.

The most polluted part of Dublin Bay is the Tolka Basin. This area of about 100 ha, was surveyed in detail in 1979 and resurveyed in 1989 using the same technique. The maps of the community types are presented in Figure 2 and the comparative data in Table 1.

The main changes between 1979 & 1989 are:-

    i) Loss of 'Stable' areas (10% to zero).

    ii) Change of Stable to Opportunistic.

    iii) Extension of Abiotic areas into Opportunistic.

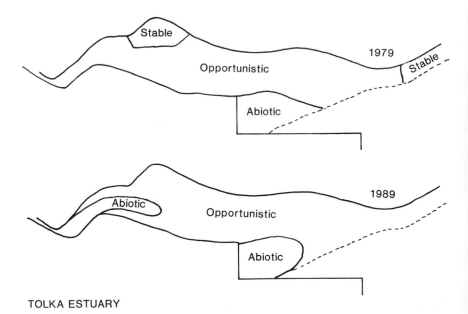

TOLKA ESTUARY

Figure 2. Map of the Tolka estuary showing extent of stable, opportunistic and abiotic communities in 1979 and 1989.

Table 1. Changes in Biological Quality Index characteristics in the Tolka Basin, 1979 - 1989.

	1979	1989	Change
C) Stable	10%	zero	-10%
B) Opportunistic	76%	68%	-8%
A) Abiotic	14%	32%	+18%
B.Q.I.	0.93	0.47	

The most probable reason for this change is a sustained loading into this area of organic particulates, probably originating from sewage and/or river water, and delivered by the rising tide.

We are convinced, that if exactly comparable areas are resurveyed at five yearly intervals, useful information will be obtained on effects of any means to reduce loading. A manager could set quantitative objectives, for example to arrest the decline in index value and ultimately to restore stable communities.

Estuarine and coastal eutrophication

The presence of substantial quantities of green macroalgae (Ulvales) has been noted in Dublin Bay since the last century (Baily 1886; Adeney 1908). A naive comment is that they are bioindicators of eutrophication. This is somewhat of a miss-application of the term. The algae are the problem, and an administrator needs to predict the size of nuisance growth. Conventional wisdom suggests that they are absorbing macronutrients from tidal water. We have evidence that this is only part of the picture.

Detailed estimates of algal biomass, coverage and nutrient content were made in Dublin Bay from June 1989 to September 1990. Twelve biomass harvests were taken during this period from 13 sampling stations throughout the bay. At each harvest up to 10 algal samples were collected at each station from quadrats along a transect. Algal samples were dried in the laboratory and dry weight was measured. Mean biomass was then calculated for each sampling station. Algal coverage was estimated from a combination of field measurements and from spectral scanning images taken from a low altitude flight over the bay in July 1989. Total amounts of algae in each area of the bay were calculated from the biomass and coverage data.

At each harvest, algal samples from each station were analysed for nitrogen and phosphorus. Total nitrogen and phosphorus were determined coulorimetrically from a single digest solution by a variant of the Kjeldahl method. After the digestion process, total nitrogen was measured using a Tecator Flow Injection Analyser, and total phosphorus was measured using a

Technicon Autoanalyser II Continuous Flow Instrument.

The nutrient requirements for algal production can thus be stated with a high degree of certainty. Nutrient measurements have also been made in water and sediments.

A first conclusion is that phosphate supply is not limiting to algal growth at any time of year in Dublin Bay. This is based particularly on the high phosphorus concentrations in algal material at all times of year. Concentrations exceeded 2000 µg g^{-1} dry weight on most occasions and frequently were above 2500 µg g^{-1}.

As a second conclusion, attention must shift to nitrogen sources. From water analyses, nitrogen ion concentrations are always of a low order, even though large quantities of total nitrogen are released from the sewage outfalls.

Sediment total nitrogen values were high in some areas (> 1000 µg g^{-1} DW), and therefore some measurements were made on nitrogen mineralization rates to assess the importance of this nutrient source to the algae. The results of these studies show:

i) Mineralization rates in late-spring are of the correct order of magnitude to supply a substantial quantity (up to 74% in some areas) of the calculated daily nitrogen requirements of the algae.

ii) Areas without algae exhibited very low mineralization rates, or even net uptake of nitrogen ions.

iii) The rate of mineralization is low when production declines later in the season.

As the pattern of algal distribution in the bay is similar from year to year, it seems that nitrogen mineralization processes may have an important bearing on the distribution of the algal mats and may explain the absence of algae from some areas.

From our detailed studies, the best hypothesis explaining intertidal green algal growths is that the mudflats are being loaded with organic nitrogen as particulates. Mineralization, mainly by the microbial population, and partly by invertebrate excretion, releases ammonium ions which are utilized by the algae. The same temperature kinetics that drive algal growth, increase nitrogen mineralization.

Hence the best predictive bioindicator for potential algal production is nitrogen mineralization rate in late spring. We measured cumulative release of nitrogen ions, from an intact column of sediment to aerated seawater, in darkness, at 15°C, over a 7 day period. The columns were collected in 10 cm diameter P.V.C. drain pipe, and capped at the base. The capped pipes were used throughout the experiment as containers, adding 1 litre of unfiltered, low nutrient seawater. As samples were taken, an equivalent volume (15 ml) to the sample was replaced. Water temperature and oxygen content were measured at each sampling. Calculations take into account the nitrogen and phosphorus content of the seawater used. These rather large experimental modules were maintained in a plant growth cabinet. It is probably worthwhile developing a more convenient module that could be contained in a smaller incubator. We will be considering this problem further.

It is worth noting that an attempt was made to use discs cut from the thallus of *Ulva lactuca* (collected in October) as a bioassay of nitrogen in seawater. This was not successful. Early winter discs, which have a very high nitrogen content (up to 7.5% dry weight), will grow in very low external concentrations, and were not responsive to experimental nitrogen addition. In order for a bioassay to work, thallus material must have a uniformly low nutrient status. This can possibly be attained if effort is placed in devising a suitable standard culture system. This, however, will increase the cost and complexity of the assay.

Nature conservation management

Management of a unit as complex as North Bull Island, and its accompanying offshore area, requires a wide range of techniques. The bird fauna, however, being at the top of the estuarine food chain, can provide valuable information on the overall environmental quality of the estuary and may be considered as a general bioindicator. We outline below the basic information that should be available to the estuarine managers.

Wintering wildfowl and waders are by far the most important groups which utilize estuaries in north-west Europe. The basic approach is to provide baseline data which should determine the diversity of species present, the size of populations, the location of roosting sites and the location of the main feeding grounds. The usual method of assessing populations is by monthly counts using standard techniques. For details and reviews of methodology see Prater (1979), Prater (1981) and Prater and Lloyd (1987). In Ireland, counts are carried out mainly by experienced amateur ornithologists working under the auspices of the Irish Wildbird Conservancy. The equivalent body in Britain for collecting such data is the British Trust for Ornithology. Counts of wintering bird populations are now available for most major estuaries in the British Isles (Prater 1981; Hutchinson 1979; Sheppard 1991). Monthly counts throughout the year are carried out in some of the more important estuaries, including Dublin Bay. From studies on feeding distribution of birds within the estuary, a map showing high priority areas can be compiled.

The bird data can provide the estuarine managers with useful information on the quality of the estuary, as there exists a general relationship between diversity and abundance of bird species and the diversity and abundance of their prey (Prater 1981). However, the data need to be interpreted with care, as a change, particularly a short-term one, in a population of a species may not necessarily indicate environmental changes within the estuary. Such changes need to be examined in the context of the local, national and even international situations. However, if a species continues to decrease, or more importantly if several species show simultaneous declines within an estuary, over a number of seasons, it may be evidence of deteriorating conditions and it would be prudent for the managers to carry out further studies. The avoidance of specific areas of the estuary by feeding birds is a good indicator of low environmental quality. In Dublin Bay the polluted Tolka estuary, which has no stable animal communities, is avoided by most birds.

In addition to waterfowl data, counts and information on other bird species that regularly utilize the estuary should be collected. Gulls can form a significant component of the bird fauna of an estuary but populations are not often assessed, partly because many gulls disperse inland to feed during the day. The size of the gull population will reflect the availability of food, and nowadays the large gull numbers in coastal populated areas usually indicates the availability of refuse. Many of the gulls that frequent Dublin Bay feed at a refuse dump at Rogerstown estuary, north of the Bay.

In recent years large number of gulls in the Dublin area have been affected by botulism during the summer months (Quinn and Crinion 1984). Proliferation of the causative organism, *Clostridium botulinum*, requires anaerobic conditions, typically associated with rotting organic matter and high temperatures. Botulism infected birds are therefore good indicators of discarded refuse, particularly domestic refuse in sealed plastic sacks. Recent decreases in the number of breeding gulls at the Dublin colonies are attributed to large losses due to botulism (Merne 1988). Data on numbers of birds infected by botulism should be collected and reviewed annually. It should be noted that reliable testing for botulism requires blood serum from live birds (Lloyd *et al.* 1976). Other incidences of mortality caused by botulism have been report from Co. Cork (Buckley and O'Halloran 1986), Northern Ireland (Radford 1979) and Britain (Lloyd *et al.* 1976).

The role of birds as bioindicators of oiling incidences is well known (Bourne and Bibby 1975; Hope-Jones *et al.* 1978; O'Keeffe 1978) and annual national "beached bird surveys" are now undertaken in many European countries. Often the presence of oiled birds may be the first indication that an oiling incident has occurred and may prompt the authorities to make further investigations. Details of all records of oiled birds should be recorded.

We recommend that the ornithological data be collected on a continuous basis, an annual review undertaken and if necessary further research be carried out or remedial actions initiated.

Management of spillage

It has been pointed out previously (Jeffrey 1990) that the attitude to adopt is to anticipate spillage with a range of measures, including the capacity to monitor their effects. Bioindicators may be used to determine the extent, the intensity and duration of a damaging incident. This is of value in directing cleanup, determining second order ecological effects and assessing liability.

Attention in Dublin needs directing to the effects of hydrocarbon spills, for which an emergency plan exists and to other possible toxic substances handled in bulk. This should include the possibility of toxic pyrolysis products being washed into tidal waters during fire fighting. Designing a simple range of bioindicators is a task that would complete the array of spillage control measures already in place.

References

Adeney, W.E. (1908) Effect of the new drainage on Dublin Harbour. *Handbook to the Dublin District British Association.*

Baily, A. (1886) *Rambles on the Irish Coast, 1. Dublin to Howth.* Carraig Book Reprints 2: 8-10, Dublin.

Bourne, W.R.P. and Bibby, C.J. (1975) Temperature and the seasonal and geographic occurrence of oiled birds on west European beaches. *Marine Pollution Bulletin,* **6,** 77-80.

Buckley, N.J. and O'Halloran, J. (1986) Mass mortality of gulls in west Cork attributed to botulism. *Irish Birds,* **3,** 283-285.

Hope-Jones, P., Monnat, J-Y, Cadbury, C.J. and Stowe, T.J. (1978) Birds oiled during the *Amoco Cadiz* incident - an interim report. *Marine Pollution Bulletin,* **9,** 307-310.

Hutchinson, C.D. (1979) *Ireland's Wetlands and their Birds.* Irish Wildbird Conservancy, Dublin.

Jeffrey, D.W. (1990) Biomonitoring of catastrophes. *Environmental Monitoring and Assessment,* **14,** 131-137.

Jeffrey, D.W. (ed.) (1977) *North Bull Island, Dublin Bay - a modern coastal natural history.* Royal Dublin Society, Dublin.

Jeffrey, D.W. and Wilson, J.G. (1985) *A Manual for the Evaluation of Estuarine Quality.* 2nd ed. National Board for Science and Technology, Dublin.

Lloyd, C.S., Thomas, G.J., McDonald, J.W., Borland, E.D., Standring, K., and Smart, J.L. (1975) Wild bird mortality caused by botulism in Britain, 1975. *Biological Conservation,* **10,** 119-129.

Merne, O.J. (1988) Recent changes in breeding seabird populations in Counties Dublin and Wicklow. *Irish East Coast Bird Report,* 1987, 69-77.

O'Keeffe, C. (1978) Oil pollution and seabirds on Irish coasts. *Irish Birds,* **1,** 206-210.

Prater, A.J. (1979) Trends in accuracy of counting birds. *Bird Study,* **26,** 198-200.

Prater, A.J. (1981) *Estuary Birds of Britain and Ireland.* Poyser, Calton.

Prater, A.J. and Lloyd, C.S. (1987) Birds. In J.M. Baker and W.J. Wolff (eds.) *Biological Surveys of Estuaries and Coasts,* 374-403. Cambridge University Press, Cambridge.

Quinn, P.J. and Crinion, R.A. (1984) A two year study of botulism in gulls in the vicinity of Dublin Bay. *Irish Veterinary Journal,* **38,** 214-219.

Radford, D.J. (1979) Gull deaths on the North Ulster coast, February 1978. *Irish Naturalist's Journal,* **19,** 391-395.

Reynolds, J.D. and Reynolds, C.P. (1990) Development and present vegetational state of Booterstown marsh, Co. Dublin, Ireland. *Bulletin of the Irish Biogeographical Society,* **13,** 173-188.

Sheppard, R. (1991) *Ireland's Wetland Wealth, the birdlife of the estuaries, lakes, coasts, rivers, bogs and turloughs of Ireland.* Irish Wildbird Conservancy, Dublin. (in press).

Wilson, J.G. and Jeffrey, D.W. (1987) Europe wide indices for monitoring estuarine quality. In D.H.S. Richardson (ed.) *Biological Indicators of Pollution*, 225-242. Royal Irish Academy, Dublin.

Benthic Macroinvertebrates as Indicators of Organic Pollution of Aquatic Ecosystems in a Semi-Arid Tropical Urban System

M. Vikram Reddy and B. Malla Rao

Environmental Biology Laboratory, Department of Zoology, Kakatiya University, Warangal - 506 009, A.P., India.

Key words: benthic macroinvertebrates, *Limnodrillus hoffmeisteri*, *Chironomus circumdatus* larvae, *Gabbia orcula*, *Eristalis tenax*, *Limnophora* spp., bioindicators, organic pollution, India.

Abstract

Seasonal sampling of benthic macroinvertebrates of different water bodies within the urban limits of Warangal revealed that the communities from the littoral zones of good quality water bodies, i.e. medium-sized water storage tanks [=reservoirs], polluted with municipal sewage, consisted of greater percentages of organic-pollution tolerant macroinvertebrates. In the areas polluted with domestic discharges located in the downstream section of a canal passing through the urban system, the benthic macroinvertebrate communities consisted of a large percentage of similar pollution tolerant communities. These communities were dominated by Oligochaetes (*Limnodrilus hoffmeisteri* Claparede, 1862 and *Pristina aequiseta* Bourne, 1891) and the larvae of Chironomidae *Chironomus circumdatus* (Kieffer). Different species of molluscs were recorded in the reservoirs and the canal, of which *Gabbia orcula* (Frauenfeld) was dominant. In a heavily polluted water body adjacent to a textile mill, the benthic macroinvertebrates consisted of species tolerant and resistant to industrial organic-effluent pollution. Individuals collected from this polluted water body consisted mainly of *Eristalis tenax* and *Limnophora* spp. The value as bioindicators of organic pollution of some of these macroinvertebrates, such as *L. hoffmeisteri*, *P. aequiseta*, *C. circumdatus*, *G. orcula*, *E. tenax* and *Limnophora* spp. are discussed with ecological justification, and their assembleges are related to water quality.

Introduction

Benthic macroinvertebrates may be defined simply as those animals caught with

Bioindicators and Environmental Management
ISBN 0-12-382590-3

a sieve of mesh size 0.5-0.6 mm. Because of their long life, constant presence, sedentary habit, comparatively large size and their endurance to external stress, they can integrate changes which reflect the characteristics of both the sediment and the water column. They present an integral measure of both autotrophic and heterotrophic processes, as well as the perturbations in these processes, occurring in the aquatic ecosystems (Wiederholm 1980).

They can also be used to complement the physico-chemical monitoring of water quality (cf. Olive et al. 1988). They have been recognized as good indicators of aquatic pollution, particularly organic pollution and eutrophication of water bodies, as by their continuous presence they reflect the long-term water quality.

In India, although studies have been conducted on the benthic macroin-vertebrates of different aquatic systems (Michael 1968; Mandal and Moitra 1975; Vasisht and Bandal 1979; Sarkar 1989; Reddy and Rao 1990), information on the use of these invertebrates as bioindicators of environmental pollution is inadequate and very much scattered (Krishnamoorthi and Sarkar 1979; Patil et al. 1983; Khan 1983; Mahadevan and Krishnaswamy 1984; Dudani et al. 1988). Moreover, there is little information available in the literature on the use of different macroinvertebrates as indicators of different levels of organic pollution at a single location. The present study identifies and recommends some macroinvertebrate species as bioindicators of different levels of organic waste pollution, originating from both domestic and industrial processes, of various aquatic ecosystems within the municipal limits of the semi-arid tropical urban system - Warangal. These species may be used as bioindicators in future pollution monitoring programmes.

Study sites

Warangal urban system (lat. 18° 0′ 31"; long. 7° 29′ 5"; altitude 263.7 m, MSL) is located in the Deccan Plateau, the semi-arid tropical region of Andhra Pradesh in South India. Different types of lentic and lotic systems, polluted with both domestic sewage and industrial organic effluents, were sampled for benthic macroinvertebrates. The sites were located in the Warangal, Kazipet and Hanamkonda areas of the urban system The water bodies were: a) two large-sized water tanks (i.e. reservoirs) called Bhadrakali and Wadepally reservoirs, the former being located in the Warangal area and the latter in the Kazipet area of the urban system; b) a canal called "Paddamori", which descends from the Wadepally reservoir and passes through the Hanamkonda area of the urban system. The canal receives directly the domestic sewage and run-off from the adjacent streets; and c) an effluent storage tank of a textile mill called Ajam Jhai Mill located in the Warangal area of the urban system. Sampling was conducted at three different sites within the littoral zones of each of the reservoirs, at four sites along the canal and at two sites in the effluent storage tank, both before and after the establishment of an Effluent Treatment Plant (E.T.P.).

Details of the different sampling sites of the Bhadrakali reservoir are given in Reddy and Rao (1990). Site-I was in a shallow zone on the southern side of the reservoir. This site received a considerable amount of municipal sewage, particularly during the dry season. Site-II was on the eastern side of the reservoir, and was perturbed by washermen washing clothes. Site-III was located on the north-east corner of the reservoir near a Hindu temple, and was comparatively deeper than the other two sites. This site received considerable amounts of discarded religious materials such as flowers, fruits and other organic materials, including various food preparations that were offered to the Goddess by the worshippers. The aquatic vegetation in the littoral zone was composed of *Vallisneria spiralis* L., *Azolla* spp., *Chara vulgaris* L. and *Aponogeton appendiculatus* Bruggen at all sites. At site-III *Eichhornia crassipes* (Mart.) Solms-Laub, *Ipomoea aquatica* Forsk. and *C. vulgaris* L. occurred.

In Wadepally reservoir, site-I was within a shallow water zone on the southern side that received a negligible amount of domestic sewage from a few nearby houses. Site-II was on the sloping side of the embankment on the eastern side of the reservoir, while site-III was in the shallow water zone on the north-western corner, where there are some perturbations by a few washermen. Site-II had no aquatic vegetation except for *I. aquatica* Forsk and a few plants on the embankment that shed leaves. Sites-I and III possessed good growths of *V. spiralis* L., *Salvinia* spp., *Cyperus papyrus* L. and *C. vulgaris* L.

The sewage canal represented the lotic system. Its site-I received less sewage compared to that of site-II and site-III, which received domestic sewage and the direct run-off from the streets. Site-IV was situated about 100 m from site-III, where the water current was very much reduced.

The effluent storage tank of the textile mill was swampy and shallow, and adjacent to the mill. It received the coloured effluents from the mill. Its site-I was the point which received the effluents directly from the mill. Site-II was separated from site-I by an embankment, and it also received domestic sewage and municipal run-off during the rainy season. However, after six months of sampling, an E.T.P. was established and the treated effluents were then released into an open area on the other side of the embankment adjacent to the effluent-storage tank of the mill.

In the sewage canal and in site-II of the effluent-storage tank before the establishment of the E.T.P., the aquatic vegetation consisted mainly of *E. crassipes* (Mart.) Solm-Laub, *Pistia stratiotes* L. and *Salvinia* spp. In addition, algal blooms of *Microcystis* spp. were recorded frequently in site-II of the effluent-storage tank. Site-I of the effluent-storage tank had no vegetation.

Materials and methods

Benthic macroinvertebrates were sampled monthly from August 1986 to February 1988, and processed using the methods described in Reddy and Rao (1987). They were collected with an iron core sampler 4.5 cm in diameter and 12.0 cm long. At each of the different water bodies five samples of benthic sediment, each separated by 2 m, were taken to 10 cm depth. Samples were placed in separate polythene bags, and preserved by adding 10 ml of 5% formaldehyde to each bag. Samples were brought to the laboratory where each was passed through a sieve with a mesh size of 0.02 mm. The residue was transferred into a petri-dish and the macroinvertebrates present were sorted with the help of a stereoscopic binocular microscope. Animals were identified to genus and species level wherever possible and stored in 80% ethanol.

Dissolved oxygen content of the water at the various sites was estimated by the Winkler modification (Alsterberg Azide modification) method.

Results and discussion

The number of species of macroinvertebrates recorded at each of the sites of the various water bodies within the municipal limits of the Warangal urban system are presented in Table 1. There was relatively more species in the Wadepally reservoir compared to that of the Bhadrakali reservoir. This was probably due to the more eutrophic conditions of the Bhadrakali reservoir owing to the various anthropogenic activities. A sewage canal that passed adjacent to the Bhadrakali reservoir leaked sewage water into the reservoir. Also, the sewage water in the upper part of the canal was diverted to the upper flat terrain of the reservoir for paddy cultivation during the winter and summer seasons, and this water ultimately entered the reservoir. Moreover, the washing of clothes in the reservoir and the dumping of large amounts of organic matter from the temple, added considerable amounts of organic matter to the system. All these activities enriched the reservoir with organic waste matter at all the sites under study. Contrary to that, the Wadepally reservoir, which received negligible amounts of domestic sewage and relatively less organic matter from clothes washing activities, was less polluted. However, there was very little difference in the dissolved oxygen content of the two water bodies (Table 2). Despite the indication that the water of the Wadepally reservoir is of relatively good quality, the higher percentage of facultative organisms occurring in the sediment of its littoral zone (Table 3), indicated substantial organic sedimentation.

The sewage canal, particularly during the rainy season, was flooded with enormous amounts of freshwater descending from the Wadepally reservoir. This water washed away most of the sediment, especially in site-II of the canal. However, during the winter and summer seasons, the canal became saturated with domestic sewage as there was no flow of freshwater from the reservoir. Of the four sites of the canal, site-I received very negligible amounts of sewage and had the highest dissolved oxygen content (Table 2). This may be the reason for the site being inhabited by a higher number of species of benthic macroinvertebrates compared to the other sites (Table 1). In accordance, Hawkes (1979) reported that domestic sewage pollution reduced the number of taxa and enhanced the density of tolerant organisms in river water.

Table 1. Number of species of benthic macroinvertebrates recorded at the various sites, and the relative organic pollution status (based on visual observations) at different sites of the various water bodies of Warangal urban system.

Water bodies	Site-I	Site-II	Site-III	Site-IV	Pollution status	Colour of the water
Wadepally	14	12	15	-	Minimum	Transparent
Bhadrakali	11	12	11	-	Medium	Turbid (muddy)
Sewage canal	18	12	11	11	Minimum (rainy season)*	Turbid (rainy season)**
	Site-I	Site-II	Site-I	Site-II		
	(Before E.T.P. establishment)		(After E.T.P. establishment)			
Effluent storage tank	3	5	6	5	High	Blackish/ blackish/brown

* nearing high pollution during remainder of year
** blackish during remainder of year

Table 2. Ranges of dissolved oxygen (mg l^{-1}) at various sites of different water bodies of Warangal urban system. The figures in parenthesis represent the mean.

Water body	Site-I	Site-II	Site-III	Site-IV
Wadepally	1.0-9.1 (5.2)	1.4-12.2 (5.8)	1.0-7.1 (5.5)	-
Bhadrakali	2.6-8.5 (5.5)	1.4-7.9 (5.2)	1.8-6.8 (5.2)	-
Sewage Canal	1.7-11.5 (4.1)	1.1-6.2 (2.7)	0.0-8.7 (2.4)	0.6-5.7 (3.2)
	Site-I	Site-II	Site-I	Site-II
	(Before E.T.P.)		(After E.T.P.)	
Effluent storage tank	Nil	Nil	0-0.7	0.4-0.6 (0.5)

Table 3. Percentage composition of benthic macroinvertebrates in Wadepally reservoir. The data represent monthly samplings from August 1986 to February 1988.

Macroinvertebrate taxa	Site-I	Site-II	Site-III
Total no. of species	14	12	15
Annelida			
Tubificidae:			
Limnodrilus hoffmeisteri Claparède 1862	29.5	26.8	33.5
Naididae:			
Pristina aequiseta Bourne 1891	17.4	11.2	19.3
Hirudinea:			
Glossiphonia Johnson 1816	-	1.5	0.5
Arthropods			
Insecta:			
Ephemeroptera:			
Indialis badia (P.& E.) (nymph)	0.7	0.5	-
Hemiptera:			
Ranatra gracilis (Dall.)	0.3	-	0.5
Neuroptera:			
Corydalus (Latreille) (larvae)	0.7	-	0.5
Diptera:			
Culicoides (Latreille) (larvae)	3.5	4.1	1.8
Chironomus circumdatus (Kieffer) (larvae)	8.3	13.6	10.0
Tabanus (Linn.) (larvae)	-	-	0.5
Odonata:			
Thithemus pallidinervis (Kirby) (nymph)	1.7	0.5	0.7
Mollusca			
Mesogastropoda:			
Bellamya dissimilis (Müller)	8.5	10.6	9.5
B. bengalensis Phase *annandalei* Kobelt	3.5	8.0	2.7
Thiara tuberculata (Müller)	5.0	6.1	3.0
Gabbia orcula (Frauenfeld)	16.8	16.1	12.9
Indoplanorbis exustus (Deshayes)	2.6	-	4.0
Lymnaea luteola f. typica Lamark	1.5	1.0	0.6

The effluent-storage tank of the textile mill was the most highly polluted body in the urban system, particularly its site-I which had no dissolved oxygen in its water (Table 2). This site possessed only three species of pollution tolerant macroinvertebrates and had a completely different species composition from the other sites. Before the establishment of the E.T.P., the number of species recorded in the sediment of sites-I and II of the effluent-storage tank was only 3 and 5, respectively. Interestingly, the numbers increased slightly to 6 and 5 in the pools (sites-I and II respectively) receiving the treated effluent after the establishment of the E.T.P., and the water acquired some dissolved oxygen.

Comparing all the water bodies in terms of species composition in relation to organic pollution and dissolved oxygen, the Wadepally reservoir was the least polluted, which was reflected by the high number of species of macroinvertebrates. The effluent-storage tank of the textile mill was highly polluted, which was probably the reason for its site-I having no aquatic plants and the lowest number of species compared to the other sites and water bodies.

In the sediment of the littoral zone of Wadepally reservoir, the annelid *Limnodrilus hoffmeisteri* (Claparède 1862) was dominant, constituting 26.8% to 33.5% of the total benthic macrofauna. This was followed by *Pristina aequiseta* (Bourne 1891) comprising 11.2% to 19.3%, and the mollusc *Gabbia orcula* (Frauenfeld) which comprised 12.9% to 16.8% of the total. The larvae of *Chironomus circumdatus* (Kieffer) comprised 8.3% to 13.3% of the total macroinvertebrates. Less than 1% of the total macroinvertebrates was constituted by the nymph of the Ephemeropteran *Indialis badia* (P. & O.), the neuropteran *Corydalus* spp., the dragonfly *Trithemus pallidinervis* (Kirby) (in site-II and III), and the hemipteran *Ranatra gracilis* (Dall) (Table 3).

In the sediment of the littoral zone of Bhadrakali reservoir *L. hoffmeisteri* constituted 22.3% to 35.4% of the total benthic macroinvertebrates, followed by the larvae of *C. circumdatus* (19.9 - 27.5%), and the mollusc *G. orcula* (8.5 - 21.2%). The hemipteran *Pantala flavescens* (Fabr.), *R. gracillis* and the coleoptera *Cybister* spp. constituted less than 1% of the total (Table 4). Compared to the previous reservoir, the chironomidae were more evident in Bhadrakali reservoir. In both reservoirs, *L. hoffmeisteri* was the dominant species, and *G. orcula* accounted for a large percentage of the total benthic invertebrates.

In the sewage canal, *L. hoffmeisteri* constituted 4.6 to 56.1% of the total benthic macroinvertebrates, followed by the larvae of *C. circumdatus* with 8.0 to 94.7% of the total. The presence of large percentages of these organisms under lotic conditions usually reflects organic enrichment (Hawkes 1979). The remaining species, except *Culicoides* spp., constituted less than 2% of the total. The ephemeropterans, hemipterans, coleopterans, the odonata and some of the Diptera (*Tabanus* spp., *Leimnophora* spp. and *Eristalis tenax* L.) and the molluscans comprised less than 1% of the total (Table 5). This may be due to the sensitivity of these organism to organic loading. Olive *et al.* (1988) reported that all coleopterans, ephemeropterans, hemipterans, lepidopterans, megalopternas, plecopterans and trichopterans were regarded as intolerant of organic pollution. Oligochaetes, hirudineans, chironomids of the genus *Chironomus* and the snail *Physella* were assigned organic pollution tolerant species.

In most of the canal sites, there was an inverse relationship between the tubificidae and the Chironomidae larvae. Bazzanti and Loret (1982) stated that the percentage composition of the Chronomidae compared to the Oligochaeta was relatively lower at most of the sites of the reservoirs and canals. In site-II of the canal, *L. hoffmeisteri* constitutes only 4.6% of the total invertebrates, while *C. circumdatus* constituted 94.7%. This was probably due to not only the texture of

Table 4. Percentage composition of benthic macroinvertebrates in Bhadrakali reservoir. The data represent monthly samplings from September 1986 to February 1988.

Macroinvertebrate taxa	Site-I	Site-II	Site-III
Total no. of species	11	12	11
Annelida			
Tubificidae:			
Limnodrilus hoffmeisteri Claparède 1862	22.3	24.3	35.4
Naididae:			
Pristina aequiseta Bourne 1891	6.5	5.5	12.9
Arthropods			
Hemiptera:			
Ranatra gracilis (Dall.)	0.7	0.8	-
Pantala flavescens (Fabr) (nymph)	-	-	0.2
Coleoptera:			
Cybister Curtis	0.9	0.3	-
Diptera:			
Aides Meigen (larvae)	1.1	2.8	0.7
Chironomus circumdatus Kieffer (larvae)	27.5	31.2	19.9
Mollusca			
Mesogastropoda:			
Bellamya dissimilis (Müller)	7.0	8.9	8.3
B. bengalensis Phase *annandalei* Kobelt	0.7	4.4	4.3
Thiara tuberculata (Müller)	2.2	0.3	7.0
Gabbia orcula (Frauenfeld)	21.2	15.8	8.5
Indoplanorbis exustus (Deshayes)	9.9	4.7	2.3
Lymnaea luteola f. *typica* Lamark	-	1.0	0.5

the sediment but also the physico-chemical characteristics of both the sediment and the water. In addition, the prey-predator relationship is important, as Chironomids are important predators of oligochaetes (Loden 1974).

The species composition of the benthic macroinvertebrate community of the effluent-storage tank was completely different from that of the rest of the water bodies of the urban system. At site-I, annelids, both adults and larvae of some arthropods and molluscs were completely absent (Table 6). The rat-tailed maggot *E. tenax* was dominant, comprising 34.0% to 50.0% of the macroinvertebrates before the establishment of the E.T.P. and 54.4% to 65.6% after its establishment. *Limnophora* spp. followed, with 41.8% to 56.5% before the establishment of the E.T.P. and 0 to 12.4% after its establishment. Although it has been reported that *E. tenax* is usually seen buried in the mud of septic pools and other highly

Table 5. Percentage composition of benthic macroinvertebrates in a sewage canal of the urban system. The data represent monthly samplings from January 1986 to December 1987.

Macroinvertebrate taxa	Site-I	Site-II	Site-III	Site-IV
Total no. of species	18	12	11	11
Annelida				
Tubificidae:				
Limnodrilus hoffmeisteri Claparède 1862	42.1	4.6	56.2	43.6
Naididae:				
Pristina aequiseta Bourne 1891	22.5	-	31.6	26.4
Earthworm	0.3	-	-	-
Arthropods				
Insecta				
Emphemeroptera				
Indialis badia (P.& E.) (nymph)	1.6	-	-	0.1
Hemiptera:				
Ranatra gracilis (Dall.)	0.9	0.1	-	0.1
Notonecta glauca (Linn.)	0.1	-	-	-
Corixa heiroglyphica (Duf.)	0.3	-	-	-
Neuroptera:				
Corydalus (Lareille) (larvae)	0.7	0.1	-	0.2
Coleoptera:				
Dytiscus	0.3	-	-	-
Odonata:				
Pantala flavescens (Fabr.) (nymph)	1.6	-	0.1	-
Diptera:				
Culicoidae (Latreille) (larvae)	2.9	0.3	1.4	1.3
Tabanus (Linn.) (larvae)	0.4	0.1	1.5	-
Limnophora (Robineau-Desovoidy)(larvae)	-	-	0.3	0.2
Eristalis tenax (Latreille) (larvae)	-	-	0.2	-
Chironomus circumdatus Kieffer (larvae)	25.4	94.5	8.2	27.8
Acari:				
Water mite	0.2	-	-	-
Mollusca				
Mesogastropoda:				
Bellamya dissimilis (Müller)	0.3	0.1	-	-
Thiara tuberculata (Müller)	-	0.1	-	0.1
Gabbia orcula (Frauenfeld)	0.1	-	0.1	0.1
Indoplanorbis exustus (Deshayes)	0.1	0.1	-	-
Lymnaea luteola f. typica Lamark	0.2	-		0.1
Vertebrate				
Amphibia:				
Tadpole larvae (*Rana* spp.)	-	-	0.4	-

organically polluted water bodies, with its breathing tube extending to the water surface (Khan 1983), there are no other relevant references available in the literature and thus, a common basis for comparison with the present data is not possible. The ephemeropteran *I. badia* and the hemipterans, viz. *Anisops breddini* Kirk and *R. gracilis*, and the molluscan, viz. *Bellamya dissimilis* Miller and *Thiara* (melanoides) *tuberculata* (Muller) colonized the pools receiving the treated effluents (Table 5). This may be due to the lower organic load and slightly better oxygenated conditions of the treated effluents (Table 2).

In lakes, many authors have shown that organic enrichment of the bottom supports a qualitative increase in Oligochaeta and a reduction in Chironomidae on the total fauna (Mylinski and Ginsburg 1977; Milbrink 1980; Seather 1980; Mastrantuono 1986). Mastrantuono (1986) stated that the application of such a concept to littoral zones may be valid. However, this concept would not be applicable to the highly organically polluted waters such as that of the effluent-

Table 6. Percentage composition of benthic macroinvertebrates in an Effluent-Storage Tank of A.J. Textile Mill. The data represent monthly samplings from February 1986 to September 1987.

Macroinvertebrate taxa	Before E.T.P.		After E.T.P.	
	Site-I	Site-II	Site-I	Site-II
Total no. of species	3	5	6	5
Arthropods				
Insecta				
Emphemeroptera				
Indialis badia (P.& E.) (nymph)	-	-	3.8	-
Hemiptera:				
Anisops breddini Kirk	-	-	5.7	-
Ranatra gracilis (Dall.)	-	-	8.2	-
Coleoptera:				
Dytiscus (Linn.)	-	3.3	-	-
Diptera:				
Eristalis tenax (Latreille)(larvae)	54.4	34.1	57.4	59.5
Culicoides (Latreille) (larvae)	3.8	-	-	8.4
Limnophora (Robineau-Descovoidy) (larvae)	41.8	56.5	-	15.5
Chironomus circumdatus (larvae)	-	91.6	-	-
Dipteran larvae (unidentified)	-	4.5	10.6	-
Mollusca				
Mesogastropoda:				
Bellamya dissimilis (Müller)	-	-	14.3	8.4
Thiara tuberculata (Müller)	-	-	-	8.2

storage tank of the textile mill where annelids were completely absent from the sediment (Table 6). Although this concept may be good for comparing different sites or situations within a single water body, it is not good for comparing different water bodies in terms of organic enrichment. Moreover, when applying the above concept to site comparisons, even in the same water body, the depth of the sites have to be taken into consideration, as the relative density and distribution of oligochaetes are strictly dependent on depth (Bazzanti and Loret 1982).

The percentage composition of some of the benthic macroinvertebrate species, particularly *L. hoffmeisteri, P. aeguiseta, C. circumdatus,* and *G. orcula* in the reservoirs and the canal, and *E. tenax* and *Limnophora* spp. in the effluent-storage tank, is interesting as they seem to be very tolerant of different levels of organic pollution. In accordance with the present study, Milbrink (1980), Saether (1980) and Mastranatuono (1986) reported the oligochaeta as the most important group, with the highest percentage occurrence at various discharge sites of the littoral zone of Lake Nemi in Central Italy. The presence of *L. hoffmeisteri* at a high percentage in littoral zones reflected the organic enrichment of the sediment (Sarkaa 1983). The presence of the Naididae, *P. aquiseta,* also indicated organic loading of the water bodies (Learner 1979). It has been reported that a higher percentage of oligochaetes to total macrobenthos can indicate conditions of polluted water (Howmiller and Beeton 1971). Wiederholm (1980) reported that the oligochaetes, mainly tubificidae, were more abundant than the chinnomidae in heavily eutrophicated lakes.

From the present investigation, it is concluded that the littoral benthic macroinvertebrate community is useful in evaluating the localized effects of organic enrichment in lentic systems, as the profundal zoo-benthos indicates the water quality as a whole (Mastrantuono 1986). The dominance of *L. hoffmeisteri,* the presence of *P. aequiseta* and the molluscan *G. orcula* in high percentages, and the negligible numbers of the developmental stages of arthropods, viz. *I. badia, R gracillis* and *Corydalus* spp., indicated water polluted with a high organic load and oxygen-depleted conditions. Furthermore, the dominance of the larvae of *C. circumdatus* followed by *L. hoffmeisteri* indicated a relatively lower organic load and less oxygen-depleted conditions. The dominance of the larvae of *E. tenax* and that of *Lemnophora* spp. indicated highly polluted conditions with little or no dissolved oxygen in the water. However, more research is needed to evaluate the different types and levels of environmental pollution, particularly organic waste pollution and eutrophication in the tropics, using benthic macroinvertebrates as bioindicators.

Acknowledgements

The present investigation was supported by a Kakatiya University Research Fellowship to Mr B. Mallo Rao. Thanks are due to Prof. M.J. Wetzel (USA) for identifying the Oligochaeta, and to Dr. P.K. Chaudhuri for identifying the Chironomidae. Other species of arthropods were identified by experts of the

Zoological Survey of India.

References

Bazzanti, M. and Loret, E. (1982) Macrobenthic community structure in a polluted lake, Lake Nemi, Central Italy. *Bollettion di Zoologica*, **49**, 79-91.

Dudani, V.K., Kumar, S., Verma, A.M. and Pandey, S. (1988) Biological indicators of paper mill waste water. *Environment and Ecology*, **6**, 600-603.

Hawkes, H.A. (1979) Invertebrates as indicators of river water quality. In A. James and L. Evison (eds.) *Biological Indicators of Water Quality*. John Wiley and Sons, New York.

Howmiller, R.D. and Beeton, A.M. (1971) Biological evaluation of environmental quality, Green Bay, Lake Michigan. *Journal of Water Pollution Control Federation*, **43**, 123-133.

Khan, K.R. (1983) Biological indicators and indices of water quality. In C.K. Varshney (ed.) *Water Pollution and Management*, 198-298. Wiley Eastern Ltd., New Delhi.

Krishnamoorthi, K.P. and Sarkar, R. (1979) Macroinvertebrates as indicators of water quality. *Proceedings of Symposium on Environmental Biology, Muzaaffarnagar*, 133-138.

Learner, M.A. (1979) The distribution and ecology of the Naididae (Oligichaeta) which inhabit the filter beds of sewage works in Britain. *Water Research*, **13**, 1291-1299.

Loden, M.S. (1974) Predation by Chironomid (Diptera) larvae on Oligochaetes. *Limnology and Oceanography*, **19**, 156-159.

Mahadevan, A. and Krishnaswamy, S. (1984) Chironomid larval population size - as index of pollution in river Vaigai. *Pollution Research*, **3**, 35-38.

Mandal, B.K. and Moitra, S.K. (1975) Studies on the bottom fauna of a fresh water pond at Burdwan. *Journal of Indian Fisheries Society*, **7**, 43-48.

Mastrantuono, L. (1986) Littoral sand zoobenthos and its relation to organic pollution in Lake Nemi, Central Italy. *Hydrobiological Bulletin*, **19**, 171-178.

Michael, R.G. (1968) Studies on the bottom fauna in a tropical freshwater fish pond. *Hydrobiologia*, **31**, 203-230.

Milbrink, G. (1980) Oligochaete community in pollution ecology : the European situation with special reference to lakes in Scandinavia. In R.O. Bsinkhurst and D.G. Cook (eds.) *Aquatic Oligochaete Biology*, 433-455. Plenum Press.

Mylinski, E. and Ginsburg, W. (1977) Macroinvertebrates as indicators of pollution. *Journal of American Water Works Association*, **69**, 538.

Olive, J.H., Jackson, J.L., Bas, J., Holland, I and Sayisky, T. (1988) Benthic macroinvertebrates as indexes of water quality in the upper Cuyahoga river. *The Ohio Journal of Science*, **88**, 91-98.

Patil, S.G., Singh, D.P. and Harshey, D.K. (1983) Ranital (Jabhalpur), a sewage polluted water body, as evidenced by chemical and biological indicators of pollution. *Journal of Environmental Biology*, **4**, 43-49.

Reddy, M.V. and Rao, B.M. (1987) Structure of benthic macmacroinvertebrate population, particularly the Tubeficidae and Chirnomid larvae, in a sewage polluted canal. *Pollution Research*, **6**, 65-68.

Reddy, M.V. and Rao, B.M. (1990) Bhadrakali tank (=reservoir) in Warangal. Seasonal community structure of benthic macroinvertebrates and some limnological characteristics of littoral zone. In P.S. Pitchaiah and B.M. Rao (eds.) *Water in Andhra Pradesh - Problems and Prospects*. Chug Publishers, New Delhi (in press).

Saether, O.A. (1980) The influence of euthrophication on deep lake benthic invertebrate communities. *Progress in Water Technology*, **12**, 161-180.

Sarkaa, J. (1983) A quantitative ecological investigation of the littoral zoobenthos of an oligotrophic Finnish lake. *Annales Zoologici Fennici*, **20**, 157-178.

Sarkar, S.K. (1989) Seasonal abundance of benthic macro-fauna in a fresh water pond. *Environment and Ecology*, **7**, 113-116.

Vasisht, H.S. and Bandal, R.S. (1979) Seasonal variation of benthic fauna in some north Indian lakes and ponds. *Indian Journal of Ecology*, **6**, 33-37.

Wiederholm, T. (1980) Use of benthos in lake monitoring. *Journal of Water Pollution Control Federation*, **52**, 537-547.

Monitoring the Effects of Agricultural Pesticides on Wildlife

H.M. Thompson and P.W. Greig-Smith

Central Science Laboratory, MAFF, Hook Rise South, Tolworth, Surbiton, Surrey, KT6 7NF, U.K.

Key words: pesticides, exposure, effects, indicators, field trials, post-approval surveillance.

Abstract

The use of agricultural chemicals in the UK is governed by controls which seek to ensure that any risks to wild species are identified and minimized before approval for use is granted. This involves a detailed, sequential programme of testing new products and is followed by post-registration monitoring of effects under field conditions.

This paper reviews the range of approaches (biological, biochemical and chemical) used by MAFF to investigate pesticide effects on wild birds and mammals in England and Wales. It includes examples of experimental field trials and discusses the national Wildlife Incident Investigation Scheme, which provides surveillance of potential environmental problems. Experience in the UK over the last 20 years has demonstrated that it is more productive to combine biomonitoring methods into an integrated multidisciplinary approach. This philosophy has ensured that potential hazards of pesticide use to wildlife may be anticipated and minimized.

Introduction

The widespread use of agricultural pesticides entails a responsibility to ensure that they do not cause harm to human health, wildlife and the environment. In the UK, this is achieved by a stepwise evaluation process which seeks to identify and minimize risks before approval for use is granted. For wildlife, this is followed by post-registration surveillance of effects under commercial use, allowing for revision of approval conditions if necessary (Hardy 1987).

In both the pre- and post-approval stages, the existence and significance of wildlife hazards can be assessed by monitoring a range of selected 'indicators'. This involves the selection of species to study, and parameters to measure in order to identify the various possible direct and indirect hazards to animals

Bioindicators and Environmental Management
ISBN 0-12-382590-3

(Fig. 1). Field monitoring may be based on biological, biochemical or chemical markers, in various combinations. This paper reviews the approaches taken by the Ministry of Agriculture, Fisheries and Food (MAFF) Central Science Laboratory, both in experimental field trials and post-approval surveillance.

Stepwise evaluation of pesticides

The initial stages in the evaluation of a pesticide involve generation of laboratory data on the intrinsic properties of the active ingredient, (e.g. solubility, volatility etc.) and basic toxicology tests on laboratory mammals (rats, mice, guinea pigs). Toxicity data collected include acute, short-term dietary, and long-term chronic tests (including reproductive effects) and data may also include information on modes of action and routes of metabolism. Although this information is primarily used in the prediction of safety to humans, it is also relevant to wildlife. In addition, toxicity tests are conducted on 'surrogate' bird species such as the mallard *Anas platyrhynchos* and Japanese quail *Coturnix coturnix japonica*. This provides a crude indication of toxicity and reproductive effects in birds and mammals. Pesticides which are likely to cause a risk to bees, due to their use in agriculture, horticulture or forestry, are tested for toxicity against honeybees (Stevenson 1968).

These laboratory data give an approximate indication of the toxicity of pesticides to animals, which may be compared to their likely exposure (Riley 1990). However, not all non-target species can be tested in the laboratory, and extrapolation between species is highly uncertain. Nor is prediction of effects under field conditions easy, because exposure is very variable. However, laboratory data allow uncertainties about the safety of a pesticide to be narrowed down to a number of specific questions about both lethal and sub-lethal effects.

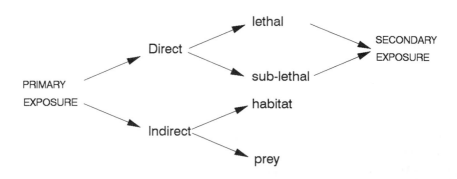

Figure 1. Possible direct and indirect hazards of pesticides to wildlife.

The next stages in evaluation are designed to answer these specific questions in order to confirm or deny a prediction of safety. They may involve further laboratory studies, or field trials tailored to individual pesticides and their proposed uses. Such trials are not essential for all products. Flexibility and expert knowledge are essential to the planning, execution and interpretation of field trials (Greig-Smith *et al*. 1990b). An integral part of this is a reliable suite of techniques to detect and measure lethal and sub-lethal effects on wildlife. Indirect effects, due to depletion of prey, for example, are rarely investigated in detail in field trials.

Once in commercial use, the undesirable side-effects of a pesticide may be detected in the course of monitoring exercises, such as surveys of residues of persistent organochlorine insecticides in predatory birds (Cooke *et al*. 1982) and incident investigations (Greig-Smith in press). Here too, a range of methods is available to identify the involvement of pesticides in the deaths or illness of wildlife.

The importance of indicators is different for pre- and post-approval studies (Table 1). In the first case, it is important to address the exposure of animals, in order to confirm, or refine previous estimates of likely contact with a chemical. Assessment of the severity of effects in experimental trials is also important, to determine whether they exceed a threshold of unacceptability. However, precise measurement of effects may not be essential because trials can only be carried out in a limited range of conditions, and the results require cautious extrapolation to predict effects under 'worst-case' conditions (Hart 1990). In contrast, the focus of post-approval monitoring is on the effects that occur in the full range of commercial use conditions, and documenting exposure is of less value.

Table 1. Comparison of need for monitors of exposure and effects in pre- and post-approval studies.

| | Need for monitoring | |
	Exposure	Effect
Pre-approval (field-trials)	Confirmation of exposure under typical conditions of use	Assess severity of adverse effects in relation to thresholds of acceptability
Post-approval (incident approval)	Not relevant	Determine level of mortality under a broad range of use conditions.

Indicator species

Species for study may be chosen for a variety of reasons, the choice of which may affect the information gathered and its interpretation. The major criteria for choice are listed by Greig-Smith (1990) (see Table 2). For field trials, several of these are generally combined to identify a small number of species that are predicted to be exposed to the pesticide and for which it is feasible to gather data. This depends on the formulation,time of application, and the crop on which the pesticide is used (Hardy 1990).

Thus seed-eating and grain-eating birds (e.g. finches *Fringilla* spp. and *Carduelis* spp., pheasants *Phasianus colchicus* and pigeons *Columbus* spp.) and certain small rodents (e.g. wood mice *Apodemus sylvaticus*) are suitable indicator species for seed treatments and granular formulations, whereas grazing species (e.g. geese *Branta* and *Anser* spp., and rabbits) are relevant for sprayed products.

In reactive investigations of incidents, when the pesticide involved, if any, is unknown, the species monitored are dictated by the biases and limitations of the arrangements for discovery, reporting and examination of carcasses (Greig-Smith, in press). For example, the poisoning of wood mice which has been the focus of many field trials (e.g. Westlake *et al.* 1980), would be unlikely to be revealed by the UK incident investigation scheme.

Biological methods

Approaches to biological monitoring of pesticide effects on wildlife range from estimates of mortality to sub-lethal effects on behaviour. The utility of each method depends on the indicator species, and certain parameters (e.g. breeding performance) are largely limited to birds.

Table 2. Criteria for the selection of 'indicator' species (from Greig-Smith, 1990).

---

1. Species that are particularly likely to be exposed.

2. Species thought to be particularly sensitive to a given exposure.

3. Species for which an adverse impact would be especially damaging.

4. Species which closely represent the typical characteristics of a large number of others in the community.

5. Abundant species permitting large sample sizes for data collection.

6. Species whose ecology and behaviour provide easy opportunities to measure effects.

The principles of each technique are discussed below and their advantages and disadvantages are compared in Table 3.

1. Changes in population density

The use of toxic pesticides may cause mortality and thus result in local population declines. Such changes in local bird populations may be monitored by well established techniques such as point counts or line transects and, during the breeding season, territory mapping (for reviews see Mineau and Peakall 1987; Bairlein 1990; Edwards *et al.* 1979). Small mammal populations, which tend to be more sedentary but less obvious than birds can be evaluated by intensive capture/mark/recapture methods (Flowerdew 1988; Greig-Smith and Westlake 1988).

Because of the high variability in population density of many species, and the difficulties of obtaining well-matched control sites, population indices are an insensitive tool. Major factors which need to be taken into consideration when using these techniques include the effect of environmental conditions on singing activity of birds, rapid replacement of certain species by animals from outside the treated area and any behavioural effects of the pesticide which may increase or reduce the conspicuousness of individuals, without causing mortality. Even major changes in numbers should not be ascribed to a pesticide effect without corroborative information of other kinds.

2. Survival and mortality of individuals

Methods to monitor the survival or death of marked individual animals (e.g. capture/recapture, and radiotracking) are of great value, if they can be combined with confirmation that exposure to the pesticide has occurred, and if the cause of disappearance can be determined. However, individual mortality may not necessarily cause any changes at the population level.

Searches to detect dead animals are often used in field trials, and casual discovery of casualties is the basis of incident investigations. Even systematic searches tend to favour larger, more conspicuous species (Fischer 1990; Mineau and Collins 1988). By implication, inability to find carcasses does not rule out the possibility that mortality has occurred; for example, animals may die outside the search area. Equally, carcasses which are found cannot, in isolation, be definitely attributed to a pesticide effect. Therefore, even with quantification of search efficiency, this method can only be a qualitative parameter. However, it does provide material for further laboratory investigations (see later).

3. Breeding performance

Both of the above parameters are primarily measures of the lethal effects of pesticides, although estimates of population levels may sometimes be confounded by sub-lethal effects. In addition, some pesticides may affect reproductive performance by direct effects on the parents or young or by indirect effects on food sources. In practice, breeding performance can only be readily evaluated in the field for birds. Parameters that can be monitored include nesting, hatching and fledging success, and growth rate of the young (Bairlein

Table 3. Advantages and limitations of biological parameters used in monitoring exposure to pesticides and its effects.

---

1. Changes in Population Density

   Advantages
   - Indicates direct and indirect effects.
   - May provide information on species abundance and diversity.
   - Indicates short and longer term effects.

   Limitations
   - Matched control sites are often difficult to find.
   - Species differences in population density.
   - Difficult to identify the cause of changes (e.g. death, emigration).
   - Relatively insensitive e.g. due to rapid replacement by some species.
   - Detectability may be affected by behavioural changes following exposure.

2. Survival and Mortality of Individuals

   Advantages
   - Focuses on direct effects.
   - Addresses severe consequences directly.
   - Carcasses provide material for other studies.
   - Tracking of known individuals removes confounding effects of replacement.

   Limitations
   - Difficult to determine cause of disappearance of individuals.
   - Carcass searches are not quantitative.
   - Favours larger,more conspicuous species.
   - Lack of carcasses does not prove an absence of mortality.
   - Only reveals lethal effects.

3. Breeding Preformance

   Advantages
   - Addresses both direct and indirect effects.
   - Allows extrapolation to population-level consequences.

   Limitations
   - Difficult to define cause of change.
   - For most species the density of breeders is low.
   - Matched control sites are often difficult to find.
   - Highly variable and often low breeding success makes detection of effects difficult.
   - There is substantial variation throughout season.

4. Behaviour

   Advantages
   - Addresses both direct and indirect effects.
   - May reveal mechanisms that alter subsequent exposure.
   - Changes are often dose-related.

Table 3. (continued)

    - Highlights species that are most at risk, by exploring routes of exposure.
    - Sub-lethal effects can be evaluated.
    Limitations
    - Behaviour changes often occur close to a lethal dose.
    - Affected birds may become less or more conspicuous.
    - Difficult to define the cause of changes.
    - Behaviour is very variable, reducing the power to detect changes.
    - Some behaviour may be difficult to interpret and quantify.

5. Autopsies
    Advantages
    - Symptoms may indicate the class of compound involved.
    - Addresses direct effects of exposure.
    Limitation
    - Effects of disease/other compound may be similar (i.e. symptoms
rarely .tb0.4"
    compound specific).

1990), which may have population-level consequences even if mortality of adult birds does not occur. In addition, longer term effects of pesticides can also be monitored, e.g. the effects of organochlorine insecticides on eggshell thickness of the peregrine falcon *Falco peregrinus* (Newton 1984).

Major considerations in using breeding performance to monitor sub-lethal effects of pesticide use include wide inter-individual variations in nesting success, and the relatively small numbers of breeding pairs that occur in most field trial areas. This results in large changes in breeding success being required to produce a significant result (Barlein 1990), unless the population can be supplemented by artificial nest-sites (Hooper *et al.* 1990).

4. Behaviour
Certain pesticides, particularly anticholinesterase compounds (i.e. organophosphorus and carbamate insecticides), affect behaviour by their interaction with the nervous system. Thus, changes in behaviour, such as singing activity, parental behaviour and territoriality in birds (Grue *et al.* 1982) may be used to indicate the effects of pesticide exposure. In addition, behaviour may also alter exposure or bias the collection of data; for example, affected birds may leave the area or become less conspicuous (Mineau and Peakall 1987; Hart 1990). Significant behavioural changes under laboratory conditions tend to occur within one order of magnitude of the lethal dose, owing to wide inter-individual variations in 'normal' behaviour (Hart 1990) and behaviour is therefore a relatively insensitive indicator of exposure, but a useful indicator of its effects.

5. Autopsies

Post-mortem examination may help to distinguish death due to pesticides from natural causes such as disease. A few chemicals produce characteristic lesions in both live-trapped animals and those collected as casualties. For example, severe internal haemorrhaging is characteristic of exposure to anticoagulant rodenticides. However, results should be interpreted with caution since symptoms are variable, and may not be specific, so that several classes of compound and/or disease may lead to similar post-mortem findings.

Biochemical methods

The selection of biochemical parameters to monitor exposure and effects of pesticides depends on the mode of action and side effects of the compound. Therefore, complementary laboratory studies are necessary to allow interpretation of changes in biochemical parameters measured in the course of field trials (Tarrant and Westlake 1988). Advantages and disadvantages of these techniques are shown in Table 4.

Measurement of enzyme inhibition is of value if compounds are known to selectively inhibit specific enzymes and if inhibition is dose-related. A well-known example is the inhibition of brain and plasma cholinesterases by organophosphate and carbamate pesticides. Inhibition of brain acetyl-cholinesterase is the primary toxic action caused by these compounds, as well as being a useful indicator of exposure (for reviews see Mineau, in press). The use of plasma esterases, which are highly sensitive to inhibition, allows a series of samples to be taken from the same individual. Thus each animal can act as its own control and problems of inter-individual variation in activity are removed. (For a review of the use of serum esterases, see Thompson, in press).

Rodenticides that act as anticoagulants can be monitored by measuring blood clotting time, which directly reflects the mode of action of these compounds (Suttie 1980).

Exposure to some compounds results in induction or release of enzymes, independent of their primary mode of action. For example, dosing with organochlorine compounds can cause induction of microsomal monooxygenase activity (Stanley *et al.* 1978) and may also cause tissue damage resulting in leakage of tissue-specific enzymes into the blood, e.g. glutamate oxaloacetate transaminase or lactate dehydrogenase (Westlake *et al.* 1979).

All these methods require comparable controls. In some cases (e.g. non-lethal monitoring of blood samples) individuals can act as their own controls if samples are taken before and after pesticide application. However, additional samples should be taken from animals in an untreated area at the same time so that any effect of environmental factors, such as temperature, can be taken into account. Many enzymes vary in activity depending on sex, age, species, season and even time of day (Hill and Murray 1987; Thompson *et al.* 1988; Hill 1989a). In addition, factors such as assay conditions and time since death (or after sample collection) and storage conditions can affect activity levels (Hill 1989b). Thus it is essential that concurrent samples are taken from the control area and are treated in exactly the same way as those from the treated area.

A major bias in the collection of samples for measurement of biochemical parameters, as for other measurements, is ensuring the collection of a representative sample of the animals affected (Mineau and Peakall 1987). Animals which are affected may become more, or less likely to be detected and caught. It may, therefore, be worth observing or radiotracking a number of known individuals in order to obtain multiple blood samples from the same animals, to reduce biases in monitoring both lethal and sub-lethal exposure (Hooper *et al.* 1990).

Table 4. Advantages and limitations of biochemical parameters used in monitoring exposure to pesticides and its effects.

---

1. Enzyme Inhibition

    Advantages
- Provides a direct indicator of exposure.
- Focuses on certain classes of compounds (e.g. organophosphorus).
- Effects are dose-related.
- Allows non-lethal sampling.
- Often very sensitive to low level exposure.
- It may be possible to perform assays 'on-site'.

    Limitations
- Wide differences between individuals ages, sexes etc.
- Effects are not compound-specific.
- Not necessarily a measure of effect depending on the mode of action of the compound.
- There may be confounding interactive effects of compounds.
- Post-mortem change and changes during storage may confuse patterns.

2. Enzyme Induction/Release

    Advantages
- Direct indicator of exposure and tissue damage.
- Non-lethal methods are available.
- It may be possible to perform assays 'on site'.

    Limitations
- Similar changes are caused by many classes of compounds.
- High doses are often needed to produce effect.
- There are wide inter-individual differences.

3. Anticoagulation

    Advantages
- Direct indicator of exposure, and of effects.
- Specific to certain classes of compound.

    Limitations
- Not readily possible for carcasses a long time post-mortem.

Chemical methods

Evaluation of chemical residues is the most direct way of estimating the distribution of pesticides in the environment, and also contributes to monitoring exposure of wildlife. Advantages and disadvantages of these techniques are shown in Table 5. They require specialized equipment and trained staff and analyses are costly. Because of the short half lives of many modern pesticides, residue levels may decline rapidly. Levels measured in animal tissues are not necessarily dose-related. However, such analyses can be invaluable in identifying the cause of death and confirming exposure of wildlife.

At the other end of the scale, the presence of low, sub-lethal levels of pesticides, such as rodenticides, is of unknown physiological significance. The development of more sensitive techniques, such as adduct formation with DNA (a highly specialized and very expensive technique), may increase the frequency with which low levels of pesticides, of which the significance is unknown, are detected.

Clearly none of the above techniques is capable, in isolation, of fully evaluating the effects of pesticides on wildlife. However, by combining several methods a fuller picture of exposure and its effects can be evolved. In this way the limitations of one method may be compensated by the advantages of another

Table 5. Advantages and limitations of chemical parameters used in monitoring exposure to pesticides and its effects.

___

1. Residue Levels
   Advantages
   - Identifies compounds involved.
   - Defines amounts present in tissues or other samples.
   - Permits direct measurement of environmental concentrations.
   Limitations
   - Some compounds have short half-lives.
   - Results are specific to the tissues analysed.
   - Often requires lethal sampling.
   - Not necessarily dose-related.
   - The significance of low residue levels is poorly understood.

2. Adduct Formation
   Advantages
   - Very sensitive.
   - Direct indicator of exposure.
   Limitations
   - Applies to limited number of compounds.
   - Very specialized technique.
   - The meaning of low levels is poorly understood.

___

The next sections discuss the use of these methods by MAFF in experimental field trials, and in the course of post-approval surveillance.

Field studies

In general, field trials aim to assess the hazard of a known pesticide to wildlife by estimating its toxicity and measuring exposure. The way in which this is approached depends on the pesticide and the specific questions raised by laboratory studies. Additional targetted laboratory studies may be undertaken to aid interpretation of biochemical, histological and residue data from field trials.

The range of parameters which have been measured and integrated in the course of published MAFF field studies is shown in Table 6. Two of these field studies will be discussed in more detail.

1. Exposure of small mammals to organophosphorus seed treatments

The withdrawal of the persistent organochlorine seed treatments in the 1970's led to the introduction of relatively less persistent but more acutely toxic organophosphorus dressings. This resulted in a need to evaluate the exposure and effects of these dressings on wildlife. The use of the compounds as seed treatments meant that granivorous birds and small mammals were most likely to be affected. Rodents are readily trapped on fields and are relatively sedentary (compared to birds). They are also easily maintained for laboratory feeding studies, which would provide the data necessary to interpret results from the field trial.

The field trials therefore aimed to evaluate the hazard of the organophosphorus seed dressings, chlorfenvinphos and carbophenothion to wood-mice (Westlake *et al.* 1980; 1982). The multidisciplinary approach consisted of measurement of residues in surface grain and in tissues of wood-mice trapped on the study field and a range of biochemical measurements including plasma and brain acetylcholinesterase and glutamate oxaloacetate transaminase (GOT). In addition complementary laboratory studies were undertaken. This approach, involving both laboratory and field investigations,showed that although exposure could be demonstrated by chemical and biochemical measurements the hazard of these organophosphorus seed treatments to wood-mice was low.

2. Effects of exposure to the aphicide demeton-S-methyl on tree sparrows (*Passer montanus*)

This study formed part of the 'Boxworth Project', a seven year multidisciplinary field study of the effects of intensive pesticide use on winter wheat on flora and fauna as well as the economics of crop production (Greig-Smith 1989).

The study aimed to investigate exposure to demeton-S-methyl (an organophosphorus aphicide) and its effects on nestling tree sparrows. The approach included studies of breeding success, effects on growth, parental visit

Table 6. Examples of the integration of measured parameters in MAFF experimental field studies.

| | | Parameters | |
Study	Biological	Biochemical	Chemical
1. Aldicarb granules on sugar beet, (Nematocide) Bunyan *et al.* (1981).	Population index insects, mammal & birds. Bird activity on field. Granule counts. Casualty searches. Post-mortem.	Brain and tissue esterases on live trapped animals and carcasses.	Residues on carcasses, soil, worms, eggs, live trapped animals, sugar beet, weeds.
2. Methiocarb as bird repellent on cherries, (Hardy *et al.* 1987).	Population index by capture/ recapture for birds. Breeding success. Casualty searches. Post-mortem. Liver histology.	Brain and blood esterase on casualities, and following destructive and non-destructive sampling of birds.	Residues in carcasses, live-trapped birds, cherries, ground deposition, drift.
3. Organo-P cereal seed treatments (Westlake *et al.* 1980, 1982).	Population index of wood mice. Grain counts	Brain and blood esterase of casualities and live-trapped. Glutamate oxaloacetate transaminase (GOT)	Residues in grain and wood mice. (Casualities and live-trapped)
4. Demeton-S-methyl aphicid on winter wheat (tree sparrows) Hart *et al.* 1990).	Tree sparrow population density. Breeding success. Nestling growth. Parental behaviour. Faecal analysis. Nestling deaths.	Blood esterase of nestlings.	
5. Molluscicide granules (Greig-Smith & Westlake 1988; Tarrant *et al.* 1990).	Population index of wood mice. Granule counts. Post-mortem.	Brain esterase on live-trapped mice and casualities.	Residues in live-trapped mice and casualities.

rates, foraging destination, diet (by faecal analysis) and biochemical measurements (serum cholinesterase and carboxylesterase).

Exposure of the nestling tree sparrows was demonstrated by serum cholinesterase inhibition. However, no significant effects were observed on parental visit rate, growth or breeding success. Thus it can be inferred that no significant hazard to the population is posed by the use of this aphicide (Hart *et al.* 1990; Greig-Smith *et al.* in press).

Post-approval surveillance

During development a pesticide can never be tested under all conditions or against all species likely to be exposed, particularly rare species. Therefore MAFF has developed a Wildlife Incident Investigation Scheme to confirm that the prediction of safety from field trials of pesticides is valid under commercial use. Over the last 25 years any suspected poisoning of wildlife thought to involve pesticide use on or near farmland has been investigated (Hardy *et al.* 1986; Greig Smith 1988).

Reports of wildlife mortality are investigated in the field both to determine the scale of the problem and record any relevant details such as local pesticide use. Casualties undergo post-mortem examination at a MAFF Veterinary Investigation Centre to screen out deaths due to disease or trauma. When cause of death cannot be determined tissue samples are sent to Tolworth Laboratory for biochemical and residue studies. Analyses are carried out for a wide range of pesticides (organochlorines, carbamates, organophosphorus, pyrethroids, rodenticides, etc.) based on field information and post-mortem findings. Brain acetylcholinesterase activity is assayed and compared to published control values (Westlake *et al.* 1983) to support residue measurements of anticholinesterase compounds. Obviously in these cases there is rarely an opportunity to collect matched control data. Unlike field trials, the pesticides involved in incident investigations may not be known and greater reliance may have to be placed on post-mortem investigations and screening for a range of pesticides. Obviously such reactive investigations can never attribute cause of death in all cases. However, it is considered adequate to allow detection of any significant problems associated with a particular pesticide or formulation. The results of these investigations are published each year in summary form (Greig-Smith *et al.* 1989, 1990a). If unacceptable hazards are revealed by this scheme then information is passed to the Advisory Committee on Pesticides, an independent body responsible for reviewing the approval status of compounds.

Conclusions

In both experimental field trials and incident investigations there is a wide range of techniques available for monitoring exposure and its effects. Experience in the UK has shown that the most productive approach is to combine the relevant methods into an integrated and multidisciplinary investigation. This ensures that the potential hazards to wildlife of pesticide use in the UK are anticipated and avoided and has resulted in few significant problems from the use of new pesticides in agriculture.

References

Bairlein, F. (1990) Estimating density and reproductive parameters for terrestrial field testing with birds. In L. Somerville and C.H. Walker (eds.). *Pesticide Effects on Terrestrial Wildlife*, 113-128. Taylor and Francis, London.

Bunyan, P.J., Van den Heuvel, H.J., Stanley, P.I. and Wright, E.N. (1981) An intensive field trial and a multi-site surveillance exercise on the use of aldicarb to investigate methods for the assessment of possible environmental hazards presented by new pesticides. *Agro-Ecosystems*, 7, 239-262.

Cooke, A.S., Bell, A.A and Haas, M.B (1982) *Predatory Birds, Pesticides and Pollution*. Institute of Terrestrial Ecology, Cambridge.

Edwards, P.J., Brown, S.M., Fletcher M.R. and Stanley P.I. (1979) The use of a bird territory mapping method for detecting mortality following pesticide application. *Agro-Ecosystems*, 5, 271-282.

Fischer, D.L (1990) Problems in the estimation of the percent mortality in carcass searching studies In L. Somerville and C.H. Walker (eds.), *Pesticide Effects on Terrrestrial Wildlife*, 285-290. Taylor and Francis, London.

Flowerdew, J.R. (1988) Methods for studying populations of wild mammals In M.P. Greaves, B.D. Smith and P.W. Greig-Smith (eds.) *Field Methods for the Study of Environmental Effects of Pesticides*, 67-76. Monograph No. 40. BCPC, Croydon.

Greig-Smith, P.W. (1988) Wildlife hazards from the use, misuse and abuse of pesticides. *Aspects of Applied Biology*, 17, 247-256.

Greig-Smith, P.W. (1989) The Boxworth Project - Environmental effects of cereal pesticides. *Journal of the Royal Agricultural Society of England*, 150, 171-187.

Greig-Smith, P.W. (1990) Intensive study versus extensive monitoring in pesticide field trials In L. Somerville and C.H. Walker (eds.) *Pesticide Effects on Terrestrial Wildlife*, 217-239. Taylor and Francis, London.

Greig-Smith, P.W. (in press) Understanding the impact of pesticides on wild birds by monitoring incidents of poisioning In R.J. Kendall and T.E. Lacher (eds.) *The Population Ecology and Wildlife Toxicology of Agricultural Pesticide Use: A Modelling Initiative for Avian Species*.

Greig-Smith, P.W. and Westlake, G.E. (1988) Approaches to hazard assessment for small mammals in cereal fields In M.P. Greaves, B.D. Smith and P.W. Greig-Smith (eds.). *Field Methods for the Study of the Environmental Effects of Pesticides*, 303-311. Monograph No. 40. BCPC, Croydon.

Greig-Smith, P.W., Fletcher, M.R., Hunter, K., Quick, M.P., Ruthven, A.D and Shaw, I.C. (1990a) *Pesticide poisoning of animals 1988; investigations of suspected incidents in Great Britain*. Report of the Environmental Panel of the Advisory Committee on Pesticides, MAFF, London.

Greig-Smith, P.W., Frampton, G.K. and Hardy, A.R. (in press) *The Boxworth*

Project: Pesticides, Cereal Farming and the Environment. HMSO, London.

Greig-Smith, P.W., Somerville, L., Walker, C.H., Hardy, A.R., Klein, W., Mogensen, B., Pfluger, W., Riley, D., Stanley, P.I and Vighi, M. (1990b) Recommendations for terrestrial field testing pesticides. In L. Somerville and C.H. Walker (eds.) *Pesticide Effects on Terrestrial Wildlife*, 353-393. Taylor and Francis, London.

Grue, C.E., Powell, G.V.N. and McChesney, M.J. (1982) Care of nestlings by wild female starlings exposed to an organophosphate pesticide. *Journal of Applied Ecology*, **19**, 327-335.

Hardy, A.R. (1987) Ecotoxicology of pesticides: the laboratory and field evaluation of the environmental hazard presented by new pesticides. In L.G. Costa (ed.) *Toxicology of Pesticides: Experimental, Clinical and Regulatory Aspects*, 185-196. NATO ASI Series Vol. H13, Springer-Verlag, Berlin.

Hardy, A.R. (1990) Estimating exposure : The identification of species at risk and routes of exposure. In L. Somerville and C.H. Walker (eds.) *Pesticide Effects on Terrestrial Wildlife*, 81-98. Taylor and Francis, London.

Hardy, A.R., Fletcher, M.R and Stanley, P.I (1986) Pesticides and Wildlife: twenty years of vertebrate wildlife incident investigations by MAFF. *State Veterinary Journal*, **40**, 182-192.

Hardy, A.R., Stanley, P.I and Greig-Smith, P.W. (1987) Birds as indicators of the intensity of use of agricultural pesticides in the UK. *ICBP, Technical Publication No. 6.* 119-132.

Hart, A.D.M. (1990) Behavioural effects in field tests of pesticides. In L. Somerville and C.H. Walker (eds.) *Pesticide Effects on Terrestrial Wildlife*, 165-180. Taylor and Francis, London.

Hart, A.D.M., Fletcher, M.R., Greig-Smith, P.W., Hardy, A.R., Jones, S.A and Thompson, H.M. (1990) Le Projet Boxworth: Effects de regimes de pesticides contradictoires sur les oiseaux de terre arable. *Proceedings 5th International Symposium, ANPP.*

Hill, E.F. (1989a) Sex and storage affect cholinesterase activity in blood plasma of Japanese quail. *Journal of Wildlife Diseases*, **25**, 580-585.

Hill, E.F. (1989b) Divergent effect of post-mortem ambient temperature on organophosphorus and carbamate inhibited brain cholinesterase activity in birds. *Pesticide Biochemistry and Physiology*, **33**, 264-275.

Hill, E.F. and Murray, H.C. (1987) Seasonal variation in diagnostic enzymes and biochemical constituents of captive northern bobwhites and passerines. *Comparative Biochemistry and Physiology*, **87B**, 933-940.

Hooper, M.J., Brewer, L.W., Cobb, G.P. and Kendall, R.J. (1990) An integrated laboratory and field approach for assessing hazards of pesticide exposure to wildlife. In L. Somerville and C.H. Walker (eds.) *Pesticide Effects on Terrestrial Wildlife*, 271-282. Taylor and Francis, London.

Mineau, P. (ed.) (in press), *Cholinesterase inhibiting insecticides: Impacts on wildlife and the environment.* Elsevier.

Mineau, P. and Collins, B.T. (1988) Avian mortality in agro-ecosystems II. Methods of detection. In M.P. Greaves, B.D. Smith and P.W. Greig-Smith (eds.) *Field Methods for the Study of Environmental Effects of Pesticides*, 13-28.

Monograph No. 40 BCPC, Croydon.

Mineau, P. and Peakall, D.B. (1987) An evaluation of avian impact assessment techniques following broad-scale forest insecticide sprays. *Environmental Toxicology and Chemistry*, **6**, 781-791.

Newton, I. (1984) Uses and effects on bird populations and organochlorine pesticides In D. Jenkins (ed.) *Agriculture and the Environment*, 80-88. ITE Symposium No. 13. ITE, Cambridge.

Riley, D. (1990) Current testing in the sequence of development of a pesticide In L. Somerville and C.H. Walker (eds.) *Pesticide Effects on Terrestrial Wildlife*, 11-24. Taylor and Francis, London.

Stanley, P.I., Bunyan, P.J., Rees, W.D., Swindon, D.M. and Westlake, G.E. (1978) Pesticide induced changes in hepatic microsomal enzyme systems: Further studies on the effects of 1,1-di(p-chlorophenyl)-2-chloroethylene (DDMU) in the Japanese quail. *Chemical Biological Interactions*, **21**, 203-213.

Stevenson, J.H. (1968) Laboratory studies on the acute contact and oral toxicity of insecticides to honeybees. *Annals of Applied Biology*, **61**, 467-472.

Suttie, J.W (1980) Mechanism of action of vitamin K: Synthesis of carboxyglutamic acid. *CRC Critical Reviews in Biochemistry*, **8**, 191-233.

Tarrant, K.A. and Westlake, G.E. (1988) Laboratory evaluation of the hazard to woodmice, *Apodemus sylvaticus*, from the agricultural use of methiocarb molluscicide pellets. *Bulletin of Environmental Contamination and Toxicology*, **40**, 147-152.

Tarrant, K.A., Johnson, I.P., Flowerdew, J.R and Greig-Smith, P.W. (1990) Effects of pesticide applications on small mammals in arable fields and the recovery of their populations. In *Proceedings 1990 British Crop Protection Conference - Pests and Diseases*, BCPC, Croydon.

Thompson, H.M., Walker, C.H. and Hardy, A.R. (1988) Avian esterases as indicators of exposure to insecticides -the factor of diurnal variation. *Bulletin of Environmental Contamination and Toxicology*, **41**, 411.

Thompson, H.M. (in press) Serum ' B' esterases as indicators of exposure to pesticides. In P. Mineau (ed) *Cholinesterase Inhibiting Insecticides: Impacts on Wildlife and the Environment*. Elsevier.

Westlake, G.E., Blunden, C.A., Brown, P.M., Martin, A.D., Sayers, P.E., Stanley, P.I. and Tarrant, K.A. (1980) Residues and effects in mice after drilling wheat treated with chlorfenvinphos and an organomercurial fungicide. *Ecotoxicology and Environmental Safety*, **4**, 1-16.

Westlake, G.E., Bunyan, P.J., Johnston, J.A., Martin A.D and Stanley, P.I. (1982) Biochemical effects in mice following exposure to wheat treated with chlorfenvinphos and carbophenothion under laboratory and field conditions. *Pesticide Biochemistry and Physiology*, **18**, 49-56.

Westlake, G.E., Bunyan, P.J., Stanley, P.I. and Walker, C.H. (1979) The effects of DDMU on plasma enzymes and blood constituents in the Japanese quail. *Chemical Biological Interactions*, **25**, 197-206.

Westlake, G.E., Martin, A.D., Stanley, P.I. and Walker, C.H. (1983) Control enzyme levels in the plasma, brain and liver from wild birds and mammals in Britain. *Comparative Biochemistry and Physiology*, **76B**, 15-24.

Aquatic Biological Early Warning Systems: An Overview

Kees J.M. Kramer[1] and Joke Botterweg[2]

[1]Laboratory for Applied Marine Research, P.O. Box 57, 1780 AB Den Helder, The Netherlands.
[2]BKH Consulting Engineers, P.O. Box 93224, 2509 AE The Hague, The Netherlands. Present address: Institute for Inland Water Management and Waste Water Treatment, DBW/RIZA, Rivers Division, P.O. Box 9072, 6800 ED Arnhem, The Netherlands.

Abstract

In recent years the development and use of Biological Early Warning Systems (BEWS) in the aquatic environment has increased. More types of BEWS have become available and managers responsible for monitoring effluents, waste or drinking water have become more interested in the application of these systems.

A number of designs have been described that may lead to automatic systems. Many other designs have been documented. Despite these different approaches, only relatively few are being tested at the moment under laboratory conditions, while even less have reached the state of operational testing and of commercial production and practical use.

This paper summarizes these designs and approaches that are or may be functioning in an aquatic BEWS, and discusses possible applications under laboratory and field conditions.

Introduction

The detection of a sudden increase of the concentration of pollutants in influent or effluent has increasingly become an important aspect of drinking water and environmental water quality management and control. Besides the chemical and physical monitoring devices, organisms are more and more applied as environmental sensors. Only a fraction of the toxic compounds that may be present in water, are covered by the routinely operated chemical monitoring networks. While the physico-chemical methods give usually well defined answers, organisms will react to a multitude of toxic compounds that may be present in natural systems. The reaction of the organism may be non-specific, but they will reflect real biological harm i.e. toxic, usually sub-lethal, effects due to

Bioindicators and Environmental Management
ISBN 0-12-382590-3

bioavailability, which can only be estimated very roughly from the total chemical concentration in the aquatic environment. For a proper understanding, no alarm does not necessarily mean that there has not been a polluting situation. Furthermore, most chemical analyses take a relatively long time (hours to days), whereas organisms may react within minutes. It is evident that there is a complementary task for both approaches.

Over the past 15 years a number of monitoring approaches and devices that incorporate organisms as sensor have been developed for the marine and freshwater environment.

An early warning monitoring system will be considered here according to the following characteristics (Cairns and Van der Schalie 1980; Diamond *et al.* 1988):

- the organisms are held either in the laboratory situation or in the field under controlled conditions;

- they are exposed in a (semi) continuous flow regime to the water that has to be monitored;

- a physiological or behavioural parameter of the organism is monitored by a recording device with the capability of responding to abnormal conditions indicated by the organism (alarm);

- the function of the monitor is primarily for the detection of short term changes in toxicity.

The choice of organism is dependent on various factors that are partly of logistic importance. They should be obtained easily or grown in the laboratory; they should live long enough without major changes in appearance; a measurable parameter should be available preferably without disturbing the organism; their maintenance (feeding, replacement) should be minimal; and the availability of ample literature for standard test organisms is advantageous.

Kenaga (1978) reviewed the various organisms that were found to be suitable for acute toxicity tests. Most groups of organisms have also been applied in a biological early warning system (BEWS). A list of aquatic organisms often used in BEWS is given in Table 1. For obvious reasons different organisms are used in different continents, but methods are usually copied without any problem, provided a suitable organism can be found. From the table it appears that fish are the most applied organisms.

For early warning a rapid response is required. Cairns and Van der Schalie (1980) presented a table indicating the response times of some organisms to various pollutants, which ranged from several minutes to over 240 hours, depending not only on the dose-response relationship, but also on the method of response detection. The requirements of the system with regard to the response times will be dependent on the application. In cases where river water monitoring aims at the warning of users further down-stream, a response time of hours or even days is acceptable, but it will be evident that much shorter response times are required, when the monitor is placed directly at the influent or effluent line in order that measures can be taken to prevent damage to the system under consideration. We suggest that in the latter situation a maximum response time is about one hour.

Table 1. Species used in various BEWS studies, with the geographical area of interest, and the measured response.

Monitoring species			
Common name	Scientific name	Area	Effect parameter
Fish			
Freshwater:			
Bluegill sunfish	*Lepomis macrochirus*	NA	Br, Ac, Av, Loc
Fathead minnow	*Pimephales promelas*	NA	Br
Goldfish	*Carassius auratus*	NA	Br, Ac, Av
Gold ide	*Leuciscus idus*	WE	Rh
Guppy	*Poecilia reticulata*	SA	Loc
Largemouth bass	*Micropterus salmoides*	SA	Br, Av, Loc
Rainbow trout	*Salmo gairdneri*	WE,NA	Br, Av, Ac
Marine:			
Sheephead minnow	*Cypridodon variegatus*	NA	Br
Salmon	*Salmo salar*	NA	Av
Arthropods			
Freshwater:			
Water flea	*Daphnia magna*	-	Ac
	Daphnia pullex	-	Ac
Bivalvia			
Freshwater:			
Zebra mussel	*Dreissena polymorpha*	WE	Vm
Swan mussel	*Anodonta cygnea*	WE	Vm
Painters' mussel	*Unio pictorum*	WE	Vm
Marine:			
Blue mussel	*Mytilus edulis*	WE,NA	Vm
Bacteria			
	Photobacterium phosphoreum	-	hv
	Synechococcus spp.	-	Ee
	Escherichia coli	-	Resp
	Sewage sludge mix	-	BOD, Resp

Area codes: NA - north American continent; WE - western Europe; SA - South Africa.

Effect parameter codes: Ac - activity; Av - avoidance; Br - breathing response; Ee - electron exchange; Loc - locomotory behaviour; hv - light emission; Ox - oxygen consumption; Resp - respiration; Rh - positive rheotaxis; Vm - valve movement.

The monitoring system should be relatively easy to operate and should produce results that are easy to interpret. For that reason appropriate methods of data analysis should be available or developed. Therefore the sensitivity of the system, which can be seen as an optimal balance between too many false alarms and no response, should preferably be adjustable depending on the local circumstances and demands. Changes in the normal environmental conditions like pH, temperature, suspended matter etc. should preferably not result in an alarm situation. False alarms may be prevented by addition of other sensing devices. To find a proper balance between high sensitivity and prevention of false alarm situations, optimization, i.e the fine tuning of the system on the detection site, is therefore unavoidable, and should be part of the installation and operation procedure.

Where biomonitoring is concerned, the term biosensor seems quite applicable. There is some confusion, however, between the use of these terms. Where in biological monitoring one uses exclusively living organisms as a sensor, biosensors are defined as devices incorporating a biological sensing element either intimately connected to or integrated within a transducer. The biological elements may range from living organisms, tissues, cells, organelles, membranes, enzymes, etc., down to nucleic acids and organic molecules (Lowe 1985; Turner *et al.* 1987; Rechnitz 1988). Essentially an overlap exists between the two fields, especially where living organisms are applied, and applications from both fields are evident (Evans *et al.* 1986a). This is, so far, almost exclusively true for biosensors based on micro-organisms. But where the applications in BEWS aim towards the detection of any harmful substance or combination of substances in the aquatic environment, the application of biosensors aims at sensor systems that can detect single components, like glucose, alcohol, methane, ammonia, acetic acid (Clarke *et al.* 1985; Karube 1987), and thus determine concentrations of these compounds, assisting the analytical chemists. In the future, however, some of their products may be used in a BEWS. In this paper the BEWS is restricted to the use of living organisms including bacteria, and the application of enzyme-electrodes and other biosensors is excluded.

Most applications of biological early warning systems are in the various fields of water intake or water discharge (see Figure 1):

Influent:
  - drinking water plants
  - aquaculture industries
  - public aquaria
  - water or sewage treatment plants
Effluent:
  - industrial discharges
  - domestic discharges
  - water and sewage treatment plants
  - cooling water
Environmental monitoring
  - rivers (spill detection, storm surges)
  - coastal waters
  - protection of oyster and mussel cultures

A BEWS will be most effective when a large gradient in the concentration of a toxic compound can be detected. The BEWS has only an indicative value, it gives signals of acute toxicity.

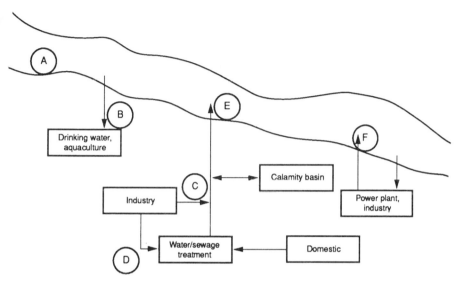

Figure 1. Various applications for biological early warning systems: environmental monitoring (A); influent monitoring (B, D); effluent monitoring (C, D, E); steering industrial processes (cooling water treatment with chlorine) (F).

A limitation to the use of BEWS in effluents can be other properties of the effluent. Often an industrial effluent has not a sufficient quality to sustain normal life, even if there is no defined toxicant present. This can be caused by: high temperature, high or low pH, content of various salts, low oxygen content, etc. In these cases a pretreatment of the effluent is necessary, e.g. cooling, pH adjustment or purging with air. Another possibility is to dilute the effluent with receiving water or tap water, at a predetermined ratio.

Applications of a biological early warning system can range over a variety of goals:

- monitoring of intake water quality (human and aquaculture industry-oriented), where an alarm may lead to closure of the intake channels.
- monitoring of potential spills to prevent damage to the natural environment (nature oriented), where an alarm may lead to the identification of the source and prosecution.
- monitoring of potential spills to detect damage or malfunctioning of the plant (industry oriented), where an alarm may lead to early detection of errors in a production process, and when possible, temporary storage of the effluent in a calamity basin.

- biomonitoring is used to guide industrial processes. The Mussel Monitor for example, placed at the exit of a cooling water line, is at present tested to adjust the amount of chlorine added to prevent fouling in the cooling system (Jenner, pers. comm.), (industrial and nature oriented).

Cairns and Gruber (1980) compared BEWSs based at different laboratories which all used a type of fish monitoring device, based on the breathing rate. Although the techniques were in general comparable, differences were clearly visible in for example, the number of fish tested, their acclamation time, the skill required for operation of the system, the level of automation, and the data collection and assessment. The increased availability of computers for automation and data manipulation and evaluation has changed a number of operations considerably.

Early work was reviewed by Cairns and Van der Schalie (1980), while Gruber and Diamond (1988) published an overview of the more recent automated biomonitoring systems, which was, however, almost entirely devoted to (mainly north American) systems using fish as sensor organisms. Botterweg (1988) and Baldwin (1990) reviewed a number of the commercially available aquatic BEWS systems, and compared their respective sensitivity.

In this paper we will not only review the commercially available systems, but also include the various approaches and types of organisms that are used, or may be used, in biological early warning systems.

Two major criteria are (i) that the response parameter is, or can be, continuously and automatically detected, and (ii) that a flow-through system has been built, or seems feasible, which is in contrast to many acute toxicity tests.

Approaches in BEWS

Means of detection
As biological early warning systems require a fast response, the detection parameters are almost automatically limited to a change in behavioural pattern and/or a physiological response.

A large number of approaches have been designed to monitor either types of response, including the occurrence of no response when a deceased organism may block either a light beam or the outlet of the system.

Behavioural responses can be summarized as swimming activity, positive rheotaxis, locomotive behaviour and preference and avoidance. Tools to follow or locate the organisms are diverse: - use of a video camera with video tracking and imaging; location detection using magnetic fields; ultrasonic sound or blocking of light beams; the actual touching of sensors placed in the aquaria. In addition, the valve movement response of bivalves, namely closure, can be classified as escape behaviour.

In the case of physiological responses we may discover suitable parameters by considering: respiration rate; ventilatory patterns (fish); heart rhythm; temperature; internal pH and ionic strength; pumping rate; bio-electric potential; photosynthesis, growth; consumption or production of chemicals; bioluminescence; and possibly others. The detection devices include for example, the use of different kinds of electrodes, thermistors and photomultipliers for light detection.

Overview of related patents and patent applications

Many of the techniques summarized above are not limited to the field of BEWS research. As automatic biological early warning systems can be of commercial interest, we scanned the patent literature to look for new techniques or combination of techniques, new developments or different approaches. The results are summarized in Table 2. In total 26 patents or patent applications were found from European countries and Japan. In almost all systems described (24) fish are the monitoring organisms. In the early part of the 20-year period covered, five patent applications describe methods to detect dead fish, a signal that even today often appears to be the earliest warning in surface waters of a harmful spill. Other principles of detection include positive rheotaxis (5), activity (13), ventilatory pattern (2), oxygen consumption (1) and current/frequency determination of electric fish (1). Other organisms involved are shellfish, crustacea (oxygen consumption) and flatworms (activity). First patent applications are covered by the following countries: Germany (10), France (7), Japan (5), UK (2) and the European patent office (2).

Only a few systems, now commercially available, appear to be covered by patent applications or patents (Aqua-Tox Control; WRc Fish Monitor; see Tables 2 and 3).

Biological early warning systems

A number of biological early warning systems, or designs that may evolve to such a system, have been developed. They are based on one or more of the means of detection described above and sometimes include designs for the entire BEWS as an integrated approach. This includes construction of aquaria or tanks and description of means of data collection and interpretation, that may lead to alarm situations. Comparisons between aquatic BEWS have been published a decade ago by Cairns and Gruber (1980) and Gruber *et al.* (1980).

In this section we will discuss the various approaches, based on the different types of organisms used.

Monitoring fish

Fish are most extensively used for early warning monitoring systems. Cairns *et al.* (1980) compared different freshwater and marine species for their suitability in biological monitoring systems. A review of the then present fish bioassay monitoring approaches is given by Van der Schalie *et al.* (1979).

In general six different approaches are applied, but not all are presently used in a BEWS (see later).

Table 2. Patents and patent applications concerning aquatic biological monitoring systems.

Inventor	Organism	Effect	System
Nicolas 1967	fish	death	blocking water effluent line
Kerren 1971	fish	rheotaxis	pressure sensor
Pauls 1974	fish	activity	reflection light on scales
Besch & Loseries 1975	fish	rheotaxis	electronic sensors
Ermisch & Juhnke 1975	fish	rheotaxis	nylon sensors
Merk 1976	fish	rheotaxis	microphone sensor
Suber 1976	fish	death	blocking rotor
Lesel & Saboureau 1977	fish	death	blocking water effluent line
Petry 1977	fish	activity	implanted magnet, inductive current measurement
Landragin 1978	fish	death	blocking water effluent line
Miller & Sandwell 1979	fish	activity	electrode chamber
Voith 1979	shellfish crustacea	activity	oxygen consumption
Anon. 1980	fish	activity	light beam detector
Rausch 1980	electric fish	activity	electric current & frequency detection, electrode
Harushige 1981	fish	death	floating fish block light beam
Tetsurou 1982	fish	activity	collision with sensor electrode
Huvé 1983	fish	activity	implanted transmitter, 2 receivers, E vs Hz correlation
Evans & Solman 1985	fish	ventilatory pattern	electrode chamber
Greaves et al. 1985	fish	activity	video registration, image processing
Schmidt 1985	crustacea	activity	blocking light beam
Takase et al. 1986	fish	activity	video camera, image processing
Onatzky & Ferrier 1986	fish	activity/ death	ultrasonic emitter, 2 receivers
Sanemitsu 1986	fish	activity/ death	TV camera, image processing, floating fish
Baba et al. 1987	fish	activity	video imaging
Evans & Solman 1988	fish	ventilatory pattern, heart beat	electrode design, electrical potential detection
Kostelecky 1988	flatworms	activity	change in electrical field, capacity measurement

Table 3. Commercially available Biological Early Warning Systems.

Monitor name	Supplier	Organism	Response
WRc MkIII Monitor	pHOX	fish	ventilation
BMI series 6000	BMI	fish	ventilation
Aqua-Tox Control	Kerren	fish	rheotaxis
Aquatest	Quantum Science Ltd	fish	rheotaxis
Züge Biotest 3	Züge	fish	rheotaxis
Toxalert	Mar.Electric	fish	avoidance
Aztec FM1000	Aztec	fish	electric pulses
Truito Sem	?	fish	activity
Unirelief	Unitika	fish	activity
Mussel Monitor	Delta Consult	mussels	valve movement
Dynamische Daphnia test	Elektron	*Daphnia*	activity
Microtox	Microbics Corps.	bacteria	light emission
Lumistox	Lange	bacteria	light emission
Stiptox	Stip	bacteria	respiration
Respiration Analyser RA-1000	Manotherm	bacteria	respiration
Toxiguard	EurControl	bacteria	respiration

Addresses:
Aztec Environmental Control Ltd, UK
Biological Monitoring Inc., P.O. Box 184, Blacksburg, Virginia 24063, USA
Delta Consult, P.O. Box 71, 4420 AC Kapelle, Netherlands
Elektron GmbH, Magdeburgerstrasse 19, 4150 Krefelt, Germany
Eur-Control GmbH, P.O. Box 2068, 4630 Bochum-Riemke, Germany
Kurt Kerren Umwellt Technik GmbH, P.O. Box 100575, 4060 Viersen, Germany
Dr Bruno Lange GmbH, Willstätterstrasse 11, 4000 Düsseldorf, Germany
Manotherm BV, P.O. Box 7050, 3000 HB Rotterdam, Netherlands
Marine Electronics Ltd, Ireland
Microbics Corp., 2232 Rutherford Road, Carlsbad, CA 92008, USA
pHOX Systems Ltd, Ivel Road, Shefford, Beds. SG17 5JU, UK
Quantum Science Ltd, UK
Stip Siepmann & Teutscher GmbH, Siemensstrasse 2, D - 6114 Gross-Umstadt
Truitosem, Japan
Unitika, Japan
Züge, Switzerland

Systems based on death detection
Spills in the aquatic environment may, even today, be detected by visual observation. The sudden appearance of hundreds or thousands of dead, floating fish may be the indication of a too low oxygen content that may, for example, be

caused by a high load of organic matter. A fish kill may also start action to find the source of the pollution. This has led to a number of designs that use the detection of dead fish as a trigger for an alarm situation. They are based on either a decrease in activity (by measuring their locomotion (Pauls 1974; Onatzky and Ferrier 1986), scanning the surface for floating objects (Harushige 1981; Sanemitsu 1986), or the use of an obstruction in the exit of the experimental tanks, causing an increase in water level (Nicolas 1967; Lesel and Saboureau 1977; Landragin 1978).

Systems based on locomotory behaviour

The use of activity patterns of fish has often been reported. A sudden introduction of a pollutant may enhance or reduce their activity. In general the systems detect the location of the fish, either continuously or semi-continuously, and compare the present activity with a previously recorded value under non-polluted conditions.

The earliest systems employed multi-chamber devices, where the presence in the different chambers was recorded (Kleerekoper 1969; cited in Gruber 1988). New techniques permits the detection in a single chamber device.

The detectors or detection systems involved are: the interruption of visible or infra red light beams (e.g. Shirer *et al.* 1968; Waller and Cairns 1972; Cairns *et al.* 1975; Morgan 1979; Anon. 1980); the detection of ultrasonic echoes (Morgan *et al.* 1982; Onatzky and Ferrier, 1986); the monitoring of the inductive current generated by implanted magnets (Petry 1977; 1982); the detection of the electrical potential that is related to the physiological condition of the fish (Wallwork and Ellison 1983); and video tracking (Lubinski *et al.* 1977; Korver and Sprague 1988). Bengtsson (1974) and Lindahl *et al.* (1977) designed fish fitness test systems, while Fisher *et al.* (1983) detected the turbulence of the water generated by the activity of the fish by a paddle sensor.

Additional literature on patent applications suggest other means of activity detection: the measurement of oxygen consumption rate which is, however, not limited to fish (Voith 1979); collision frequency with an immersed sensor (Tetsurou 1982); and application of an implanted transmitter with two receivers, resulting in potential frequency relations (Huvé 1983). During the last few years the use of video detection and image processing has also been documented (Greaves *et al.* 1985; Takase *et al.* 1986; Sanemitsu 1986; Baba *et al.* 1987).

Systems based on positive rheotaxis

These systems record a special type of locomotory behaviour and use the habit of fish living in running waters always to swim upstream, positive rheotaxis. Two possible reactions can happen when a sudden increase in a pollutant concentration occurs. The fish try to escape from the source water, thus swim downstream (loss of positive rheotaxis), or they rapidly lose sufficient condition to be able to swim upstream, and they are carried away downstream by the flow-through current.

The detection systems indicate the number of incidents that the fish are observed in a downstream section, blocking a light beam (Waller and Cairns

1972; Hasselrot 1975; Poels 1975; Van Hoof 1980) or infra red light (Morgan 1977) by photo-electric cells. The fish are sometimes stimulated to swim back to the upstream end of the system to prevent the accidental registration of healthy fish, e.g. by a mild electric shock (Poels 1975) or a strong light area (Hasselrot 1975). This reduces the incidence of false alarm situations. Another approach intermittently increases the speed of the water current forcing the fish to respond (Juhnke and Besch 1971, Besch *et al.* 1977). Touching of pressure sensitive bars at the down-stream end of their circular test tank (Arena basin, Kerren, 1971) detects the displacement of the fish (Ermisch and Juhnke 1975). This kinetic screen also serves as a stimulator to remain up-stream. Other detectors have been based on this principle (Besch and Loseries 1975; Merk 1976). An example is given in Figure 2. Sharf (1979) built a conveyer belt to transport exhausted fish out of the system, before detection.

Systems based on preference/avoidance
The ability of fish to avoid toxic compounds has been used in tests applying slowly running water in so-called fluvarium test systems. The fish have the possibility to choose between "good" and "bad" water quality, thus permitting preference and avoidance tests (Sprague 1964; Scherer and Nowak 1973; Poels 1975; Van Hoof 1980).

These systems have been reviewed by Cherry and Cairns (1982) and Cherry *et al.* (1982). In all systems special care has been taken to guide the fish to the tested water in such a way that undesired interference with the good quality water is minimized. Mixing of the waters produces controlled gradients, which enables an estimate of the water quality in between the extremes (e.g. Ishio 1964; Giattina *et al.* 1982).

Another approach includes the use of the ampullary electro-receptor organs of catfish *Ictalurus nebulosus*, which detect weak natural electric fields. The fish are trained to perform certain actions and positive responses are awarded by the supply of food. A deteriorated aquatic condition may result in an adverse response (Neuman *et al.* 1990). As the location of the fish, rather than overall activity, must be detected, optical methods are mostly applied. A number of techniques mentioned in the section "Systems based on locomotory behaviour" may also be used.

Systems based on breathing pattern, heart rhythm
The respiration of fish is closely related to the breathing rate or the movement of the gill covers. It is usually found that the breathing rate under pollution stress is faster and more erratic. This change in response is used in the monitoring devices.

Basically two different detection mechanisms are used. Some older systems apply detector electrodes connected directly to the fish (Sparks *et al.* 1972; Heath 1972). Other systems allow the fish a maximum amount of freedom under test conditions and use dual electrodes placed on or in the walls of the containers (Spoor *et al.* 1971; Waller and Cairns 1972; Morgan 1977; Miller and Sandwell 1979; Evans and Solman 1985; 1988; Evans *et al.* 1986b). Not only can the

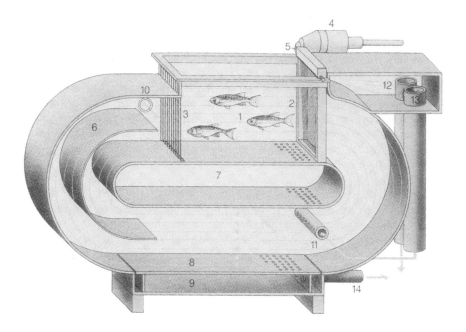

Figure 2. The Arena basin, a system based on positive rheotaxis: 1) test room; 2) contact detector; 3) screen; 4) pressure detector; 5) elastic tubing; 10) inlet; 11) laminar flow; 12) overflow; 13) recirculation; 14) outlet (after Kerren Umwelt Technik GmbH).

ventilatory rate be determined, but additional information can be obtained from the electrical signal on the strength of the movement (ventilatory depth) and the coughing rate (Drummond and Carlson 1977). This technique also allows the detection of heart beat, though the signal is less strong than that of the ventilatory pattern (Evans and Walwork 1988).

Systems based on weak electrical pulses of fish
Several fish produce weak electrical pulses periodically. The frequency of the pulses is species dependent and varies between 1 and 1600 s^{-1} at a temperature range between 20-30°C (Rausch 1980). Most important species belong to the families *Mormyriden* and *Gymnotiden*. The Nile pike *Gnatonemus petersi* especially seems to be promising with this respect. On this measuring method several operational systems have been based, using the Nile pike. Based on a normal pulse frequency of *c*. 18 s^{-1}, the fish respond to pollution by a decrease in activity and in pulse frequency (Geller 1984; Ewen 1987).

Monitoring water fleas (*Daphnia*)

Water fleas (*Cladocera*) are small crustacea. *Daphnia* species appear to be especially sensitive to a large number of chemical compounds and have therefore been used in toxicity testing. In addition, the "dynamic Daphnia-test" has been developed to detect sub-lethal effects (Knie 1978). The system is based on the detection of the activity of the water fleas.

This flow-through test consists of a vertical tube, the influent entering from above. Light (IR) sources and detectors are mounted opposite each other along the tube. The principle of detection is based on the (total) number of light beam blocking events per unit of time caused by the Daphnids in the water column. A decrease in the number of pulses indicates a deterioration of environmental conditions.

The only system known, the Daphnia test (see Figure 3), has been tested under laboratory and field conditions (Botterweg 1989; Smith and Bailey 1988).

Monitoring bivalves

Bivalves correspond very well with the requirements that should be met when selecting a suitable monitoring organism. They are sedentary, abundant and available throughout the year; they have a manageable size and are hardy enough to be handled in the laboratory (Phillips 1977). They are common both in freshwater and the marine environment.

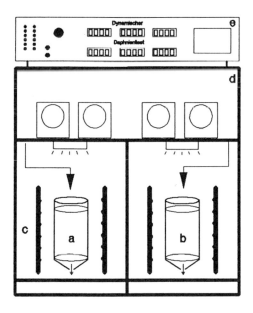

Figure 3. Scheme of the dynamic *Daphnia* test: a,b) test chambers; c) light emitters, detectors; d) pumps; e) electronics (after Elektron GmbH).

Systems using valve movement response

The valve movement response method is based on the fact that mussels have their shells open for respiration and feeding most of the time. It has been shown that they close their shells for an extended period when under stress. This response was used to study a number of natural and anthropogenic effects, including a series of toxicants, like trace metals (Salánki and Varanka 1976; Davenport 1977; Manley and Davenport 1979), pesticides (Salánki and Varanka 1978) and other trace organics (Sabourin and Tullis 1981; Slooff *et al.* 1983).

Traditionally the valve movement was recorded by mechanical means. Schuring and Geense (1972) were the first to use electromagnetic induction to measure the valve displacement. This system was improved into a high frequency (HF) electromagnetic induction system, EMIS (Jenner *et al.* 1989). The electronic sensor consists of two small coils, glued to opposite shell halves of the mussel. One coil acts as a transmitter, the other coil as a receiver. A linear response is thus obtained.

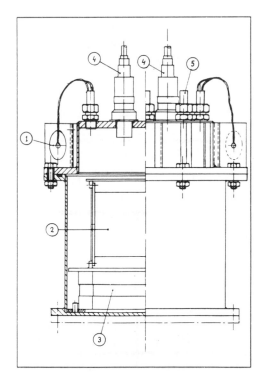

Figure 4. Scheme of the Mussel Monitor: 1) mussel with sensors; 2) electronics; 3) batteries; 4) underwater connectors; 5) sensor connectors (after Delta Consult).

Based on laboratory experience, an early warning system has been developed using this valve movement response (De Zwart and Slooff 1987; Kramer *et al.* 1989). This system consists of a waterproof housing, enabling *in situ* operation. Eight mussels are attached to the housing, and followed individually (Kramer 1989, Figure 4).

Most applied bivalves, so far, are freshwater species - the zebra mussel *Dreissena polymorpha* and the blue or common mussel *Mytilus* spp., but other species have responded very well in laboratory test systems.

Systems using pumping rate, oxygen demand, etc.
Other physiological parameters of bivalve molluscs have been used to detect environmental changes (Akberali and Trueman 1985). Examples that offer possible use in a BEWS are the heart beat (Akberali and Black 1980; Sabourin and Tullis 1981), burrowing activity (Barnes 1955; Salánki and Varanka 1976), and the respiration and pumping or filtration rate (Abel 1976; Manley and Davenport 1979; Manley 1983; Slooff *et al.* 1983).

Voith (1979) described a system that was based on the activity measurement, which detects oxygen consumption. In principle this system is not limited to shellfish only. The pumping rate detection (Famme *et al.* 1986; Salánki *et al.*, this volume) offers possibilities for use in a BEWS.

Monitoring bacteria

In general two approaches are applied to the use of bacteria as early warning sensors. In the first approach, free living bacteria as single or mixed culture, are present in a test cell. They may be on a substrate, or biofilter, or freely suspended. The second method immobilizes the bacteria, usually monocultures, in electrodes, where they act as an intermediate for instrumental analysis e.g. potentiometric or amperometric.

The systems described here are limited to those that react to general environmental conditions rather than to one specific compound. Several systems are presently used as sentinels, mainly of sewage treatment plants (see table 3).

Systems based on incorporation in biofilters
The inhibition of nitrification by a mixture of *Nitrosomas* and *Nitrobacter* on a substrate of granite chips in a biofilter was one of the first attempts to use bacteria as a tool in biomonitoring studies (Holland and Green 1975). The oxygen consumption of activated sludge has been applied as a parameter by Shieh and Yee (1985), while Martin (1988) used activated sludge in an on-line respirometer to detect the bacterial biomass respiration. In a small, automated active sludge unit, the oxygen concentration is followed continuously to monitor the influent of a sewage treatment plant (Clarke *et al.* 1977). The measurement of biological oxygen demand can also be used in automated early detection systems. Another continuously operated system (toxiguard) monitors the oxygen production of a culture of bacteria on a filter (Solyom 1977).

Systems based on immobilization in electrodes

New developments in biosensor technology allow incorporation of bacteria in electrodes.

An electrode to determine the general microbial activity, as measured by the electron transport system, was developed by several groups (Turner *et al.* 1983; Ramsay *et al.* 1986; Dobbs and Briers 1988).

The immobilization of *Escherichia coli* at the surface of a CO_2 sensing electrode, allowed the potentiometric detection of CO_2 production by the bacteria. The net rate of CO_2 production reflects the complex series of biochemical reactions which constitute the respiratory processes of the cells (Dorward and Barisas 1984).

Systems based on light emission

Luminescent bacteria, such as *Photobacterium phosphoreum*, have been found to be useful in acute bioassay tests. The principle is based on the reduction of light produced under stress of pollutants (Bulich 1979). The detection of fluorescent light by a photomultiplier in a biophotometer is very sensitive. As the bacteria, which are supplied in freeze dried form originate from the marine environment, an electrolyte has to be added (e.g. 2% NaCl) for freshwater applications.

The method has been developed into standard toxicity screening systems, like the "Microtox Assay System" (Beckman 1978) or "Lumis Tox" (Link 1988). An automated version of the detection system has been described (Levi *et al.* 1989).

Monitoring insect larvae, worms and phototrophic organisms

A number of other types of organisms have been used in (acute) toxicity studies. Consequently these techniques could be evaluated for use in BEWS. So far, this has not led to a system that is automated and operates under continuous flow conditions. A number of techniques have been included here as they may develop into such a system.

Techniques for the monitoring of aquatic insect larvae

Several techniques have been developed that use the response of midge larvae (chironomids) to pollutants. Batac-Catalan and White (1983) used the "optical-fibre light interruption biomonitoring system" to detect effects upon the ventilation system of *Chironomus* spp. For example, they found a distinct reaction even at low chromium concentrations.

An impedance-conversion technique was applied to measure the short term activity of the chironomids *Glyptotendipes pallens*, as a function of cadmium concentration (Heinis and Swain 1986; Heinis *et al.* 1990). The response of larvae of *Chironomus riparius* to organic toxicants has also been documented in Heinis and Crommentuijn (1989). This method is at present working under laboratory conditions, and an operational field system is under development.

Techniques for the monitoring of worms

So far, only two approaches have been reported of sub-lethal effect detection in worms, which may be used in early warning systems.

A protocol was developed to rapidly identify sub-acute environmentally limiting factors, by measuring the adaptive responses of sewage worms (tubificids) in a bioassay flow-through test chamber (Coler *et al.* 1988). The detection system is based on the increased rate of respiration under stress. The respiration rate is determined by analysing upstream and downstream oxygen content, together with flow rate, to estimate uptake per unit time.

In another approach, flatworms (e.g. *Tubularia*) are kept in a test chamber that consists of an inner and outer conductor. An electrical field is generated and the capacitance is determined as a function of the activity of the organisms (Kostelecky 1988).

Techniques for the monitoring of zooplankton

In toxicity studies zooplankton species are often found to have a relatively high sensitivity to toxicants. Behavioural responses, especially their locomotory activity, may offer a tool to study this group of organisms as biological sensors. The marine tintinnid (*Favella* spp.) has been studied for its swimming behaviour under different environmental conditions (Buskey and Stoecker 1989). The system uses a video system with dark-field IR illumination and subsequent automated motion analysis. Quantity and quality of the food seems an important factor, but no details are known about the response to pollutants.

Techniques for the monitoring of phototrophic organisms

For the detection of herbicides the use of phototrophic organisms seems a logic choice. Several approaches have been followed: detection of fluorescence, the inhibition of growth rate, oxygen production and morphological changes.

To be able to rapidly detect the quantitative and qualitative effect of eutrophication, Noack *et al.* (1985) developed a fluorescence system that estimates the phytoplankton density and their distribution. In another approach the alga *Scenedesmus subspicatus*, cultured in a continuous culture is grown at a constant density. By pumping around flow-through cuvettes, the fluorescence is detected as a function of toxicant concentration (Benecke *et al.* 1982). The system showed a good performance in a series of toxicity tests (Schmidt 1987).

Cyano-bacteria *Synechococcus* spp., immobilized on an electrode, produce photosynthetic intermediates that are re-oxidized at the electrode and amperometrically detected (Rawson *et al.* 1987; 1989).

Application of a so-called "cage culture turbidostat" (Skipnes *et al.* 1980) is proposed for the monitoring of surface waters or effluents. It is supposed that this flow-through system is more sensitive then batch culture systems (Wangersky and Maass 1990), and has been applied to test the toxicity of oil (Ostgaard *et al.* 1984). The technique establishes a rate of growth for the test organism, e.g. the diatom *Phaeodactylum tricornutum*, in continuous culture with unpolluted medium, and then examines the changes in population growth rate with the introduction of the pollutant. Since the cultures are monitored continuously, immediate or short-term effects may readily be seen. A monitoring design is now under construction (Wangerski, pers. comm.).

Cairns *et al.* (1973) designed a system for the automatic determination of diatoms, for use in estimating water quality. Microscope images are stored and compared using an image processor (laser holography), with pre-selected species on slides (Cairns *et al.* 1982). Besides taxonomic differences, sub-lethal morphological changes, like cell wall thickness, can be determined.

Availability of biological early warning systems

Of all the methods that have been developed over the last twenty years, many have remained at a laboratory stage. Few are at the moment tested under experimental conditions and even fewer are really commercialized and for sale. It appears that a system which works well in a laboratory environment is not easy to transform to a system that can be used in the natural environment or near effluent lines. Besides the already mentioned qualities, like response time, easy usage, type of organism, false alarm prevention, data treatment, etc., it should be robust and low in maintenance. Finally it should be designed, built, tested and marketed for field use. From the list of possible approaches few methods have surfaced that are beyond the laboratory stage, and operate as a BEWS.

We have summarized the biological early warning systems that to our knowledge are commercially available (Table 3; see also Baldwin (1990) for a more extensive description of several of these systems). Most abundant are the systems that use the response of fish, either by positive rheotaxis (3), breathing response (2), activity (2), avoidance (1) or the use of electrical fish (1). The other organisms involve the valve movement of mussels (1), the activity of *Daphnia* (1) and systems using the respiration of bacteria in various approaches (3) or their bioluminescence (2). All other proposed methods are at various stages of laboratory development.

The cost of the systems vary considerably, even between systems based on the same operating principle. But often it is not clear to what extent the system is complete, and what are the operating costs (man-hours for maintenance). The complexity of the data treatment will certainly also play an important role in estimating the price of the complete system

Detection limits

A comparison between different systems regarding their sensitivity for specific toxicants, even between systems using the same principle of detection, is difficult. Apart from the use of different fish species, too many factors may influence the conditional settings of the systems. Some efforts have been made to compare systems for their use as BEWS (Heath 1972; Cairns and Gruber 1980), but without comparisons of the respective sensitivity. It is well known that different groups of organisms may have different sensitivities towards toxicants (Diamond *et al.* 1988).

A general picture of the sensitivity, however, may be abstracted from the literature. A number of systems were selected based on the availability of response data (detection limits), and the possibility to monitor the water or effluent continuously and automatically. Only systems with real early warning approaches, i.e. response within one or a few hours, are given. Botterweg (1988) compared 13 systems and reviewed the toxicity data for 65 compounds (where applicable). A real comparison of their sensitivity was only possible for 18 toxicants. In this comparison the electrode chamber system of Spoor *et al.* (1971) using ventilation rate of trout was most sensitive, but failed when tested in a river environment because of the occurrence of too many false alarms. When this system is not included, a final 13 compounds could be compared. The system using phototrophic organisms was most sensitive for six of the toxicants, while the remaining seven were distributed evenly over six systems for their sensitivity. A summary of these results are presented in Table 4.

The most sensitive organism or monitoring system suitable for the detection of all possible toxic substances does not exist. Therefore it is recommended that when a wide range of toxicants need to be detected, several BEW systems be incorporated, that are based on the response of different organisms.

Data retrieval, manipulation and evaluation

New activities in the field of BEWS have certainly been induced by the availability of PC-computers. The large number of data that were in former days recorded on strip-chart recorders, and interpreted by measuring with a ruler did not stimulate this type of research. The method of data retrieval is usually not a question of how to store the data, but what data, and how fast is an interpretation needed. The amount of time this process will take will depend on the BEWS system. Some responses are reliably detectable within minutes, and require not 15, 30 or more minutes of data-averages before a decision can be reached.

Gruber and Cairns (1981) made an effort to compare various data acquisition methods, using for example, a method that compares statistically the present finding with the findings at the same time the day before (static baseline approach), or a method where the present finding statistically deviates from the average of say the last hour (time series approach). Usually data reduction can be achieved, and is necessary, for example by averaging results - either running averages or by using time blocks; by storage of data only for the last 24 or 48 hours, and over-writing the storage medium, if essential, with storage of some key data, or simply storing only the alarm events with date and time. In the latter case it is especially convenient for the system to check its functioning and to signal every few hours that all functions are operating satisfactorily. No alarm does not necessarily mean that there has not been a polluting situation. The observation of several organisms together in one tank, or the summing up of the data for all individual results, is a reduction of data but, in our opinion, may lead to a less sensitive BEWS. Individual observation will give information on the specific behaviour of each organism, and the evaluation may take this specific

Table 4. Comparison of the compound specific sensitivity of various biological monitoring systems. Concentrations in mg l^{-1}, unless otherwise stated (based on Botterweg 1988; Botterweg *et al.* 1990).

System	A	B	C	D	E	F	G
effect parameter:	WRC breathing rate	Arena rheo- taxis	Poels rheo- taxis	Spoor ventil. pattern	Petry acti- vity	Daphnia activity	bivalves valve move.
compound							
ammonia	0.11	1.2	-	5	0.021	-	-
cadmium	-	0.3	-	-	0.025	0.2	0.37
chlorine	-	-	-	-	-	-	0.005
chloroform	-	-	-	20	-	-	106.0
chromium	-	-	-	5	-	-	-
copper	-	-	0.5	0.06-0.5	-	-	0.005
cyanide	-	-	0.5	0.01-0.13	-	-	0.53
endosulphan (thiodan)	-	0.1	-	-	-	-	-
hexachlorobutadiene	-	-	-	0.05	5.1	-	0.26
lead	-	-	-	1.0	-	-	0.25
lindane	0.003	-	0.06	0.04	-	-	0.11
malathion	-	-	2.0	-	-	-	-
mercury	-	-	1.0	0.01	-	-	-
paraquat	0.8	-	-	-	-	-	-
pentachlorophenol	0.14	-	-	0.06	-	-	0.34
phenol	2.0	-	-	1-4.0	-	1.0	26.4
tetrachloroethylene (μl l^{-1})	-	-	4.0	-	-	-	-
TBT	-	-	-	-	-	-	0.005
xylene	-	-	-	2.0	-	-	16.2
isophorone	-	0.95	-	-	-	-	-
diesel oil	-	>50μgl^{-1}	-	-	-	-	-
3,4 DCA	-	0.5	-	-	-	-	-
disulfotone	-	7.0	-	-	-	-	-

A: WRC Mk(III) Fish Monitor (Evans *et al.* 1986b; Evans & Walwork 1988)
B: Arena Basin (Besch *et al.* 1977)
C: KIWA System (Poels 1975; 1977; Van Hoof 1980)
D: Electrode Chamber (Spoor *et al.* 1971; Morgan 1977; Slooff 1979)
E: Motility test (Petry 1982)
F: Dynamic Daphnia Test (Knie *et al.* 1983; Smith & Bailey 1988)
G: Mussel Monitor (Slooff *et al.* 1983; Slooff & De Zwart 1986; Doherty *et al.* 1987; Kramer *et al.* (1989)

Table 4. (continued).

System	H	I	J	K	L	M	N
effect parameter	insect activity	bact. nitrific.	bact. O_2-cons	Photo-trophic	bact. CO_2 elect.	bact. light emiss.	bivalve filtration rate
compound							
ammonia	-	-	-	-	-	-	-
cadmium	10.0	2.0	-	0.007	0.3	20-50	-
chlorine	-	-	-	-	-	-	-
chloroform	-	-	-	-	-	-	-
chromium	0.1	5.0	25.0	-	-	-	-
copper	-	4.0	-	0.003	1.1	0.8	0.15
cyanide	-	0.04	5.0	0.007	3.5	4.0	0.3
endosulphan (thiodan)	-	0.1	-	-	-	-	-
hexachlorobutadiene	-	-	-	-	-	-	-
lead	-	-	50.0	-	0.13	0.6	-
lindane	-	sat.	-	-	-	-	-
malathion	-	50	-	-	-	-	-
mercury	-	-	0.5	0.037	-	0.02	0.04
paraquat	-	4	-	-	-	-	-
pentachlorophenol	-	-	-	-	-	-	-
phenol	-	1.5	-	7.8	1209	28.0	-
tetrachloroethylene (μl l^{-1})	-	-	-	▪			▪
TBT	-	-	-	-	-	-	-
xylene	-	-	-	-	-	-	-
isophorone	-	-	-	-	-	-	-
diesel oil	-	-	-	▪	-	-	-
3,4 DCA	-	-	-	▪		▪	-
disulfotone	-	-	-	-	-	-	-

H: Aquasensor (Batac-Catalan & White 1983; Leuchs 1986; Heinis *et al.* 1990)
I: Bacteria nitrification (Holland & Green 1975)
J: Bacteria O_2 consumption (Clarke *et al.* 1977; Solyom 1977; Shieh & Yee 1985)
K: Bacteria *E. coli* electrode, CO_2 production (Dorward & Barisas 1984)
L: Phototrophic organisms (Schmidt 1987; Rawson *et al.* 1987)
M: Microtox (Beckman 1981; De Zwart & Slooff 1983)
N: Bivalves filtration rate (Abel 1976)

behaviour into account. When the activity is normalized for each individual, e.g. to percentage, and the alarm criterion is set as 'double activity', the system will be more sensitive and more reliable. More fine tuning of the BEWS becomes possible through a selection of the number of organisms that should react. It will be clear that a statistical treatment of the data is inevitable to optimize the system, and to minimize the chance of too many future false alarms. When a system is optimized with respect to the critical limit for alarm detection - the number of organisms that should react, the time they need to react, the increase or decrease in 'activity' required for response etc., it is not always necessary to treat all data extensively by statistical procedures in an operational BEWS. This contrasts with, for example, acute toxicity tests, where statistics are used to prove that a certain dose is causing or is not causing an effect. After an alarm, a stored data set may statistically be evaluated, as a tool to evaluate the alarm criteria and to improve, when necessary, future operation e.g. by determination of 95% confidence limits, one or two-way analysis of variance, etc. A BEWS is, though often quite a complex system, essentially only one sensor. Often the data evaluation process will involve the inclusion of data from other environmental sensors, such as temperature, pH, suspended matter content, dissolved oxygen concentration. Alarm situations may thus be interpreted using a number of environmental parameters that effect organisms, but may not necessarily be harmful. This evaluation process, involving additional parameters, will help to eliminate false alarms. An example of such a system was developed at the WRc, and called the 'intake protection system' (Dobbs and Briers 1988).

New developments include remote operation and data transmission by modem telephone line, and if applicable radio transceivers or even satellite communication can be used for remote locations (Morgan *et al.* 1988).

Conclusions

Numerous approaches are described to measure behavioural or physiological parameters of organisms. Many of these approaches have been usefully applied in acute toxicity studies. Important criteria for application of these techniques in a biological early warning system (BEWS) are the possibility of automated detection, either *in situ* or in a flow-through system.

It requires quite some effort before the detection system, that works well under laboratory conditions, is transformed into a system that can reliably operate under field conditions.

Traditionally only fish were used in BEWS, but there is now a greater variety of organisms used. Until now four groups of organisms were applied as sensing organisms in commercially available BEWS. They include fish (9 systems, based on ventilation pattern, positive rheotaxis, avoidance and weak electrical pulses), *Daphnia* (1 system, based on activity), bivalves (1 system, based on valve movement response) and bacteria (5 systems based on light emission and various respiration parameters), totally 16 systems.

Different organisms demonstrate different sensitivities for different toxicants. A wide spectrum of organisms in an integrated biological early warning system will be more reliable than the use of one species only. It is therefore seen as an advantage that more groups of organisms are available for application in a BEWS. It is expected that several more groups will be included in the next few years in practical BEWS's, e.g. aquatic insects and several bacteria-based biosensors. For the specific detection of a given compound, or a group of compounds, one species may be suitable, like bivalves, for the detection of chlorine, or algae, for the detection of herbicides. What system or systems are to be advised for a specific situation, will strongly depend on the application.

In the future, the list of sensitivities of a given system for various toxic compounds will be extended, but never completed. The real advantage of BEWS is the possibility of detection, by a given organism, that there is something wrong with their local environment.

A major activity for the near future is the practical incorporation of such systems, in the various fields of application. This will be in addition to physico-chemical monitoring systems. If the environmental and/or industrial oriented managers can be convinced of the advantages, and realize the disadvantages, of living biota as sensors/indicators for environmental quality, biological early warning systems will become an important sensing tool in the near future.

References

Abel, P.D. (1976) Effect of some pollutants on the filtration rate of *Mytilus. Marine Pollution Bulletin*, **7**, 228-231.

Akberali, H.B. and Trueman, E.R. (1985) Effects of the environmental stress on marine molluscs. *Advances in Marine Biology*, **22**, 102-198.

Akberali, H.B. and Black, J.E. (1980) Behavioural responses of the bivalve *Scrobicularia plana* (da Costa) subjected to short-term copper (Cu II) concentrations. *Marine Environmental Research*, **4**, 97-107.

Anon. (1980) Verfahren und Vorrichtung zur Ueberwachen von Gewässern und Abwässern. Patent application, publication nr. and date: DE 29.06.194, 28.08.1980.

Baba, K., Watanabe, S. and Yoda, M. (1987) An apparatus to monitor the inflow of toxicants. Patent application, publication nr. and date: JP 62-83663, 17.04.1987.

Baldwin, I.G. (1990) Review of fish monitors and other whole organisms monitoring systems. *Water Research Centre Publication UM 1109*, Medmenham, U.K. pp.62.

Barnes, G.E. (1955) The behaviour of *Anodonta cygnea* I.., and its neurophysiological basis. *Journal of Experimental Biology*, **32**, 158-174.

Batac-Catalan, Z. and White, D.S. (1983) Effect of chromium on larval Chironomidae as determined by the optical-fiber light interruption biomonitoring system. In W.E. Bishop, R.D. Cardwell and B.B. Heidolf (eds.) *Aquatic Toxicology and Hazard Assessment: Sixth Symposium*, 469-481.

ASTM, Philadelphia.

Beckman Instruments (1981) *Microtox application notes.* Carlsbad, CA, USA.

Beckman Instruments (1978) *Microtox model 2055 toxicity analyzer system.* Carlsbad, CA, USA.

Benecke, G., Falke, W. and Schmidt, C. (1982) Use of algal fluorescence for an automated biological monitoring system. *Bulletin of Environmental Contamination and Toxicology,* **28,** 385-395.

Bengtsson, B.E. (1974) The effect of zinc on the availability of minnow to compensate for torque in a rotating water current. *Bulletin of Environmental Contamination and Toxicology,* **12,** 654-658.

Besch, W.K., Kembal, A., Meyer-Waarden, K. and Scharf, B. (1977) A biological monitoring system employing rheotaxis of fish. In J. Cairns, K.L. Dickson and G.F. Westlake (eds.) *Biological Monitoring of Water and Effluent Quality,* 56-74. ASTM, Special Technical Publication 607.

Besch, W.K. and Loseries, H.G. (1975) Verfahren zur Früherkennung von toxischen Verunreinigungen in Wasser mit Hilfe von Testfishen. Patent application, publication nr. and date: DT 23.62.084.

Botterwig, J. (1988) Using biological early warning systems for continuous monitoring of toxic substances in the aquatic environment. Literature survey. BKH consulting engineers in assignment of DBW/RIZA. *Publications and reports of the project "Ecological Rehabilitation of the River Rhine",* 5-88. DBW/RIZA nr 89.002 (in Dutch with English abstract), 24.

Botterwig, J. (1989) Evaluation of biological early warning systems (BEWS) in the river Rhine at Lobith (the Netherlands). Bioalarm project. Part 1. BKH consulting engineers in assignment of DBW/RIZA. *Publications and reports of the project "Ecological Rehabilitation of the River Rhine",* 8-89. DBW/RIZA nr 89.045 (in Dutch with English abstract).

Botterwig, J., Van de Guchte, C. and De la Hay, M.A.A. (1990) Sensitivity and reliability of two Biological Early Warning Systems (BEWS), in prep.

Bulich, A.A. (1979) Use of luminescent bacteria for determining toxicity in aquatic environments. In L.L. Marking and R.A. Kimerle (eds.) *Aquatic Toxicology,* ASTM STP 667. 98-106.

Buskey, E.J. and Stoecker, D.K. (1989) Behavioural responses of the marine tintinnid *Favella* spp. to phytoplankton: influence of chemical, mechanical and photic stimuli. *Journal of Experimental Marine Biology and Ecology,* **132,** 1-16.

Cairns, J. and Gruber, D. (1980) A comparison of methods and instrumentation of biological early warning systems. *Water Resources Bulletin,* **16,** 261-266.

Cairns, J. and Van der Schalie, W.H. (1980) Biological monitoring. Part I. Early warning systems. *Water Research,* **14,** 1179-1196.

Cairns, J., Dickson, K.L. and Westlake, G.F. (1975) Continuous biological monitoring to establish parameters for water pollution control. *Progress in Water Technology,* **9,** 829-841.

Cairns, J., Dickson, K.L. and Lanza, G. (1973) Rapid biological monitoring systems for determining aquatic community structure in receiving systems. In *Biological Methods for the Assessment of Water Quality.* ASTM STP 528. 148-

163.

Cairns, J., Thompson, K.W., Landers, J.D., McKee, M.J. and Hendricks, A.C. (1980) Suitability of some freshwater and marine fishes for use with a minicomputer interfaced biological monitoring system. *Water Resources Bulletin*, **16**, 421-427.

Cairns, J., Almeida, S. and Fujii, H. (1982) Automated identification of diatoms. *Bioscience*, **32**, 98-102.

Cherry, D.S. and Cairns, J. (1982) Biological monitoring. Part V. Preference and avoidance studies. *Water Research*, **16**, 263-301.

Cherry, D.S., Larrick, S.R., Giattina, J.D., Cairns, J. and Van Hassel, J. (1982) Influence of temperature selection upon the chlorine avoidance of cold and warm water fish. *Canadian Journal of Fisheries and Aquatic Sciences*, **39**, 162-173.

Clarke, D.J., Calder, M.R., Carr, R.J.G., Blake-Coleman, B.C., Moody, S.C. and Collinge, T.A. (1985) The development and application of biosensing devices for bioreactor monitoring and control. *Biosensors*, **1**, 213-320.

Clarke, A.N., Eckenfelder, W.W. and Roth, J.A. (1977) The development of an influent monitor for biological treatment systems. *Progress in Water Technology*, **9**, 103-107.

Coler, R.A., Coler, M.S. and Kostecki, P.T. (1988) Tubificid behaviour as a stress indicator. *Water Research*, **22**, 263-267.

Davenport, J. (1977). A study of the effect of copper applied continuously and discontinuously to specimens of *Mytilus edulis* (L.) exposed to steady and fluctuating salinity levels. *Journal of the Marine Biological Association of the United Kingdom*, **57**, 63-74.

De Zwart, D. and Slooff, W. (1987) Continuous effluent biomonitoring with an early warning system. In Bengtsson, B.E., Norbert-King, T.J. and Mount, D.I. (eds.) *Effluent and Ambient Toxicity Testing in the GötaÄlv and Viskan Rivers, Sweden.* Naturvardsverket report 3275.

De Zwart, D. and Slooff, W. (1983) The MICROTOX as an alternative assay in the acute toxicity assessment of water pollutants. *Aquatic Toxicology*, **4**, 129-138.

Diamond, J.M., Collins, M. and Gruber, D. (1988) An overview of automated biomonitoring - past developments and future needs. In D.S. Gruber and J.M. Diamond (eds.) *Automated Biomonitoring: Living Sensors as Environmental Monitors*, 23-29. Ellis Horwood, Chichester.

Dobbs, A.J. and Briers, M.G. (1988) Water quality monitoring using chemical and biological sensors. *Analytical Proceedings*, **25**, 278-279.

Doherty, F.G., Cherry, D.S. and Cairns, J. (1987) Valve closure responses of the asiatic clam *Corbicula fluminea* exposed to cadmium and zinc. *Hydrobiologia*, **153**, 159-167.

Dorward, E.J. and Barisas, B.G. (1984) Acute toxicity screening of water pollutants using a bacterial electrode. *Environmental Science and Technology*, **18**, 967-972.

Drummond, R.A. and Carlson, R.W. (1977) *Procedures for measuring the cough (gill purge) rates of fish.* EPA-600/3-77-133. U.S. EPA, Washington DC.

Ermisch, R. and Juhnke, I. (1975) Vorrichtung zur automatischen qualitativen

Erfassung toxischer Stoffe im Wasser. Patent application, publication nr. and date: DT 23.65.214, 17.07.1975.

Evans, G.P. and Solman, A.J. (1985) Continuous monitoring of water quality. Patent application, publication nr. and date: EP 0.158.522, 16.10.1985.

Evans, G.P. and Solman, A.J. (1988) Electrodes for continuous monitoring water quality. Patent application, publication nr. and date: GB 2.195.543, 13.04.1988.

Evans, G.P. and Walwork, J.F. (1988) The WRc fish monitor and other biomonitoring methods. In D.S. Gruber and J.M. Diamond (eds.) *Automated Biomonitoring: Living Sensors as Environmental Monitors*, 75-90. Ellis Horwood, Chichester.

Evans, G.P., Briers, M.G. and Rawson, D.M. (1986a) Can biosensors help to protect drinking water? *Biosensors*, **2**, 287-300.

Evans, G.P., Johnson, D. and Whithell, C. (1986b) Development of the WRc Mk 3 fish monitor: description of the system and its response to some commonly encountered pollutants. Water Research Centre. Environment TR 233.

Ewen, R. (1987) *Biological Testing for Toxicity Control in Open Waters*. Endress and Hauser, Germany.

Famme, P., Riisgard, H.U. and Jorgensen, C.B. (1986) On direct measurements of pumping rates in the mussel *Mytilus edulis*. *Marine Biology*, **92**, 323-327.

Fisher, J.W., Putnam, M.E., Dilege, R.A., Livingstone, J.M. and Geiger, D.L. (1983) Biological monitoring of bluegill activity. *Water Resources Bulletin*, **19**, 211-215.

Geller, W. (1984) A toxicity warning monitor using weakly electric fish, *Gnathonemus petrsii*. *Water Research*, **18**, 1285-1290.

Giattina, J.D., Gorton, R.R. and Steven, D.G. (1982) The avoidance of copper and nickel by rainbow trout as monitored by a computer-based acquisition system. *Transactions of the American Fisheries Society*, **111**, 491-504.

Greaves, J.O.B., Smith, E.H. and Wilson, R.S. (1985) Environmental monitoring system. Patent application, publication nr. and date: EP 0.162.688, 27.11.1985.

Gruber, D. (1988) A historical perspective. In D.S. Gruber and J.M. Diamond (eds.) *Automated Biomonitoring: Living Sensors as Environmental Monitors*, 15-20. Ellis Horwood, Chichester.

Gruber, D. and Cairns, J. (1981) Data acquisition and evaluation in biological monitoring systems. *Hydrobiologia*, **83**, 387-393.

Gruber, D.S. and Diamond, J.M. (eds.) (1988) *Automated Biomonitoring: Living Sensors as Environmental Monitors*. Ellis Horwood, Chichester.

Gruber, D., Cairns, J., Dickson, K.L., Hendricks, A.C. and Miller, W.R. (1980) Recent concepts and developments of an automated biological monitoring system. *Journal Water Pollution Control Federation*, **52**, 465-471.

Harushige, U. (1981) Monitoring method for quality of water. Patent application, publication nr. and date: JP 56-128459, 07.10.1981.

Hasselrot, T.B. (1975) Bioassay methods of the national Swedish environment protection board. *Journal Water Pollution Control Federation*, **47**, 851-857.

Heath, A.G. (1972) A critical comparison of methods for measuring fish respiratory movements. *Water Research*, **6**, 1-7.

Heinis, F. and Swain, W. (1986) Impedance conversion as a method of research for assessing behaviourial responses of aquatic invertebrates. *Hydrobiological Bulletin*, **19**, 183-192.

Heinis, F., Timmermans, K.R. and Swain, W.R. (1990) Short-term lethal effects of cadmium on the filter feeding chironomid larva *Glyptotendipes pallens* (Meigen) (Diptera). *Aquatic Toxicology*, **16**, 73-86.

Heinis, F. and Crommentuijn, T. (1989) Development of a biological monitoring system with water sediment inhabiting insect larvae as test organisms. In E.C.L. Marteijn (ed.) *Project Ecological Rehabilitation of the River Rhine*. Publications and reports no.9. DBW/RIZA. Lelystad.

Holland, G.J. and Green, A. (1975) Development of a gross pollution detector: laboratory studies. *Water Treatment Examination*, **4**, 81-99.

Huvé, J.-L. (1983) Dispositif métrologique d'analyse et d'alarme automatique en cas de pollution basé sur l'utilisation d'un biocapteur et de sa chaine de télémesure. Patent application, publication nr. and date: FR 25.18.265, 17.06.1983.

Ishio, S. (1964) Behaviour of fish exposed to toxic substances. In *Advances in Water Pollution Research*, Vol.1., 19-33. Pergamon, London.

Jenner, H.A., Noppert, F. and Sikking, T. (1989) A new system for the detection of valve movement response of bivalves. KEMA scientific and technical report 1989-7-2.

Juhnke, I. and Besch, W.K. (1971) Eine neue Testmethode zur Früherkennung akut toxischer Inhaltsstoffe im Wasser. *Gewässern und Abwässern*, **50/51**, 107-114.

Karube, I. (1987) Micro-organism based sensors. In A.P.F. Turner, I. Karube and G.S. Wilson (eds.) *Biosensors: Fundamentals and Applications*, 13-29. Oxford University Press, Oxford.

Kenaga, E.E. (1978) Test organisms and methods useful for early assessment of acute toxicity of chemicals. *Environmental Science and Technology*, **12**, 1322-1329.

Kerren, K. (1971) Method for the automated qualitative determination of toxic compounds in water. Patent application, publication nr. and date: DE P.21.64.702, 27.12.1971; NL 72.17.467, 29.06.1973; FR 2.170.514, 14.09.1973.

Kleerekoper, H. (1969) *Olfaction in Fishes*. Indiana University Press, Bloomington; cited in Gruber, 1988.

Knie, J. (1978) Der dynamischen Daphnientest - ein automatischer Biomonitor zur Überwachung von Gewässern und Abwässern. *Wasser und Boden*, **12**, 310-312.

Knie, J., Halke, A., Juhnke, I. and Schiller, W. (1983) Ergebnisse der Untersuchungen von chemischen Stoffen mit vier Biotests. Deutsche Gewässerkündl. *Mitt.*, **27**, 77-79.

Korver, R.M. and Sprague, J.B. (1988) A real-time computerized video tracking system to monitor locomotor behaviour. In D.S. Gruber and J.M. Diamond (eds.) *Automated Biomonitoring: Living Sensors as Environmental Monitors*, 157-171. Ellis Horwood, Chichester.

Kostelecky, J. (1988) Verfahren und Vorrichtung zur Bestimmung von Schadstoffen in Gewässern und Abwässern. Patent application, publication nr. and date: DE 37.08.753, 29.09.1988.

Kramer, K.J.M. (1989) The mussel monitor: mussels as early warning systems. Information leaflet, MT-TNO, KEMA, RIVM, Delta consult. Delft, pp.4.

Kramer, K.J.M., Jenner, H.A. and De Zwart, D. (1989) The valve movement response of mussels: a tool in biological monitoring. *Hydrobiologia*, **188/189**, 433-443.

Landragin, G. (1978) Dispositif de détection biologique de la pollution des eaux. Patent application, publication nr. and date: FR 23.56.141, 20.01.1978.

Lesel, R. and Saboureau, J.-L. (1977) Procédé et appareillage pour la détection des pollutions chimiques dans les eaux. Patent application, publication nr. and date: FR 23.31.024, 03.06.1977.

Levi, Y., Henriet, C., Coutant, J.P., Lucas, M. and Leger, G. (1989) Monitoring acute toxicity in rivers with the help of the Microtox Test. *Water Supply*, **7**, 25-31.

Lindahl, P.E., Olofsson, S. and Schwanbom, E. (1977) Rotary flow technique for testing fitness of fish. In J. Cairns, K.L. Dickson and G.F. Westlake (eds.) *Biological Monitoring of Water and Effluent Quality*, 75-84. ASTM, Special Technical Publication 607.

Link, M. (1988) Das Dr.Lange-System "LUMIStox". Bestimmung der biologischen Toxizität mit dem Leuchtbakterientest. Anwendungsbericht Bio Nr. 101, Dr Lange GmbH, Düsseldorf.

Lowe, C.R. (1985) An introduction to the concepts and technology of biosensors. *Biosensors*, **1**, 3-16.

Lubinski, K.S., Dickson, K.L. and Cairns, J. (1977) Microprocessor-based interface converts video signals for object tracking. *Computer Design*, **Dec.**, 81-87.

Manley, A.R. and Davenport, J. (1979) Behavioural responses of some marine bivalves to heigtened seawater copper concentrations. *Bulletin of Environmental Contamination and Toxicology*, **22**, 739-744.

Manley, A.R. (1983) The effects of copper on the behaviour, respiration, filtration and ventilation activity of *Mytilus edulis*. *Journal of the Marine Biological Association of the United Kingdom*, **63**, 205-222.

Martin, J.V. (1988) Biomonitoring of polluted waters: three systems. In D.S. Gruber and J.M. Diamond (eds.) *Automated Biomonitoring: Living Sensors as Environmental Monitors*, 172-181. Ellis Horwood, Chichester.

Merk, H.-D. (1976) Automatische Nachweisvorrichtungen akuter Intoxikationen in Gewässern und Abwässern. Patent application, publication nr. and date: DT 24.55.621, 26.05.1976.

Miller, W.F. and Sandwell, F.W. (1979) Continuous monitoring of water quality using fish. Patent application, publication nr. and date: GB 1.555.683, 14.11.1979.

Morgan, W.S.G. (1977) An electronic system to monitor the effects of changes in water quality on fish operulum rhythms. In J. Cairns, K.L. Dickson and G.F. Westlake (eds.) *Biological Monitoring of Water and Effluent Quality*, 38-55. ASTM, Special Technical Publication 607.

Morgan, W.S.G. (1979) Fish locomotor behaviour patterns as a monitoring tool. *Journal Water Pollution Control Federation*, **51**, 580-589.

Morgan, W.S.G., Kühn, P.C., Allais, B. and Wallis, G. (1982) An appraisal of the performance of a continuous automatic fish biomonitoring system at an industrial site. *Water Science Technology*, **14**, 151-161.

Morgan, E.L., Young, R.C. and Wright, J.R. (1988) Developing portable computer-automated biomonitoring for a regional water quality surveillance network. In D.S. Gruber and J.M. Diamond (eds.) *Automated Biomonitoring: Living Sensors as Environmental Monitors*, 127-144. Ellis Horwood, Chichester.

Neuman, I.S.A., Van Rossum, C., Peters, R.C. and Teunis, P.F.M. (1990) Cadium deteriorates electro-navigation performance in the catfish *Ictalurus nebulosus*. *Proceedings of ESCPB conference, August 27-31, Utrecht, The Netherlands.* (in press).

Nicolas, C. (1967) Protection des rivières et piscicultures, contre les empoisonnement avec possibilité d'identifier la nature du poison meurtrier. Patent application, publication nr. and date: FR 14.76.724, 06.03.1967.

Noack, U., Herden, N., Löffler, J., Warcup, C. and Gorsler, M. (1985) Kontinueliche erfassung der Algen-Biomasse mittels Chlorophyll-Fluoreszenz in Gütemessstationen des Gewässerüberwachungssystems in Niedersachsen. *Z. Wasser-Abwasser Forschung*, **18**, 177-182.

Onatzky, J.-P. and Ferrier, G. (1986) Procédés et dispositifs pour détecter la pollution de l'eau. Patent application, publication nr. and date: FR 25.73.875, 30.05.1986.

Ostgaard, K., Eide, I. and Jensen, A. (1984) Exposure of phytoplankton to Ekofisk crude oil. *Marine Environmental Research*, **11**, 183-200.

Pauls, J. (1974) Einrichtung und Verfahren zur laufenden Kontrolle von Brauchwasser auf giftige Stoffe. Patent application, publication nr. and date: DT 23.04.315, 01.08.1974.

Petry, H. (1982) The "Motility test": an early warning system for the biological control of waters. *Zentralbalt Bakteriologie Mikrobiologie und Hygiene Series (B)*, **176**, 391-412.

Petry, H. (1977) Vorrichtung zur Bewegungsmessung von Tieren mit Hilfe von induzierten Spannungen. Patent application, publication nr. and date: DT 26.13.713, 13.10.1977.

Phillips, D.J.H. (1977) The use of biological indicator organisms to monitor trace metal pollution in marine and estuarine environments - a review. *Environmental Pollution*, **13**, 281-317.

Poels, C.L.M. (1975) Continuous automatic monitoring of surface water with fish. *Water Treatment Examination.*, **24**, 46-56.

Poels, C.L.M. (1977) An automatic system for rapid detection of acute high concentrations of toxic substances in surface water using trout. In J. Cairns, K.L. Dickson and G.F. Westlake (eds.) *Biological Monitoring of Water and Effluent Quality, 85-95.* ASTM, Special Technical Publication 607.

Ramsey, G., Turner, A.P.F., Franklin, A. and Higgins, I.J. (1986) Rapid bioelectrochemical methods for the detection of living microorganisms. In

A. Johnson (ed.) *Modelling and Control of Biotechnological Processes*, Proceedings of the 1st IFAC Symposium, December 11-13 1985, Noordwijkerhout, The Netherlands. Pergamon, Oxford.

Rausch, E. (1980) Einrichtung zur Ueberwachung der Wasserqualität mithilfe elektrischer Fische. Patent application, publication nr. and date: DE 29.06.884, 28.08.1980.

Rawson, D.M., Willmer, A.J. and Cardosi, M.F. (1987) The development of whole cell biosensors for on-line screening of herbicide pollution in surface waters. *Tox. Assessment*, 2, 325-340.

Rawson, D.M., Willmer, A.J. and Turner, A.P.F. (1989) Whole cell biosensors for environmnetal monitoring. *Biosensors*, 4, 299-311.

Rechnitz, G.A. (1988) Biosensors. *Chemical Engineering News*, 66, 24-36.

Sabourin, T.D. and Tullis, R.E. (1981) Effect of three aromatic hydrocarbons on respiration and heart rates of the mussel *Mytilus californianus*. *Bulletin of Environmental Contamination and Toxicology*, 26, 729-736.

Salánki, J. and Varanka, L. (1976) Effect of copper and lead compounds on the activity of the freshwater mussel. *Annals of Biology (Tihany)*, 43, 21-27.

Salánki, J. and Varanka, L. (1978) Effect of some insecticides on the periodic activity of the freshwater mussel (*Anodonta cygnea* L.) *Acta Biologica Academiae Scientiarum Hungaricae*, 29, 173-180.

Sanemitsu, T. (1986) Water quality examination apparatus. Patent application, publication nr. and date: JP 61-59260, 26.03.1986.

Scherer, E. and Nowak, J. (1973) Apparatus for recording avoidance movements of fish. *Journal of the Fisheries Research Board of Canada*, 30, 1594-1596.

Schmidt, Ch. (1987) Anwendingsbereiche und Ergebnisse des Algen fluoreszenztests. *Archiv für Hydrobiologie, Beiheft Ergebnisse Limnologie*, 29, 107-116.

Schmidt, H.R. (1985) Verfahren und optisches Messgerät zum Nachweis von toxischen Verbindungen durch Motilitätsmessung. Patent application, publication nr. and date: DE 33.45.196, 04.07.1984.

Schuring, B.J. and Geense, M.J. (1972) Een electronische schakeling voor het registreren van openingshoek van de mossel *Mytilus edulis* L. TNO-Rapport CL 72/47.

Sharf, B.W. (1979) A fish test alarm device for the continual recording of acute toxic substances in water. *Archive für Hydrobiologie*, 85, 250-256.

Shieh, W.K. and Yee, C.J. (1985) Microbial toxcity monitor for *in situ* continuous applications. *Biotechnology and Bioengineering*, 27, 1500-1506.

Shirer, H.W., Cairns, J. and Waller, W.T. (1968) A simple apparatus for measuring activity patterns of fishes. *Water Resources Bulletin*, 4, 27-43.

Skipnes, O., Eide, I. and Jensen, A. (1980) Cage culture turbidostat: a device for rapid determination of algal growth rate. *Applied and Environmental Microbiology*, 40, 318-325.

Slooff, W., DeZwart, D. and Marquenie, J.M. (1983) Detection limits of a biological monitoring system for chemical water pollution based on mussel activity. *Bulletin of Environmental Contamination and Toxicology*, 30, 400-405.

Slooff, W. and DeZwart, D. (1986) Continuous effluent biomonitoring with an early warning system, In B.E. Bengtsson, T.J. Norbert-King and D.I. Mount (eds.) *Effluent and Ambient Toxicity Testing in the Göta Älv and Viskan Rivers, Sweden.* Naturvardverket Report 3275.

Slooff, W. (1979) Detection limits of a biological monitoring system based on fish respiration. *Bulletin of Environmental Contamination and Toxicology,* **23,** 517-523.

Smith, E.H. and Bailey, H.C. (1988) Development of a system for continuous biomonitoring of a domestic water source for early warning of contaminants. In D.S. Gruber and J.M. Diamond (eds.) *Automated Biomonitoring: Living Sensors as Environmental Monitors,* 182-205. Ellis Horwood, Chichester.

Solyom, P. (1977) Industrial experiences with Toxiguard, a toxicity monitoring system. *Progress in Water Technology,* **9,** 193-198.

Sparks, R.E., Cairns, J., McNabb, R.A. and Suter, G. (1972) Monitoring zinc concentrations in water using the respiratory response of bluegills (*Lepomis macrochirus* Rafinesque). *Hydrobiologia,* **40,** 361-369.

Spoor, W.A., Neiheisel, T.W. and Drummond, R.A. (1971) An electrode chamber for recording respiratory and other movements of free swimming animals. *Transactions of the American Fisheries Society,* **1,** 22-28.

Sprague, J.B. (1964) Avoidance of copper-zinc solutions by young salmon in the laboratory. *Journal Water Pollution Control Federation,* **36,** 990-1004.

Suber, Soc. (1976) Système permettant de détecter la mort par intoxication d'un poisson témoin installé dans un bac alimenté par de l'eau dont on contrôle la viabilité. Patent application, publication nr. and date: FR 22.91.556, 11.06.1976.

Takase, I., Nishioka, K. and Taraoka, I. (1986) An instrument to detect toxicants in water sources. Patent application, publication nr. and date: JP 61-224690, 06.10.1986.

Tetsurou, H. (1982) Detector for changes in water quality. Patent application, publication nr. and date: JP 57-1969, 07.01.1982.

Turner, A.P.F., Ramsey, G. and Higgins, I.J. (1983) Applications of electron transfer between biological systems and electrodes. *Biochemical Society Transactions,* **11,** 445-448.

Turner, A.P.F., Karube, I. and Wilson, G.S. (eds.) (1987) *Biosensors: Fundamentals and Applications.* Oxford University Press, Oxford.

Van der Schalie, W.H., Dickson, K.L., Westlake, G.F. and Cairns, J. (1979) Fish bioassay monitoring of waste using trout. *Environmental Management,* **3,** 217-235.

Van Hoof, F. (1980) Evaluation of an automatic system for detection of toxic substances in surface water using trout. *Bulletin of Environmental Contamination and Toxicology,* **25,** 221-225.

Voith, J.M. (1979) Procédé et installation de mesure de la consommation d'oxygène par des animaux de rivière ou de mer, par example pour un test avertisseur à poissons. Patent application, publication nr. and date: FR 24.27.604, 28.12.1979; DE 28.24.435, 03.06.1978.

Waller, W.T. and Cairns, J. (1972) The use of fish movement patterns to monitor zinc in water. *Water Research,* **6,** 257-269.

Wallwork, J.F. and Ellison, G. (1983) Automated sensor systems for water recource pollution warning and treatment process control. *Aqua,* 6, 313-320.
Wangerersky, P.J. and Maass, R.L. (1990) Bioavailability: the organism as sensor. In *Proceedings of 10th International Symposium on Chemistry of the Mediterranean,* May 9-16, Primosten, Yugoslavia, (in press).

The Application of Biological Monitoring to Urban Streams: A System Designed for Environmental Health Professionals

David Atkin and Paul Birch

Environment and Industry Research Unit, The Polytechnic of East London, Romford Road, Stratford, London E15 4LZ, England.

Key words: biotic index, macroinvertebrate, urban stream, environmental health, London.

Abstract

Biological monitoring has provided a valuable and convenient basis for determining river water quality in the London Borough of Bromley. Due to increased awareness of the requirement for information on river water quality, the Polytechnic of East London devised a biologically based monitoring programme, to assist with appraisal of waterways in the Borough. The project was designed to optimize the information gained, whilst allowing for a minimum of taxonomic skill, so that non-biologists could undertake the work, if required.

Introduction

The streams, ponds and rivers of Bromley form an important part of the character of the Borough, linking the more open, rural southern region with the urban northern part. They are an integral part of many open areas, including parks and are fished and paddled in by children.

Although the rivers pass through much of the Borough, very little work had been carried out on their nature or water quality. The value of these waterways as havens for wildlife and as a recreational resource for the public makes them very important from the environmental health point of view.

The aims of the project were as follows:
1. Assess the present condition of the waterways.
2. Indicate areas where a significant degree of pollution exists.
3. Identify the sources of such pollution.
4. Provide a comprehensive database against which to compare future variations in water quality.
5. Allow 'problem areas', where water quality could be improved, to be identified.

Bioindicators and Environmental Management
ISBN 0-12-382590-3

The best river and stream management requires water quality to be based on physical, chemical and biological characteristics. Traditionally, measurements of water quality have used chemical criteria or certain key fish species. However, with increasing amounts of research into the susceptibility of aquatic biota to pollution has come increasing awareness of their importance as indicators of pollution. The value of information gained from monitoring invertebrates is based on their continual presence in the water. This makes them susceptible to any factors altering the quality of their environment over a period of time and so compensates for the 'snap-shot' nature of chemical monitoring.

As biological methods are not pollutant specific, the pollutant does not have to be identified before surveillance begins. Biological studies make important contributions under conditions of toxic, intermittent or mild organic pollution, whereas changes in water quality are not easily detected by chemical means alone. Biological methods show the integral effects of mixed pollutants on the biota. As pollution control measures continue to reduce gross point-source pollution, sensitive biological techniques will be required to detect the more subtle disruptions as well as non-point source pollution.

Furthermore, criteria restricted to chemical, physical and bacteriological parameters no longer suffice when the value of water extends beyond its agricultural, domestic and industrial use to include aesthetic, recreational, and ecological dimensions. Biological methods of water quality assessment actually measure the biota. All chemical methods must eventually be interpreted on a biological basis, hence, the trend in river monitoring now is to combine chemical and biological methods. Chemical monitoring provides measurements of the levels of materials considered to be environmental hazards, whilst biological methods indicate long-term, total water quality. A combination of the two allows both the identification of those pollutants having the most serious effect and the overall water quality.

Most biologists concentrate on a particular area of the ecosystem, such as plankton, periphyton, macrobenthos or fish. A clear preference for using macroinvertebrates has emerged for the following reasons:

1. They are differentially sensitive to pollutants of various types.
2. They react quickly, and are capable of a graded response to a broad spectrum of kinds and degrees of stress.
3. They are ubiquitous, abundant and easy to collect.
4. Identification and enumeration is not as tedious as with some of the other groups.
5. They are mainly sedentary and therefore reflect local conditions
6. They have life spans long enough to provide a record of environmental quality.
7. They are very heterogeneous, consisting of representatives of several phyla.

Other groups of organisms have some, but not all of these attributes. It is this kind of combined approach that has been employed by the Bromley Environmental Health Department in monitoring the main river systems in the area, using selected macroinvertebrates.

Methodology

The survey was undertaken within the boundary of the Borough. The sites were set at intervals along the length of each of the river systems to reflect local areas of concern. Invertebrate sampling, however, was restricted to areas where the rivers are shallow, fairly rapid and have gravel bases. These conditions provide the necessary habitat for those species selected in the biotic index. These requirements, however, precluded some stretches of the rivers, especially those that have been canalized. This is clearly an important problem in urban monitoring.

There have been a variety of sampling regimes devised to assess the invertebrates community but the 'kick sampling' technique remains the most popular. This sampling method was undertaken, with a minimum of training, by the Environmental Health Officers. Sample time was standardized to allow quantification and comparison of the invertebrate communities obtained. Kick sampling involves moving upstream, whilst continually disturbing the substrate with one foot. Animals thus dislodged are carried by the current into a 1 mm mesh net, held close to the substrate, downstream of the disturbance.

The animals collected are sorted in the field into taxonomic groups and their numbers estimated. This information is then used in calculating the biotic index values and the animals are returned.

The Biotic Index

There are two main problems in the use of biotic indices for those unfamiliar with freshwater biology. Firstly, although sampling presents no major problems, the identification and classification of organisms often appears to be a daunting task. Secondly, the scoring schemes used in biotic indices are often complex and the interpretation of results may require some biological experience.

To enable those Environmental Health Officers with little experience of freshwater biology to undertake this type of assessment, a simple easy to use biotic index was devised by the Polytechnic of East London. The original Basic Biotic Index (BBI) displayed in Figure 1 was developed specifically for the rivers in the London Borough of Bromley, based upon the natural distribution of groups of invertebrates in the area and their relative tolerance to pollution. Like all such biotic indices it is based upon the differing susceptibility of benthic invertebrates to organic pollution. Each group is awarded a score, those less tolerant to pollution are awarded the higher scores.

The BBI measures the presence and abundance of a "core" of organisms which show a range of responses to pollution. The data presented by Aston and Andrews (1978) was used to identify suitable organisms for this index, and the scoring system is derived from both the British Monitoring Working Party score system and the Revised Chandler score developed by Bryce *et al.*(1978).

ORIGINAL BASIC BIOTIC INDEX (BBI) - SCORE SHEET
London Borough of Bromley version

STREAM: SITE NAME:

SITE CODE: GRID REF.(   )   DATE: /   /

HABITAT: WORKER:

NAME GROUP	1	2	3	4	5	6	TOTAL	BBI SCORE	SCORING
			SUB-SAMPLES						
STONEFLIES PLECOPTERA (Nemouridae)									<50 = 7 >50 = 8
FRESHWATER MOLLUSCA LIMPETS (Ancylidae)									<50 = 6 >50 = 7
MAYFLIES EPHEMEROPTERA (Baetidae)									<50 = 5 >50 = 6
FRESHWATER CRUSTACEA SHRIMPS (Gammaridae)									<50 = 5 >50 = 5
FLATWORM TRICLADIA									<50 = 4 >50 = 3
COILED MOLLUSCA GASTROPODS									<50 = 3 >50 = 2
WATER CRUSTACEA HOGLOUSE (Asellidae)									<50 = 3 >50 = 1
LEECHES HIRUDINEA									<50 = 3 >50 = 1
MIDGE CHIRONOMIDAE LARVAE									<50 = 2 >50 = 1
WORMS OLIGOCHAETA									<50 = 1 >50 = 1
OTHER GROUPS									No score
FISH									No score

1   2   3   4   5   6   TOTAL
Number of individuals

BBI SCORE

NOTES:

Figure 1. The original basic biotic index score sheet.

The simplification of a biotic index reduces its reliability but only to a very limited extent when it is designed for use in a localized area. Since only a similar set of waterways were considered, it was possible to remove confusing or uncommon organisms from the BBI without unduly affecting its sensitivity. The invertebrates chosen in the calculation of the BBI were all previously recorded in the area, are relatively easy to identify and act as reliable indicators of water quality.

A "key" was produced for use by the Environmental Health Officers, consisting of both large scale and life size illustrations of the commonest organisms which occur in the streams. The identification of an organism, as shown in Figure 2, relies on both the illustrations and a list of diagnostic features of the group.

The BBI score for each group is given at the base of each page of the key. The number of individuals recorded in each group is taken into account by the index so that some elements of the community structure are used in the calculation of the index score. For example, a large number of midge larvae or leeches might indicate organic enrichment, while many freshwater limpets or mayflies in a sample would suggest relatively unpolluted conditions. The type of habitat and occurrence of each group is also described on each page of the key.

In many samples, especially those from "clean" sites, organisms that do not fall in any key group will be found. These may include beetles (Coleoptera), water mites (Hydracarina) and non-chironomid fly larvae (Diptera). The presence of these and other non-scoring groups are simply noted.

It was decided on the basis of data obtained over the first six months to alter the original index to include further species of invertebrates and to make more precise estimations of abundance. The initial practical usage of the index highlighted two main weaknesses. Firstly, the splitting of numbers and individuals into categories of more or less than 50 took no account of very low or very high numbers. Whenever such figures were recorded they would have had a disproportionate effect on the score for the site. Secondly, several groups of invertebrates were being recorded but not scored, which would show marked differences between sites. These groups, particularly cranefly and caddisfly larvae, were easily incorporated into the existing key and index.

The most satisfactory method of estimating the effects of abundance proved to be a three-way split into:

a) numbers less than 10,

b) numbers between 10 and 100, and

c) numbers greater than 100;

with each being awarded different scores as for the earlier BBI. These categories take account of the most common and biologically significant differences in numbers. The Revised Basic Biotic Index score sheet is displayed in Figure 3. The BBI's performance was assessed by comparing the scores obtained with that of the BMWP.

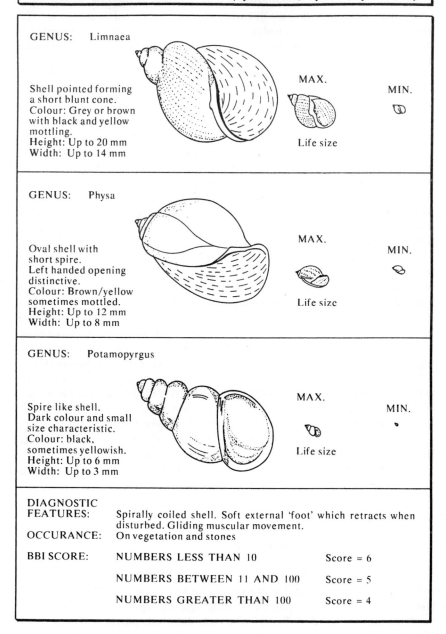

COILED GASTROPODS - MOLLUSCA (Lymnacidae, Physidae, Hydrobidae)

GENUS: Limnaea

Shell pointed forming
a short blunt cone.
Colour: Grey or brown
with black and yellow
mottling.
Height: Up to 20 mm
Width: Up to 14 mm

MAX.

MIN.

Life size

GENUS: Physa

Oval shell with
short spire.
Left handed opening
distinctive.
Colour: Brown/yellow
sometimes mottled.
Height: Up to 12 mm
Width: Up to 8 mm

MAX.

MIN.

Life size

GENUS: Potamopyrgus

Spire like shell.
Dark colour and small
size characteristic.
Colour: black,
sometimes yellowish.
Height: Up to 6 mm
Width: Up to 3 mm

MAX.

MIN.

Life size

DIAGNOSTIC
FEATURES: Spirally coiled shell. Soft external 'foot' which retracts when
                 disturbed. Gliding muscular movement.
OCCURANCE: On vegetation and stones

BBI SCORE: NUMBERS LESS THAN 10   Score = 6

                 NUMBERS BETWEEN 11 AND 100   Score = 5

                 NUMBERS GREATER THAN 100   Score = 4

Figure 2. A section of the Bromley Key.

REVISED BASIC BIOTIC INDEX (BBI) - SCORE SHEET
London Borough of Bromley version

STREAM: SITE NAME:

SITE CODE: GRID REF.(   ) DATE: /   /

HABITAT: WORKER:

NAME GROUP	SUB-SAMPLES 1	2	3	4	TOTAL	BBI SCORE	BBI SCORING < 10	< 100	> 100
STONEFLIES PLECOPTERA (Nemouridae)							16	17	18
CASED TRICHOPTERA CADDISFLY (Limnephilidae)							14	15	16
FRESHWATER MOLLUSCA LIMPETS (Ancylidae)							13	14	15
CRANEFLIES DIPTERA (Tipulidae)							12	13	14
MAYFLIES EPHEMEROPTERA (Baetidae)							9	10	10
FRESHWATER CRUSTACEA SHRIMPS (Gammaridae)							9	9	9
FLATWORM TRICLADIA							8	7	5
UNCASED TRICHOPTERA CADDISFLY (Hydropsychidae)							8	7	5
COILED MOLLUSCA GASTROPODS							6	5	4
WATER CRUSTACEA HOGLOUSE (Asellidae)							5	4	3
LEECHES HIRUDINEA							5	4	3
MIDGE CHIRONOMIDAE LARVAE							3	2	1
WORMS OLIGOCHAETA							2	1	1

Number of individuals

BBI SCORE

NOTES

Figure 3. The revised basic biotic score index sheet.

Correlations were made between the BBI and BMWP scores based on 135 macroinvertebrate samples from a wide range of sites, not just within the Borough. In the absence of non-scoring organisms, the BBI score was found to be closely related to that of the BMWP score. However, since the data was from many different streams, the BBI performs poorly where high scores occur because the greater diversity increases the number of non-scoring groups. In the Bromley catchment, there was less divergence between the two indices due to the design of the BBI for use in this particular area. The BBI score for a site can therefore be used to indicate the minimum BMWP score and its probable maximum, as a measure of water quality.

One of the functions of the BBI is to reduce the effects of seasonality of invertebrates. Those groups used in the index were chosen because they are represented in one form or another for most of the year. The differences in the monthly BBI scores are therefore more likely to be an effect of the environment, rather than the natural cycles of the invertebrates.

Conclusions

The work undertaken since 1986 has supplied a wealth of information on the quality of the Borough's waterways. The four major river systems have now been intensively surveyed and documented and in future it will be relatively easy to periodically update this Borough wide database.

As a baseline investigation, the information obtained using the basic biotic index is invaluable, proving sufficiently comprehensive to allow detailed comparisons with future data while being easily collected and collated.

Feedback from the Environmental Health Officers has been good. They have found the BBI system easy to use and interpretate. This biotic index has proven a valuable and convenient means for Environmental Health Officers to determining the water quality of local rivers.

Acknowledgement

The authors would like to acknowledge the support and enthusiasm of staff of the London Borough of Bromley, particularly the Environmental Health Department, during this project.

References

Aston, K.A. and Andrews M.J. (1978) Freshwater macroinvertebrates in Londons rivers 1970-77. *The London Naturalist,* **57,** 34-52.

Bryce, D., Cafoor, I.M., Dale, C.R. and Jarret, A.F. (1978) *Macro-invertebrates and the bioassay of water quality: A report based on a survey of the River Lee.* Nelpress.

Soil Fauna as Bioindicators of Biological After-Effects of the Chernobyl Atomic Power Station Accident

Dmitrii A. Krivolutzkii and Andrei D. Pokarzhevskii

Institute of Evolutionary Animal Morphology and Ecology of the USSR Academy of Sciences, 117971, Moscow, Leninsky Prospect, 33 USSR.

Key words: radioactive contamination, soil fauna, population effects, Chernobyl.

Abstract

Litter dwelling populations decreased sharply at contaminated plots during the first year after the Chernobyl atomic power station accident. Effects on soil dwellers were not so marked due to shielding by the soil. 30 Gy doses did not directly affect adult animals but their eggs and juveniles. Population recovery processes were continued 2 - 2.5 years after the accident into the contaminated zone. Data on the Chernobyl contaminated plots confirmed results of experimental manipulation with soil fauna at artificially contaminated plots.

Introduction

The effects of ionizing radiation on soil animal populations have been the subject of many experimental and field studies at naturally and artificially contaminated plots. It has been revealed that insects, mites and earthworms respond to radioactive pollution, although they are more radio-resistant than vertebrates (Krivolutzkii 1983; Krivolutzkii et al. 1988). In some cases, this resistance is explainable by the shielding effect of the soil. Some changes in the community structure after irradiation were noted, as were changes in population numbers. These are due to not only primary radio-biological effects but to secondary ones, which result from biotic relationship disruptions in polluted ecosystems. Hence soil animals could be useful for estimations of radioactive contamination impacts after the Chernobyl atomic power station accident, at least in a 30 km contaminated zone around the station.

Materials and methods

This study was carried out in the framework of ecological research of short and long term after-effects of the Chernobyl accident. Areas were sampled during July and September 1986, April 1987 and October 1988. The sites were within the 30 km protection area of the Chernobyl APS accident, and on relative control plots 70 km south of the station.

In July 1986 samples were collected from soil and litter of pine forest stands 50-60 years old on sandy soil with a thick needle litter and moss layer and with a humus layer 1-2 cm thick. The plots were 3 km from the station (Izumrudnoe), 30 km south of the station at the edge of the protection zone and 70 km south of the station near the village of Lutezh. Animals from these samples (225 cm^2) were collected by Tullgren funnels. On the same plots, samples (0.25 m^2) were taken in April 1987 and in October 1988, with hand sorting used for collection.

In September 1986 samples (0.25 m^2) were collected in potato fields near the village of Kopachi (3 km from the station) and in the village of Demidov (70 km south of the station).

In July and September 1986, 10 samples were taken in each plot, and in April 1987 and October 1988 twenty four cores were taken in each plot.

Results and discussion

In July 1986 only small soil and litter animals were found in the samples from the heavily contaminated area, and earthworms were absent. One specimen of earthworm was found in the cores from the plot of relative control. Adult oribatid mites and their juvenile stages were dominant among soil animals in that period, but the juvenile stadia proved to be more sensitive to radioactive contamination than adult animals. Differences between numbers of juvenile oribatids in control and heavily contaminated plots varied more sharply in comparison with adult oribatids. For mesofauna populations there were no differences between the control plot and the plot at the edge of the 30 km zone (Fig. 1).

In Table 1 data are presented on differences in oribatid mite numbers for litter, and soil pore dwellers. Although differences in species composition are found for both groups of oribatids, and radioactive contamination led to decreases in population numbers of both litter and soil dwellers, litter dwellers were influenced by contamination significantly more than soil ones.

In September 1986, soil populations of earthworms from contaminated and control plots within the potato fields differed in numbers to a lower degree than populations of mesofauna in July, but differences were significant (2-3 times) (Fig. 2). There were marked differences in the immature/mature ratio for earthworms: in Kopachi the ratio was 1.0; in Chernobyl (17 km from the station) it was 0.8; and in Demidov it was 0.47. Among immature specimens in Kopachi the mass of animals was close to 200 mg but in Dimidov was only 30 mg. Hence in the contaminated zone, young earthworms had not survived after the accident, although earthworms had bred in this period (cocoons were found in Kopachi soil). Viktorov (pers. comm.) found chromosome abberations in 19% of primary spermatocytes of *Apporectodea caliginosa caliginosa* from contaminated plots.

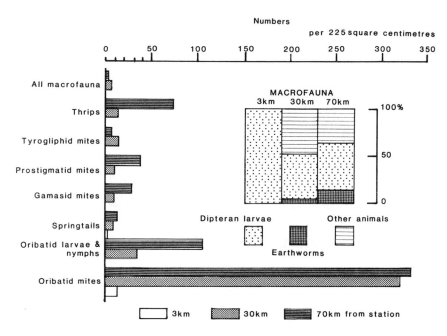

Figure 1. Soil fauna population number in the pine stand soils of the accident area in July 1986.

Table 1. Oribatid mites in soil at the pine stand plots, July 1986. (Figures are mean ± S.E. number of animals m⁻².)

Species and groups	Distance from the station		
	3 km	30 km	70 km
Litter dwellers			
Species in 10 cores	5	22	23
Density	97.8 ± 84.4^a	6222 ± 1333^d	4622 ± 1244da
Species bound in all cores	none	5	6
Soil pore dwellers			
Oppiella nova	44.4 ± 17.8^a	3244 ± 1022^d	1866 ± 311da
Quadroppia quadricarinata	57.8 ± 44.4^a	667 ± 356^a	1022 ± 222da
Suctobelbella spp.	160.0 ± 84.4^a	2311 ± 1022^b	2800 ± 889^{ca}
Tectocepheus velatus	62.2 ± 26.7^a	1200 ± 311^c	2500 ± 490^{cb}
Species bound in all cores	none	4	4

a - no diff.; b - P<0.05; c - P<0.01; d - P<0.001

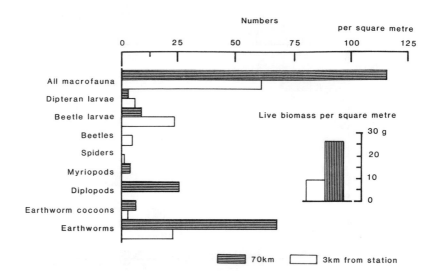

Figure 2. Soil fauna population density and biomass in the potato field soils of the accident area in September 1986.

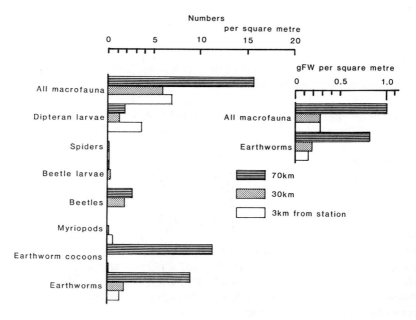

Figure 3. Soil fauna population density and biomass in the pine stand soil in April 1987.

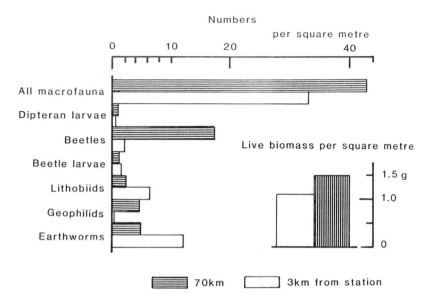

Figure 4. Soil fauna population density and biomass in the pine stand soils in October 1988.

In April 1987, the censuses in the pine stands showed that the differences in population numbers of resident groups and earthworms between the contaminated and control plots were still considerable (Fig. 3). It is obvious that recovery processes in the resident populations were very slow after the first year after the accident. However, it is also clear that the animals were capable of reproduction in contaminated soil, since their cocoons were found there.

In October 1988 soil animal populations had recovered and for earthworms the number in contaminated soils were slightly greater than in the control soil (Fig. 4).

The main reason for crucial differences in earthworm population number, on contaminated plots and relative control plots during the first year after the accident, was high doses of ionizing radiation from radioactive fallout and "hot particles" (Table 2). Hot particles induced drastic effects in litter dweller populations, especially oribatid mites and other microarthropods because the "hot particles" size was comparable with soil microarthropod size. Soil dwellers, such as earthworms, or soil pore dwellers were affected to a smaller extent because soil served as a screen for radiation from "hot particles" and beta-radiation. An absorbed dose in litter was 3 times higher than in soil after two and a half months after the accident. Early in the experimental investigations the same phenomena was noted but at much higher levels of contamination, up to 65-125 MBq m^2 in the low range to 2^{10} MBq m^2 at the highest ranges (Krivolutzkii 1983; Krivolutzkii *et al.* 1988). It is important to note that Crossley and Shanks

(1967) found disturbances to arthropod communities at absorbed doses near 80-95 Gy, i.e. at higher doses than in the Chernobyl contaminated area. An explanation is that in ours and Crossley and Shanks experiments, solutions of radioactive salts were used (in our experiments Sr-90, Cs-137, Zn-93, Ru-106, Ce-144, Pu-239, in Crossley and Shanks experiments Cs-137 on sand particles) but in the Chernobyl soil "hot particles" influenced on soil dweller populations.

But the main after-effects of ionizing radiation were manifested in reproduction processes of soil fauna. Among mesofauna the first instars were almost totally lacking in soil populations in polluted soil. The recovery processes of the earthworm *Dendrobaena octaedra* began only at a dose rate of 7.2 mr h^{-1} for gamma-radiation and 130-170 Bq cm^2. This earthworm species is represented in the Chernobyl area by a hexaploid form, which reproduces by obligate apomictic parthenogenesis (Viktorov, pers. comm.). Effects on juvenile stadia of earthworms were noted early in investigations on plots with high levels of natural radioactivity at doses of only 50-150 mkr h^{-1} (Krivolutzkii *et al.* 1988).

Table 2. Dose characters of the sample plots on soil surface.

Plots and distance from the station	Gamma radiation mr h^{-1}	Beta radiation Bq cm^2	Total absorbed dose Gy
July 1986			
Pine stands, 3 km	49.7		29.4
Pine stands, 30 km	6.1		9.2
September 1986			
Potato fields, 3 km	8.0	170-830	86.0
Potato fields, 30 km	0.2		
April 1987			
Pine stands, 3 km			
litter	7.2	130-170	
soil surface	7.2	33	
soil 5 cm	7.2	15	
Pine stands, 30 km			
litter	below det. level	33-60	
soil surface	below det. level	2.3-2.5	
soil 5 cm	below det. level	2.3	
Pine stands,70 km			
litter	below det. level	4.0	
soil surface	below det. level	0.2	
soil 5cm		0.2	
October 1988			
Pine stands, 3 km			
litter	0.9-1.4		
Pine stands , 70 km			
litter	0.01		

In these investigations growth rate decrease was noted for all species of earthworms and some species were absent in the radioactive plots. In the Chernobyl contaminated zone in pine stands *Aporrectodea caliginosa caliginosa* disappeared after the first year. This may be a result of differences in reproductive patterns for the two species of earthworms met here. *A. c. caliginosa* reproduces amphimictically and *D. octaedra* apomictically. The latter has an advantage due to parthenogenesis. The higher population density of earthworms in contaminated soil in comparison with the control in the autumn of 1988 possibly results from much slower recovery processes of myriapoda - *Geophilida* (like other amphimictic invertebrates), which mainly feed on earthworms.

Hence soil faunal studies in the contaminated zone after the Chernobyl atomic power station accident confirmed that soil animal populations are among the sensitive bioindicators of radioactive contamination in the early period after this type of contamination. They also confirmed the results of early investigations under experimental conditions.

References

Crossley, D.A. and Shanks, M.H. (1967) Survival of irradiated insects in field environments. *Report of Oak Ridge National Laboratory, N 4168*, 65-67.

Krivolutzkii, D.A. (1983) Radioecology of terrestrial animal communities. *Energoatomizdat*, Moscow (in Russian).

Krivolutzkii, D.A., Tikhomirov, F.A., Fedorov, E.A., Pokarzhevskii, A.D. and Taskaev, A.I. (1988) Ionizing radiation influence on biogeocoenosis. *Nauka*, Moscow (in Russian).

The Chernobyl Accident: Fallout and Possible Effects in Norway

Per Oftedal

Department of Biology, Division of General Genetics, University of Oslo, PO Box 1031 Blindern, 0315 Oslo 3, Norway.

Key words: radioactive fallout, effects on humans, grazing animals, milk, Chernobyl, Norway.

Abstract

Radioactive fallout from the Chernobyl accident was registered in Norway from April 28, 1986, onwards. The extent of contamination became clear within 2-3 weeks, but detailed measurements during several months proved necessary to identify areas and intensities of different degrees of contamination, mainly by 134-Cs and 137-Cs. Aerial surveys from helicopter proved more useful than from ordinary plane, due to the irregular features of the terrain. Contamination proved to be very variable and closely related to the occurrence of local rain showers during the ten-day period of high aerial radioactivity over Norway.

The districts most heavily contaminated are in relatively high ground, with oligothrophic lakes and rather meagre grazing lands utilized mainly by sheep and reindeer.

Contamination levels, countermeasures, and various interpretations of the situation and possible ecological effects are presented. Similarities with and differences from the fallout situation following the atmospheric bomb tests in the fifties and early sixties are discussed.

Introduction

The Chernobyl reactor failure constitutes the greatest industrial accident ever, in terms of widespread and long-lasting radioactive contamination. Other accidents, in particular the Bhopal chemical catastrophe in India, have led immediately to greater losses of human life. The time scale is, however, of a different order than in that caused by the releases from Chernobyl. The situation is possibly comparable to the areas immediately surrounding the atomic bomb test sites. However, a large reactor like the Chernobyl unit contains several

Bioindicators and Environmental Management
ISBN 0-12-382590-3

orders of magnitude more uranium, transuraniums, and fission products than even the largest bombs, and the uncontrolled release during the reactor fire thus had gigantic dimensions, lasting ten days.

Obviously, the major part of the contamination is found in the regions surrounding the site. However, although the debris was not lifted beyond the tropopause, there were significant amounts of radioactivity deposited at distances of 1000-2000 kilometres. The present account is concerned with the situation in Norway. The post-Chernobyl conditions are to some extent compared with the experiences resulting from the atmospheric bomb testing during the period 1954-1962.

The pattern of contamination

A clear increase in the air activity was registered in Southern and Central Norway within three days of the start of the releases, reaching Cs-134 levels of 1-10 Bq m^3 on the first day, followed by lower levels, and finally a second peak of 0.01-0.1 Bq m^3 about 8-10 days later. Levels on the west coast were lower, and in the far north (Tromso), the activity hardly reached 0.01 Bq m^3. Levels of I-131 were consistently higher, on some days and at some locations by almost two orders of magnitude. This was mainly due to the I-131 content being less variable than the Cs-134 content.

As was to be expected, ground contamination level was dominantly determined by the pattern of precipitation during the cloud passages. Rain fell over certain areas in the valleys and the high ground of central Norway, and in two regions in mid-Norway. The latter areas form an extension of the band of high deposition crossing the Scandinavian peninsula. It is fortunate that the rainfall over Norway was limited to few and relatively local showers.

Ground contamination was identified initially by the readings from mobile instruments, in cars and helicopter. Due to the nature of the terrain, with relatively deep and narrow valleys, survey by ordinary aeroplane was of limited value. Subsequently, areas identified as the most highly contaminated were surveyed by way of soil samples, and a coherent picture of the contamination pattern had been established by early June, i.e. within a month after the period of deposition (Fig. 1). Subsequent more detailed surveys confirmed the initial results (Backe et al. 1987).

The areas most contaminated had activities in excess of 200 kBq m^2, with reference to Cs isotopes. Subsequent detailed measurements by way of soil samples indicate that the deposition varies a great deal at any given locality. Within an area of 100-200 m^2 values differ by more than an order of magnitude. This shows the need for standardized and systematic sampling procedures in order to obtain representative and comparable observations (R.A. Olsen, pers. comm.).

Initially, about one-third of the total radioactivity measured was Cs-134 and two-thirds was Cs-137. The level of Sr-90 was not extensively registered, mainly because it demands a more refined technical effort, but it is in general considered

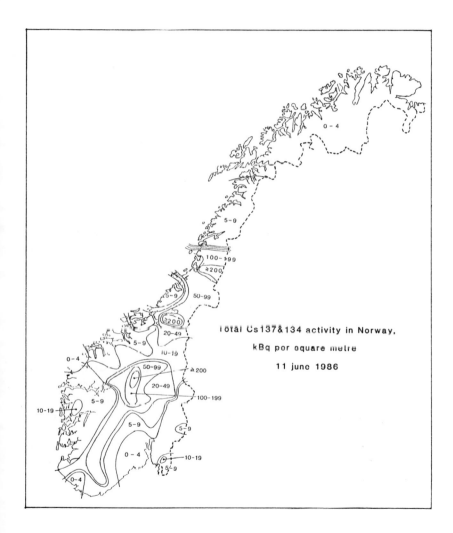

Figure 1. Contamination levels in Norway as determined during the first month following the Chernobyl accident. Courtesy Health Directorate/National Institute of Radiation Hygiene.

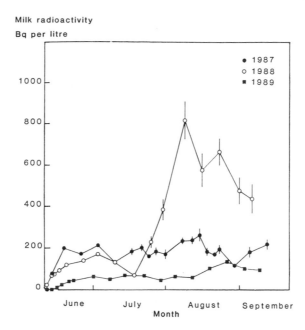

Figure 2. The importance of mushroom contamination for Cs-137 content in goat milk. 1988 was a good mushroom year, 1987 and 1989 were not (K. Hove *et al.*, 1990).

to be low. The contaminated areas are in general sparsely populated and are utilized for summer grazing mainly of sheep, and for year-round grazing by wild or semi-domesticated reindeer. Soil cover is thin with very little humus. Cultivated areas are relatively small. The total population of the highly contaminated areas is about 200 000.

The first summer, a major fraction of the fallout was seen as surface contamination of vegetation. In the following years, the upper few centimetres of soil of uncultivated land held most of the contamination, and vegetation was increasingly contaminated through root uptake. The most characteristic pathway of transfer to animals proved to be by way of mushrooms, in particular *Rozites caperata*. Uptake by sheep of Cs-137 is highly dependent upon whether it is a good year for mushrooms or not. Mushroom content of Cs-137 was found up to 300 kBq kg^{-1} dry matter, being 60-80x the content of grass in the same area (Fig. 2) (Hove *et al.*, 1990).

Contamination of animals

Simply because of the economic importance, the predominant investigations have been centred around sheep, goats, cattle and reindeer, and in addition wild trout.

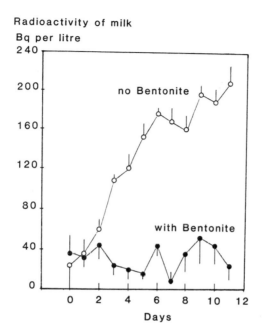

Figure 3. Effects of bentonite on transfer of radioactivity from contaminated fodder to milk in goats (K. Hove, pers. comm.).

During the summer of 1986, action levels of 370 Bq l^{-1} for milk, and 600 Bq kg^{-1} for meat were instituted. During the summer and during the slaughter period in the fall of 1986, milk and meat produced in relatively large districts were discarded as non-edible. Of Norway's more than two million sheep, there was nearly 0.5 million in the highly contaminated districts. Evaluation of contamination was made on the basis of samples of animals measured from each grazing herd. On this basis, 160 000 sheep were destroyed as unfit for human consumption. Farmers were fully compensated for all produce that could not be marketed.

Since 1986, the contamination has been met with quite efficient counter-measures. In the main, this consisted of keeping the sheep on lowland cultivated pastures for a suitable period prior to slaughter. In 1988 this period was up to 16 weeks in some herds. At present, the period is 4-6 weeks. In addition to lowland grazing, the sheep are treated with Cs-binders (zeolite, bentonite, or Prussian Blue (ferric-ferro-hexocyanate)). The remedials have been administered as salt-licks, or as slowly-dissolving bolus implanted in the rumen. The latter technique has proved feasible and efficient even for reindeer and goats, in the field, reducing the content by 80% to 90% relative to controls (Fig. 3). In 1987, about 280 000 sheep, and in 1988 about 430 000 needed one or other of the above treatments. Because of these measures very few animals had to be discarded.

A year after the accident, it was realized that the action levels chosen had serious socio-economic effects on groups of mountain farmers and reindeer herding families. In view of the economic and cultural impact the loss of their traditional way of life would have for these groups, the action level for reindeer meat, wild game and fish was raised to 6000 Bq kg^{-1}. For the general population, this would constitute a very small additional dose burden, since the contribution to the normal diet from these sources would be almost negligible. For those groups in the highly contaminated districts normally having a large fraction of their food intake from fish, game and their own farm production, advice was given as to how often meals of given contamination levels could be eaten (once a week, once a month, for children, for pregnant women). This advice has been followed and is to a large extent effective, as has been shown by regular whole-body radioactivity measurements of members of those groups (Strand *et al.* 1989).

By monitoring external radiation as well as the level of dietary intake it could be shown that during the first year after the releases, the importance of internal contamination gradually increased, while that of external irradiation decreased (Fig. 4).

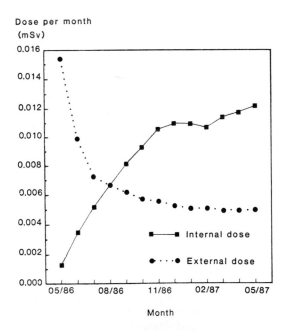

Figure 4. Average external and internal dose equivalents in Norway by month during the first year after the deposition of radioactivity from Chernobyl (R.T. Lie, pers. comm.).

Effects of present fallout

The doses received by animals due to the content of radioactive caesium are such that it may be possible to recognize radiation effects. A constant body content of 10 kBq kg^{-1} body weight gives a dose of 0.562 milliSivert per year. This corresponds to one half of the natural background dose, approximately. Thus, one would not expect dramatic and immediate effects.

Reindeer is the species with the highest body contents, in individuals values up to about 100 kBq kg^{-1} have been measured. There is considerable seasonal variation, related to differences in diet during winter and summer, and the calculation of accumulated dose is difficult.

Semi-quantitative observations of reproductive success in several "domestic" reindeer herds give indications of a reduction in number of surviving calves in 1987 by 10% to 25% relative to preceding and following years. Correspondingly, increases in chromosomal aberrations in lymphocytes of calves and adults have been observed, but again interpretation of the data is difficult due to uncertainties regarding control values, normal inter-herd variability, and average caesium content.

Man is himself the most closely observed and best registered organism (Strand *et al.* 1990). The Norwegian Birth Registry in Bergen receives obligatory reports concerning all births and newborn in the country. In a careful collation of data according to municipality and month of birth, the frequency of malformations seen during each of the first 13 months after the fallout was compared with the mean frequency in the corresponding month in the preceding three years, and with the calculated fetal dose according to local external and dietary contamination. For this period of 13 months, a statistically significant positive correlation was seen between dose and relative frequency of total CNS (central nervous system) anomalies. However, this survey has been continued, and no significant deviations from the pre-Chernobyl pattern is observable in subsequent periods. Thus, the interpretation in uncertain, with regional changes in dosimetry, diagnostic acuity and other factors being important variables that are difficult to evaluate (R.T. Lie, pers. comm.).

Effects of earlier fallout

Norway received - as most West-facing coastlines - above average amounts of radioactive fallout from the atmospheric atom bomb tests 1954-1962 (Westerlund *et al.* 1987; Hove and Strand 1990). The fallout pattern is largely determined by the precipitation pattern, and in Norway there is a clear difference between on the one hand the West and North West coast, and on the other the South and Central inland districts. In the West there is also a strong seasonal variation (Fig. 4).

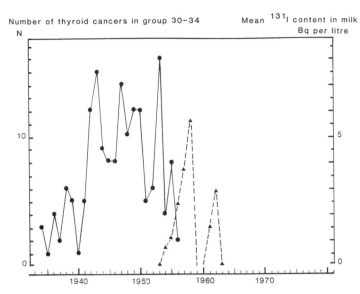

Figure 5. Number of thyroid cancer in 30-34 year old women by year of birth, and mean annual content of I-131 in Norwegian milk. Data from Cancer Registry of Norway, and the Norwegian Defence Research Establishment. Updated and redrawn from Oftedal and Lund 1983.

Two effects have been studied with a view to possible connection with this earlier fallout, namely thyroid cancer frequencies presumably related to fallout of radioactive iodine I-131, and fetal brain damage expressed as reduced scholastic ability, presumably related to external and internal irradiation during the 8th to 15th week of fetal development, shown in Hiroshima/Nagasaki to be a period of high radiation sensitivity of the brain (Otake and Schull 1984).

The doses received by the thyroid could be roughly estimated for the various cohorts on the basis of annual mean content of I-131 in milk. All cancer cases are reported by law to the Cancer Registry of Norway (Oslo). The number of thyroid cancers in the age group of 27-38 years showed a sudden increase by 3-4 fold for the cohorts born in 1941 and later (Fig. 5). These patients were from 12 years of age and less when the contamination started, and 21 years or less when it came to an end. The doses received by these cohorts were of the order of 5 milliSivert, and the only common biological characteristic of the group easily recognized is that it went through puberty during the period of exposure. The significance of this coincidence is not known. However, there is no known reason other than radiation that might account for this abrupt increase. As seen normally for this type of cancer, the rate in males is about one-third of that in females. The pattern of increase is closely similar in the two sexes.

Scholastic ability is measured in Norway by the Basic School Council, which is the governmental organ for overseeing teaching in public schools in the first nine years. Norm-related tests are offered to all teachers, of which about one half

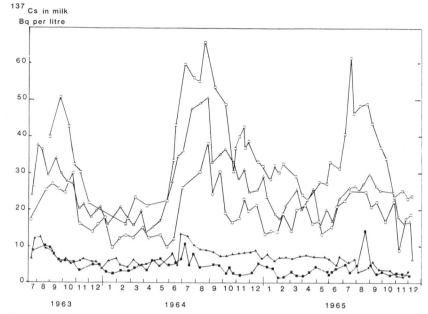

Figure 6. Content of Cs-137 in milk from five Norwegian dairies 1963-65. The three upper curves are from West and North coast dairies, and the two lower ones from inland East and Central Norway. Unpublished data from the Norwegian Defence Research Establishment.

uses the tests, and 10% of the test results are returned to the Council. Tests are offered in the 7th and 9th year, in Norwegian, English and Mathematics. A total of more than 12 000 tests of children born in 1965 have been analysed (Oftedal 1989a; 1989b) - Concentration of Cs-137 was measured bi-weekly in milk from ten dairies suitably located in different regions of the country, and the regional means have been used to indicate the relative variation in dose over the year in the West/North (i.e. municipalities with more than 1000 mm rainfall per year) and in the East (less than this value) (Fig. 6.). The milk content of Cs-137 in all probability form an important contribution to the fetal dose, but effects from other radio-isotopes and pathways of exposure cannot be excluded.

Children born in 1965 were chosen for study, since 1964 showed the peak content of Cs-137 in milk. A rather broad peak occurred during July-August of that year.

Children enter school the year they become seven years of age, and those born in December 1965 were on the average some 2% less capable in the test than those born in January 1965. In the material from schools in the East, scholastic ability in children showed a steady increase in mean value with age, with no characteristic deviations. In the material from schools on the West/North coast, the same development over the year is seen, except that in the months of January-March there is a clear depression. This depression corresponds to the peak of milk Cs-137 content coinciding with the fetal period of maximum

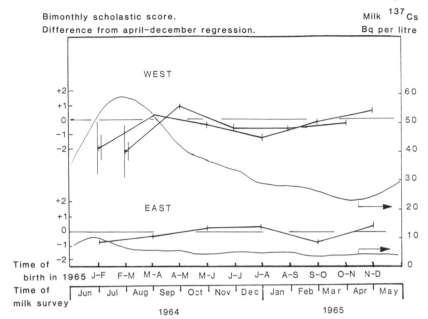

Figure 7. Bimonthly mean scholastic success relative to April-December regression, in West/North and East schools, according to time of birth in 1965. Mean milk content of Cs-137 shifted 6 months to make maximum content in 1964 coincide with maximum deviation in scholastic achievement in 1965 cohort (from Oftedal 1990).

sensitivity. The depression corresponds to about one year's development (Fig. 7). Again, the doses are small and probably amount to no more than a doubling of the natural background dose rate. In comparison with the effects on IQ seen in Hiroshima/Nagasaki, the effects are one to two orders of magnitude greater than expected.

Conclusions

The studies of Chernobyl and bomb fallout in Norway have led to many interesting findingings.

a) In oligotrophic areas - terrestrial and aquatic - available nutrients/minerals are recirculated effectively, leading to biological half-lives of radioactivity approaching the physical half-lives. The problems of contamination of mountain-grazing sheep and reindeer may last for fifty years. On cultivated agricultural land the dilution and sequestration of radioactivity is much more rapid (Hove and Strand 1990).

b) With caesium-binders - e.g. bentonite or Prussian Blue - it is possible to reduce caesium body burdens effectively and with acceptable economy (K. Hove, pers. comm.).

c) Given sufficiently refined methods of observation and sufficiently large materials - as in human epidemiology covering a significant fraction of the country's population - it is possible to demonstrate the probable presence of effects of very low doses. The effects registered following Chernobyl or bomb test fallout do not constitute significant risks for the individual, but serve to show that the concept of "population dose" is a valid concern even when the doses appear insignificant and the consequences for any single individual cannot be identified.

References

Backe, S., Bjerke, H., Rudjord, A.L. and Ugeltveit, F. (1987) The fallout of Caesium in Norway after the Chernobyl accident. *Radiation Protection Dosimetry*, **18**, 105-107.

Hove, K., Pedersen, O., Garmo, T.H., Hansen, H.S. and Staaland, H. (1990) Fungi: A major source of radiocaesium contamination of grazing ruminants in Norway. *Health Physics*, **59**, 189-192.

Hove, K. and Strand, P. (1990) Predictions for the duration of the Chernobyl radiocaesium problem in non-cultivated areas based on a reassessment of the behaviour of fallout from nuclear bomb tests. In *Environmental Contamination Following a Major Nuclear Accident*, 215-223. IAEA, Vienna.

Oftedal, P. and Lund, E. (1983) Cancer of the thyroid and I-131 fallout in Norway. In *Biological Effects of Low-level Radiation*, 231-239. IAEA, Vienna.

Oftedal, P. (1989a) Scholastic achievement in relation to fetal exposure to radioactive fallout in Norway. In K.F. Baverstock and J.W. Stather (eds.) *Low Dose Radiation*, 345-353. Taylor and Francis, London.

Oftedal, P. (1989b) Radiation fallout in Norway in 1964. In *Berzelius Symposium, XV, Umeå, Sweden*, 75-81.

Oftedal, P. (1990) A holistic view of low level radiation effects in biological systems. *Canadian Journal of Physics*, (in press).

Otake, M. and Schull, W.J. (1984) *In utero* exposure to A-bomb radiation and mental retardation: a reassessment. *British Journal of Radiology*, **57**, 409-414.

Strand, P., Boe, E., Berteig, L., Berthelsen, T., Strand, T., Trygg, K. and Harbitz, O. (1989) Whole-body counting and dietary surveys in Norway during the first year after the Chernobyl accident. *Radiation Protection Dosimetry*, **27**, 163-171.

Strand, P., Selnæs, T.D. and Andersson, A. (1990) Total-body measurements in Southern Norway after the Chernobyl accident 1987-1989. *National Institute of Radiation Hygiene 1990*, **4**, 1-20 (in Norwegian).

Westerlund, E.A., Berthelsen, T. and Berteig, L. (1987) Body burdens in Norwegian Lapps 1965-1983. *Health Physics*, **52 (2)**.

Prediction of Radiocaesium Levels in Vegetation and Herbivores Using Bioindicators

E.J. McGee, P.A. Colgan and H.J. Synott

Nuclear Energy Board, 3 Clonskeagh Square, Dublin 14, Ireland.

Key words: radiocaesium, uplands, predictions, plant uptake, sheep, Chernobyl, Ireland.

Abstract

Radiocaesium deposition and activity levels in soils on four farms in Ireland are presented and shown to produce conflicting patterns of radioisotope content. The data show considerable variability to exist for both deposition and activity at adjacent sites in upland situations.

Soil deposition and activities are compared with associated plant activities using concentration ratios and transfer factors. An alternative predictive method using *Calluna vulgaris*, a common component of these systems, as a baseline to which activity levels in other species may be compared, is proposed.

Measurement of radiocaesium activity levels in faecal pellets from sheep are compared with *in-vivo* measurement of flocks and may prove to be a simple and economical alternative to live monitoring in the determination of mean flock activity.

Introduction

The Chernobyl accident in April 1986 resulted in significant deposition of caesium-137, caesium-134 and iodine-131. The iodine-131 fallout decayed rapidly but the radiocaesium components of the fallout are of more long term significance (Horrill 1983) due primarily to the long physical half-lives of 30.2 years and 2.1 years for caesium-137 and caesium-134 respectively. Deposition was largely governed by climatic conditions. Heavier deposition tended to occur in areas of higher altitude (Belli *et al.* 1989) and where rainfall was greatest at the time of passing of the Chernobyl plume (Smith and Clark 1989).

Mountainous regions tended to be the most seriously affected areas where the combination of altitude and higher precipitation resulted in enhanced fallout. Since most of these areas in north-western Europe are dominated by montane blanket bogs it could have been predicted, on the basis of previous studies (Grueter 1971; Fredriksson 1970; Bunzl and Kracke 1986), that the transfer of deposited radioactivity from peat soils into the food chain would be high, and that the ecological half-life of fallout in these areas would be greater than in mineral soils where radiocaesium becomes rapidly fixed to the clay minerals (Schultz *et al.* 1959; Livens and Loveland 1988). Models in use at the time of Chernobyl underestimated the duration of radiocaesium availability in upland areas and data presented in this paper show that radiocaesium activity levels in upland sheep remain high.

The measurement of transfer and accumulation of radionuclides in ecosystems is essential to the assessment and prediction of risk to biota, and in the formulation of effective emergency plans. Soil-to-plant concentration ratios have been extensively used in the study of radionuclide transfer from soil to plants (Sheppard and Evenden 1988; Horrill *et al.* 1989; Bunzl and Kracke 1987). However, there is considerable variability in the methodologies applied. Values can show remarkable variations for similar species on similar soil types using the same experimental procedures. This is in part due to large differences in deposition values over small areas. For example, Oughton *et al.* (1990) found deposition to upland soils to vary by a factor of 20 within 1 m^2. The soil-to-plant concentration ratio has not proved to be a reliable method for predicting plant activity concentrations since the relationship between soil and plant concentrations is seldom linear (Simon and Ibrahim 1987). Oughton *et al.* (1990) use an alternative to the concentration ratio, which they refer to as a transfer factor, defined as activity in vegetation (Bq kg^{-1}) divided by soil deposition (Bq m^{-2}).

In this study a comparison between concentration ratios, transfer factors and an alternative ratio, suggested from the presented results, is carried out. The alternative method is based on the ratio between radiocaesium content of *Calluna vulgaris* and activities in different plant species: it is referred to as the *Calluna vulgaris* to plant ratio (CPR).

The majority of Irish uplands are grazed by sheep, which are the principal agricultural product of these areas. Consequently, radioactivity levels in sheep are monitored and assessed on an ongoing basis in order to ensure that radioactivity in meat remains within acceptable limits (Colgan 1988; Colgan and Scully 1989). Standard methods for the *in-vivo* measurement of radiocaesium in sheep have been developed (Meredith *et al.* 1988; Sherlock *et al.* 1988) but are labour intensive and expensive. Prediction of concentrations in animals may be possible based on sampling and measurement of concentrations in faeces. Preliminary results of an investigation carried out on this topic are also presented.

Materials and methods

Four sites were studied in Ireland at the following locations: farm-A, County Donegal (National Grid Reference C 209430); farm-B, County Donegal (H 207392); farm-C, County Cavan (H 128207); and farm-D, County Waterford (S 050070). Soils and vegetation samples were taken on all farms, and data are presented for live-monitored ewes, lambs and for faeces collected from the grazing areas of these flocks on the farm-C site.

Soil sampling was carried out on two occasions on all farms. The first set of samples were collected in June/July 1989 and the second set during the following October/November. Within each farm a number of areas were defined by subjective ecological assessment of soils and vegetation. From each such area between one and five soil monoliths of surface area 30 x 30 cm were extracted, with larger numbers of samples used to represent the more extensive sampling areas. Each monolith was then sub-divided in 5 cm sections and these sub-samples were bulked for each sampling area at each depth horizon.

Samples of individual plant species were selectively gathered from around the soil sampling sites on each occasion. *Calluna vulgaris, Erica tetralix* and *Juncus squarrosus* were collected on all four farms. *Erica cinerea* was collected on all except the farm-C site where it did not occur.

Soils were dried to constant weight at 108°C and ground to less than 1mm particle size. Plant samples were sorted to remove dead material and contamination. Only young growing shoots of Ericoid species were analyzed. Samples of *Juncus squarrosus* were sorted to separate live leaves for subsequent analysis. Plants were dried for 24 hours at 108°C and ground to less than 1mm particle size prior to counting.

Soils and plants were analyzed for radiocaesium (caesium-137 and caesium-134) by high resolution gamma spectrometry using an Ortex GMX-series germanium detector (resolution 1.7 keV at 1.33 MeV; relative efficiency 28%). Activity concentrations were determined by automatic peak fitting using the Nuclear Data VAX/VMS spectroscopy applications software package, and are reported at the 95% confidence level.

Results and discussion

Table 1 compares the values calculated for deposition (Bq m^2), to those for the same sites on an activity (Bq kg^{-1}) basis for the first sampling round carried out in June 1989. Table 2 presents similar data for the second sampling round for comparison purposes. Deposition is an expression which considers activity in terms of the density of the soil per square metre, down to a depth of 10 cm. The classification of areas used in the table approximate to a simple ecological description of each area. Mineral soils, for example, had higher mineral contents than the other soil types, but the percentage mineral content varied between farms.

Both Table 1 and Table 2 clearly show that interpretation may be seriously altered depending on how results are presented. Table 1 shows that on farm-A deposition to mineral soils was 3.6 times greater than on the heather knolls, but comparing the same results based solely on activity levels, a 2.3 times difference is observed. A more serious situation arises in the results from farm-C; here the activity levels were highest on the deep peats, but deposition to the mineral soils was more than twice that to the deep peats. The discrepancy between deposition

Table 1. Radiocaesium deposition* (Bq m^2) and activity concentrations (Bq kg^{-1})* for soils at 4 upland sites in Ireland.

Farm		Mineral Soils	Deep Peat	Heather Knolls	Burnt Peat
		June/July 1989			
A	Deposition	17,998	5,416	4,940	-
	Activity	1908	1,204	846	-
B	Deposition	-	3,450	5,745	-
	Activity	-	772	903	-
C	Deposition	9,614	4,165	-	3,634
	Activity	861	957	-	844
D	Deposition	7,186	6,256	5,571	-
	Activity	743	674	679	-

* integrated to 10cm and decay-corrected to the date of sampling

Table 2. Radiocaesium deposition* (Bq m^2) and activity concentrations (Bq kg^{-1})* for soils at 4 upland sites in Ireland.

Farm		Mineral Soils	Deep Peat	Heather Knolls	Burnt Peat
		October/November 1989			
A	Deposition	9,672	6,425	4,468	-
	Activity	996	1,342	786	-
B	Deposition	-	3,591	4,158	-
	Activity	-	781	653	-
C	Deposition	21,404	2,841	-	4,048
	Activity	1,744	613	-	976
D	Deposition	8,084	7,015	6,421	-
	Activity	851	755	771	-

* integrated to 10cm and decay-corrected to the date of sampling

Table 3. Radiocaesium deposition profiles (Bq m^2) for upland mineral soils at farm-A and farm-C.

Farm	Section (cm)	Chernobyl Cs-137	Weapons Cs-137	Bulk Density
A	0-5	12,405	2,687	0.18
	5-10	431	2,475	0.26
	10-20	132	701	0.19
	20-30	0	161	0.12
	Total	12,968	6,024	
C	0-5	1,891	6,542	0.22
	5-10	25	1,156	0.31
	10-20	0	130	0.50
	20-30	0	0	0.56
	Total	1,916	7,828	

values and the activity values are especially obvious for mineral soils on all of the farms and this is due to the greater density of these soils compared to soils of higher organic content. In Table 2 it is seen that activity was higher in deep peat soils than in mineral soils, but deposition values show the reverse to be true.

A further consideration is the very significant difference between the results for the first and second round of soil samples. It should be remembered that the samples were taken from directly adjacent sites and the same procedures were used in sample preparation. Bulk density values used in the calculation of deposition were similar and cannot be responsible for the differences observed. Neither set of results is consistently higher or lower than the other, and therefore it is very unlikely that any systematic error has caused the variations observed. It seems most likely that these differences are the result of large field variations in deposition, as have previously been reported for studies on heavy metal deposition to upland peatlands (Oldfield *et al.* 1979).

The data presented also show that the deposition to mineral soils on each farm greatly exceeded deposition to the other soil types, and this is evident on farms-A and B where the organic matter contents were lowest of all the mineral soil sites surveyed. Both of these sites were also situated in topographical locations which received run-off from adjacent slopes. It is possible that these areas acted as sinks for radiocaesium in run-off either at the time of Chernobyl deposition, or through remobilization of previously deposited radioisotopes from adjacent peaty soils.

Table 3 presents detailed data for the mineral soil sites at farms-A and C. It can be seen that on farm-A, the principal source of Cs-137 deposition is from Chernobyl fallout, but on farm-C the situation is reversed and the weapons component greatly exceeds that of Chernobyl. The pattern of weapons deposition is probably quite different to the Chernobyl pattern since the two occurred over different time periods and under different climatic circumstances. However, run-off may have accentuated the deposition at both sites. Farm-A is at the base of a mountain slope while the farm-C site is one that floods occasionally: it is located in the valley of a small river and soils have perhaps scavenged radiocaesium from solution in floodwaters. Since such floods are intermittent it is likely that there have been few such events since the Chernobyl incident, perhaps explaining the higher weapons deposition at this site.

Bulk density is greater in the deeper sections of the farm-C mineral soils (Table 2) and it can be seen that Chernobyl Cs-137 has only penetrated to the 5-10 cm horizon on this site, while it has penetrated to the 10-20 cm horizon in the less dense (higher organic content, lower mineral content) farm-A soils. The pattern of weapons Cs-137 penetration is similar but has reached deeper levels at both sites than has the Chernobyl component.

Table 4 presents total radiocaesium analyses of plant samples collected on each of the four research farms. Sample results are presented for two sampling occasions which coincided with soil collection and are referred to as sample 1 and sample 2 in the Tables. Highest concentrations were found in *Calluna vulgaris* at between 1174 and 2699 Bq kg^{-1}. Lowest concentrations were found in samples of *Erica cinerea* at between 120 and 228 Bq kg^{-1} with other species showing intermediate activity levels.

Table 4. Total radiocaesium content of *Erica tetralix, Erica cinerea, Juncus squarrosus* and *Calluna vulgaris* taken from the four research farms on two occasions in 1989

Farm		Radiocaesium (Bq kg^{-1})*			
		E. tetralix	*E. cinerea*	*J. squarrosus*	*C. vulgaris*
A	Sample 1	127	120	557	1174
	Sample 2	220	159	570	1210
B	Sample 1	360	191	514	1480
	Sample 2	294	122	463	1245
C	Sample 1	235	-	639	1723
	Sample 2	306	-	361	1693
D	Sample 1	698	228	537	2699
	Sample 2	456	131	622	2110

* dry weight, decay-corrected to the date of sampling

The soil activity and soil deposition data presented in Tables 1 and 2 are used in conjunction with the plant analysis data of Table 4 to calculate concentration ratios (CR) and transfer factors (TF). These are presented in Table 5 together with *Calluna vulgaris* to plant ratios (CPR) calculated from data in Table 4. The difference between sample 1 and sample 2 is presented as a percentage of the larger value. This percentage difference is then used to assess the reliability of the measurement. Ideally, since each calculation is an attempt to find a predictive value, there should be no difference between sample 1 and sample 2. The closeness between the two figures is therefore a measure of the reliability of the method.

It was found in eight out of eleven cases that the CPR showed a smaller percentage difference between the first and second samples than either the TF or the CR. This suggests that a plant to plant ratio may be a more reliable predictive method than either of the other two systems. All plant samples are collected and processed in exactly the same manner and this reduces the possibility of error in calculating the CPR, in contrast to a comparison between soil and plant where each must be sampled and processed using different methods. An advantage of the CPR is that sampling is faster and easier than when soil collection is necessary, sample preparation is also made simpler.

Table 6 presents data for radiocaesium analysis of ewes, lambs and faecal samples taken from farm-C during the summer of 1990. It should be noted that results of the faecal samples were on a dry weight basis, while the animal data

Table 5. Comparison between the reproducibility of concentration ratio (CR), transfer factor (TF) and *Calluna vulgaris*-plant ratio (CPR) as a means of predicting radiocaesium activity in vegetation

Farm		E. tetralix			E. cinerea			J. squarrosus		
		CR	TF	CPR	CR	TF	CPR	CR	TF	CPR
A	Sample 1	0.11	0.02	0.11	0.10	0.02	0.21	0.46	0.10	0.47
	Sample 2	0.16	0.03	0.18	0.12	0.03	0.27	0.42	0.09	0.47
	%difference	35%	33%	39%	17%	33%	22%	9%	10%	0
B	Sample 1	0.47	0.10	0.24	0.25	0.06	0.13	0.67	0.15	0.35
	Sample 2	0.37	0.08	0.24	0.16	0.03	0.10	0.59	0.13	0.37
	%difference	21%	20%	0	36%	50%	23%	8%	13%	6%
C	Sample 1	0.25	0.06	0.14	-	-	-	0.67	0.15	0.37
	Sample 2	0.50	0.11	0.18	-	-	-	0.58	0.12	0.21
	%difference	50%	45%	22%	-	-	-	14%	20%	43%
D	Sample 1	1.04	0.11	0.26	0.34	0.04	0.08	0.80	0.09	0.20
	Sample 2	0.60	0.07	0.22	0.17	0.02	0.06	0.82	0.08	0.29
	%difference	42%	36%	15%	50%	50%	25%	3%	12%	31%

Table 6. Radiocaesium activity in ewes, lambs and faeces from farm-C.

Sample	Date (1990)	Minimum (Bq kg⁻¹)	Maximum (Bq kg⁻¹)	Mean (Bq kg⁻¹)	Standard Deviation	Sample Size
Ewes	May	138	1053	398	193	70
Lambs	May	29	652	201	133	26
Faeces	May	510	1489	1043	406	4
Ewes	Jun	208	1613	923	257	56
Lambs	Jun	218	1171	617	244	29
Faeces	Jun	777	1572	1194	351	4
Ewes	Jul	401	1974	1126	384	54
Lambs	Jul	416	1956	1199	354	26
Faeces	Jul	1202	1760	1509	230	4
Ewes	Aug	572	1842	1150	310	57
Lambs	Aug	513	1674	1147	281	22
Faeces	Aug	873	1496	1134	230	5

Activities for lambs and ewes are reported on a live weight basis
Activities in faeces are reported on a dry weight basis

are reported as live weight. Mean concentration in the flesh of lambs was lowest in May, considerably lower than in the ewes or in the faecal samples, but mean concentrations increased during the season until the levels in ewes and lambs were approximately equivalent in July and August. Concentrations in ewes were relatively high in May and also increased during the season, but this increase was not as marked as in the lambs. There was a large difference between concentrations in faeces, ewes and lambs in May, but the difference lessened in June and concentrations were very similar in July and August. The data set suggests that a comparison on an absolute basis between faeces and animal measurements would overestimate the concentrations in animals in May, June and July, but might yield a more accurate estimate in August. Overestimation, however, might be considered the safer predictive situation when risk evaluation is necessary.

Conclusions

The data clearly illustrate some of the problems relating to presentation of radioisotope analyses of soils and it is recommended that deposition values be quoted as standard rather than activity concentrations. Large field variations in activity and deposition were found through replication of a sampling programme. Further study is required in order to determine the optimum number of samples required for reliable estimation of deposition in such upland ecosystems where highly non-uniform deposition seems to be prevalent.

It appears that areas of mineral soils may act as sinks for mobilized or remobilized radioisotopes originating in surrounding peat soils, and depth profiles reveal that penetration of Chernobyl caesium has been less than the weapons component, but that penetration of both is inhibited in soils of higher mineral content.

A comparison between concentration ratios, transfer factors and a *Calluna vulgaris* to plant ratio (CPR) reveals that the CPR is a more reliable parameter than the other two in eight out of eleven presented cases. The sole use of plant materials also means faster sampling and easier analysis than when soils are analyzed in conjunction with plant samples. Further study of the method is required and it is recommended that other plant to plant ratios might be investigated for other agricultural and semi-natural ecosystems based on a reliable and common representative of the relevant vegetation association.

Monitoring of ewes and lambs shows a seasonal pattern with lambs initially lower than ewes, concentrations converge during the summer season. A comparison of the absolute concentration in collected faeces with those of the flocks shows that the mean faecal concentration is greater than either the ewes or lambs in May and June, but all were very similar later in the season. The results show that analysis of faeces, which are easily collected and analyzed with a minimum of input from farmers or hardship on animals, is a promising alternative to live monitoring of animals in the prediction of mean flock activity. Work on the utility of faeces as a bioindicator is continuing and will be reported in the literature in due course.

Acknowledgements

The authors would like to thank the Irish Department of Agriculture, The Ministry of Agriculture, Fisheries and Food in the U.K. (contract 663) and the CEC (contract B17*0044-C) for their financial assistance. Field and laboratory work was provided by Niamh Mulvany, and Dr. David Jeffrey (Trinity College, Dublin) gave valuable assistance in project design.

References

Belli, M., Blasi, M., Borgia, A., Desiato, F., Poggi, M., Sansone, U., Menegon, S. and Nazzi, P. (1989) First results of a radioecological research on the agricultural environment on a north-eastern region of Italy (Friuli-Venezia Giulia). In *Proceedings of a Conference on the Transfer of Radionuclides in Semi-natural Ecosystems.* Udine, Italy, 11-15 September 1989.

Bunzl, K. and Kracke, W. (1986) Accumulation of fallout ^{137}Cs in some plants and berries of the family Ericaceae. *Health Physics,* 50, 540-542.

Bunzl, K. and Kracke, W. (1987) Soil to plant transfer of [239+240]Pu, [238]Pu, [241]Am, ^{137}Cs and ^{90}Sr from global fallout in flour and bran from wheat, rye, barley and oats, as obtained by field measurements. *The Science of the Total*

Environment, **63**, 111-124.

Colgan, P.A. (1988) *A report on the levels of radiocaesium activity in mountain sheep. October - December 1987*. Nuclear Energy Board, Dublin.

Colgan, P.A. and Scully, B.J. (1989) *Sheep Monitoring Programme, January - December 1988*. Nuclear Energy Board, Dublin.

Fredriksson, L. (1970) Plant uptake of fission products III. Uptake of ^{137}Cs by *Trifolium pratense* as influenced by the potassium and calcium level in the soil. *Lantbrukshogskolans Annaler*, **36**, 41-60.

Grueter, H. (1971) Radioactive fission product ^{137}Cs in mushrooms in W. Germany during 1963-1970. *Health Physics*, **20**, 655-656.

Horrill, A.D. (1983) Radioactivity in a saltmarsh. In P.J. Coughtrey, J.N.B. Bell and T.M.Roberts (eds.) *Ecological Aspects of Radionuclide Release*. Special Publications Series of the British Ecological Society Number 3. Blackwell Scientific Publications.

Horrill, A.D., Kennedy,V.H. and Harwood, T.R. (1989) The concentration of Chernobyl derived radionuclides in species characteristic of natural and semi-natural ecosystems. In *Proceedings of a Conference on the Transfer of Radionuclides in Semi-natural Ecosystems*. Udine, Italy 11-15 September 1989.

Livens, F.R. and Loveland, P.J. (1988) The influence of soil properties on the environmental mobility of Cs in Cumbria. *Soil Use and Management*, **4**, 69-75.

Meredith, R.C., Mondon, K.J. and Sherlock, J.C. (1988) A rapid method for the *in vivo* monitoring of radiocaesium activity in sheep. *Journal of Environmental Radioactivity*, **7**, 209-214.

Oldfield, F., Thompson, R. and Barber, K.E. (1979) The effect of microtopography and vegetation on the catchment of airborne particulates measured by remanent magnetism. *Quaternary Research*, **12**, 326-332.

Oughton, D.H., Salbu, B. and Strand, P. (1990) Mobility factors for estimating the bioavailability of radionuclides in soil. *Proceedings of Symposium and Workshop on the Validity of Environmental Transfer Models*. October 8-12, Uppsala, Sweden.

Sherlock, J., Andrews, D., Dunderdale, J., Lally, A. and Shaw, P. (1988) The *in vivo* measurement of radiocaesium activity in lambs. *Journal of Environmental Radioactivity*, **7**, 215-220.

Sheppard, S.C. and Evenden, W.G. (1988) The assumption of linearity in soil and plant concentration ratios: an experimental evaluation. *Journal of Environmental Radioactivity*, **7**, 221-247.

Schulz, R.K., Overstreet, R. and Barshad, I. (1959) On the soil chemistry of caesium-137. *Soil Science*, **89**, 16-27.

Simon, S.L. and Ibrahim, S.A. (1987) The plant/soil concentration ratio for calcium, radium, lead, and polonium: evidence for non-linearity with reference to substrate concentration. *Journal of Environmental Radioactivity*, **5**, 123-142.

Smith, F.B. and Clark, M.J. (1989) *The transport and deposition of airborne debris from the Chernobyl nuclear power plant accident with special emphasis to the United Kingdom*. Meterological Office, Scientific Paper No. 42. Her Majesty's Stationery Office, London.

The Structure of Animal Communities as Bioindicators of Landscape Deterioration

J. Boháč and R. Fuchs

Institute of Landscape Ecology, Czechoslovak Academy of Sciences, Ceské Budejovice.

Key words: ecological classification, square mapping, environmental factors, invertebrates, birds, Czechoslovakia.

Abstract

Biodiagnostic evaluation methods were developed during the five years 1986-1990. Biodiagnostic investigations are made on a range of life forms and at various scales. They are performed both at population and community level. The data on plants and animals must be obtained at the appropriate scale of the investigation (local, regional, fluvial).

Biomonitoring at regional and fluvial scales is mainly by using populations of small mammals and communities of birds. These groups, owing to their large size and migratory habits, are optimal bioindicators of the complex of factors which degrade the landscape and lead to air pollution.

Populations and communities of invertebrates are mainly used for biomonitoring at a local scale. These groups have small body sizes and a lower tendency for migration, and are therefore suitable for indicating local environmental factors such as unsuitable application of fertilizers and pesticides, unsuitable methods of landscape management, overdrainage and the consequential desiccation of the landscape.

Biomonitoring at the local scale is often the sole method for the investigation of the biotic conditions in reserves and national parks. The investigation of key organisms and their communities in portions of the landscape is important for the research at the regional scale. The mosaic of various well-preserved landscape units increases the stability of the landscape.

Introduction

The subjective statement that a landscape is diversified, has a high natural value

Bioindicators and Environmental Management
ISBN 0-12-382590-3

and steadily exhibits a relatively low degree of adverse effects from man, is insufficient for those who are deeply involved in nature and its problems. Simple statements are equally inadequate when choosing sites for protection as reserves or biocorridors on the basis of their high natural value. Such general statements must be supported in an adequate manner, preferably by quantitative data. The main obstacle to this is the difficulty of expressing natural value in numerical form. However, methods exist allowing us to evaluate landscape sites and larger areas in a more exact manner than by a general description in words.

These methods are largely based on the evaluation of the relative proportions (frequencies) of some components of the biota within the region of interest. Analysis of the frequency of various ecological groups of species can reveal the major factors contributing to changes in the ecosystems studied. Sometimes the conclusions are quite simple; but their formulation would not be grounded if the ecological analysis of the communities was lacking.

This kind of biological research can be less expensive than instrumental research. For instance, it is not feasible to make daily measurements over a period of years of micro-meteorological factors or the concentration of toxic substances over a large area. Neither do single environmental measurements mirror the situation faithfully because of variations in seasonal climate and emission of pollutants etc. On the other hand, animals live for years on the territory and mirror the natural conditions in a cumulative manner.

Bird communities are preferred for large-area monitoring. Their size and migrating ability make birds suitable for monitoring complexes of effects associated with devastation of the landscape over large areas.

Invertebrates, smaller in size and possessing lower migrating ability, are well suited to shorter-range biomonitoring (poorly-controlled use of fertilizers and pesticides, inadequate landscape management, land drainage, etc.)

Invertebrate communities - methods of study

When investigating a smaller area, we can evaluate the frequency of individuals of species having certain ecological requirements in the community. For instance, we can examine the abundances of hygrophilic species, or the abundances of relict, rare species of invertebrates. Classifications which can be employed for this purpose have been worked out. Although the results of such ecological analysis can be affected by various factors (e.g. professional skill and experience of the researcher), we consider this approach very suitable for biodiagnostic purposes. Its results are very illustrative and can be well interpreted.

As a promising indicator of landscape disturbance, use can be made of species with known responses to the activities of man. For instance, the Czech arachnologist Buchar (1983) suggested the classification of spiders into three groups: relicts of the 1st rank, relicts of the 2nd rank, and expansive species. Expansive species are those which are capable of forming viable populations on artificially deforested areas. Relicts of the 1st rank are species occurring predominantly in protected regions and regions unaffected by man. Relicts of the 2nd rank include species that occur in man-made forest ecosystems and do not penetrate into non-forest ecosystems. The extensive data available concerning the occurrence of some groups of invertebrates in Czechoslovakia form a basis for classification of other invertebrates as well. If such a classification exists, it is necessary to evaluate data from rather extensive territories and establish the frequencies of individuals from the various categories of sites affected by human activity (Fig. 1). Thus, from the ecological analysis of beetle communities on different biotopes, we were able to set up a scale for comparing any other sample of the community. A low frequency of expansive species is indicative of a high natural value of the community (Ruzicka 1986). The frequency of expansive species is usually increased in communities of biotopes more extensively affected by man. On the other hand, a high frequency of individuals of expansive species does not imply that the locality is naturally valueless. This proviso applies particularly to sites which are the original sites of occurrence of these species (littoral biotopes, meadows).

   Based on the classification of beetles according to the relictness of occurrence, an index mirroring the degree of anthropogenic stress has been suggested (Bohác 1988). The basis for the calculation is the frequency of all the three groups in the community. The index takes values from 0, when only expansive species are present (the stability of the community is lowest due to the strongest anthropogenic effect), to 100, when only relicts of the 1st rank are found in the community and the ecosystem is disturbed by the activity of man to the least extent.

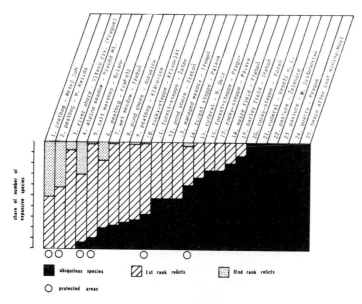

Figure 1. Structure of the relictness of the occurrence of species of examined localities.

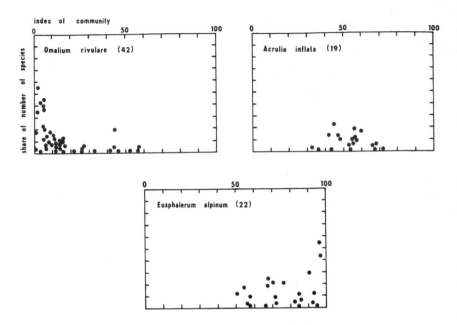

Figure 2. Examples of the relationship between the frequency of the species in the community and index of community by anthropophilic, anthropoindifferent and anthropophobic species (the number of studied biotopes in parenthesis).

From a quantitative evaluation of the degree to which communities of beetles are affected in various ecosystems, we established three groups with respect to the anthropogenic effect on biocoenoses: anthropophilic, anthropoindifferent, and anthropophobic (Fig. 2). The picture demonstrates that, for anthropophilic species, the index lies largely between 0 and 50, for anthropoindifferent species, between 30 and 70, and for anthropophobic species, between 70 and 100. This classification can serve as a audit re-check of the correctness of the species classification according to the relictness of occurrence (Bohác 1988).

In addition to the classification of invertebrates into groups according to the relictness of occurrence, other approaches can also be applied to the ecological analysis of communities. These include the percentage abundance of life forms, of size groups, of groups of different thermopreference and hygropreference, of zoogeographical groups, and sex ratio values. These criteria will help us to gain a deeper insight into the mechanisms of changes taking place in communities affected by man. The research into a territory should be complete within one or two years. Repetition of the research in a suitable period of time can bring valuable data concerning changes in the ecosystem.

Invertebrate communities - use of research results

The proposed procedures of ecological analysis for evaluation of the community structure were employed in the study of beetle communities in biotopes with different degrees of anthropogenic effects (Fig. 3). Increased effect of man was found to bring about increase in the frequency of expansive species, increase in the frequency of species with summer activity of the imagoes, and decrease in the proportion of species with winter activity of the imagoes. Furthermore, the proportions of winged species and individuals possessing a higher migrating ability, large-body-size species (size groups IV and V), species with higher thermopreference and lower hygropreference, and species with area of occurrence wider than European are also increased. Whereas a decrease is noted in the number of life forms and fraction of species with limited area of occurrence (narrower than European). A decrease in also observed in the beetle community state index, which is an integral index of the degree of anthropogenic stress of the ecosystem. More extensive human activity brings about an alteration of the sex ratio index from zero.

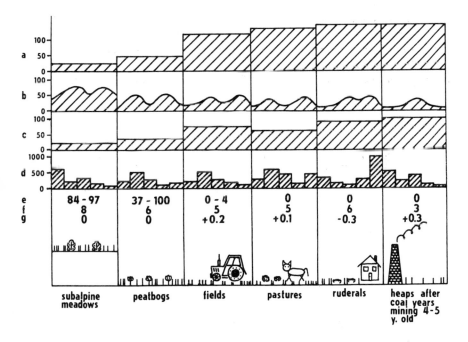

Figure 3. The influence of man on commuities of staphylinid beetles of non-forest landscape (a) percentage share of expansive species, (b) seasonal dynamics (spec. m^{-2}), (c) percentage of species with good migrating ability, (d) distribution of specimens in dependence of the body size (spec.), (e) index of community, (f) number of life forms (g) sex ratio index (female - male/N).

From our research, it was possible to determine the qualitative and quantitative parameters of beetle communities mirroring a critical state brought about by the activity of man (Boháč 1988a): the number of species in the community is lower than 10; only anthropophilic species occur; the community state index value is lower than 35; the number of life forms is lower than 4; the fraction of individuals of large-size species (size groups IV and V) exceeds 20%; the fraction of species with summer activity of imagoes exceeds 40%; non-flying species are absent; the fraction of widespread species is higher than 90%; that of species with enhanced thermopreference is higher than 70% and of species with reduced thermopreference is higher than 60%; the sex ratio in the community is higher than 0.1 (agrosystems) or lower than -0.1 (urban ecosystems).

Bird communities - methods of study

The use of bird communities for bioindication of landscape deterioration is limited by methodological and information factors. Only a limited set of methods exists for the investigation of the structure of bird communities, and this structure only mirrors some of the environmental parameters.

The methodological potential of examining bird communities is considerably wider than for other groups of animals; however, information that can be extracted from such study is poorer than in the case of some groups of invertebrates. Thus, the use of bird communities for this purpose is aimed at the investigation of extensive, topographically heterogeneous landscape complexes.

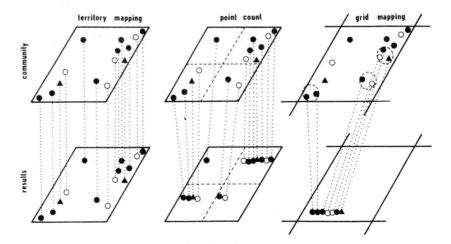

Figure 4. The information contents of the results of the basic methods for studying the structure of the breeding bird communities.

The following three methods are employed for studying the structure of breeding bird communities: spot mapping (IBCC 1970); point counts (Blondel *et al.* 1970); and grid mapping (Sharrock 1974; Bezzel and Rantfl 1974). These methods differ in the actual field work, time consumption, and information content of the results.

The spot mapping method allows the community to be treated as a whole (Fig. 4). Moreover, it mirrors its actual spatial structure, i.e. territories of individual pairs of individual species. The point counts method also deals with the community as a whole (Fig. 4), but only mirrors its schematised structure, i.e. the occurrence of all individuals of all species within squares 200 x 200 m to 300 x 300 m area. The grid mapping method usually only enables representative samples from the community to be handled (Fig. 4). It also mirrors its schematised structure, i.e. the species abundances within plots 1 x 1 m to 10 x 10 km area.

The spot mapping method is highly time-consuming. A researcher can examine an area of 0.01 to 0.2 km^2 per day. The point counts method is one order of magnitude faster, the area examined by a researcher in a day being 0.5 to 5 km^2. The time consumed by grid mapping depends on the number and area of the partial plots. For each of them, 1 to 4 days of work must be considered.

Bird communities - use of research results

The breeding bird community structure mirrors a complex of factors of the environment determining the occurrence and abundance of the species. They include topographic, trophic and climatic conditions, interspecies relationships and direct anthropogenic influence.

The factors that determine the structure of bird communities also determine their potential in the landscape deterioration bioindication. Bird communities mirror rather sensitively any intervention into the landscape structure. These include: replacement of natural forest ecosystems by seminatural or artificial ones; change in woodland age composition; loss of dispersed vegetation; watercourse regulation; drainage of waterlogged meadows, etc. More difficult to establish are interventions into trophic chains, e.g. as a consequence of intensification of agriculture or air pollution. Direct effects of toxic substances are apparent, particularly in some predators and owls.

The opportunities offered by bird communities for the bioindication of landscape deterioration are substantially affected by the methods employed for evaluating the community structure. These can be divided into autoecological or synecolgoical characteristics of the communities and characteristics of the partial synusies determined with reference to certain links with the environment. Exploration statistics methods, e.g. cluster or ordination analysis, also find wide application.

The basic autecological characteristics of the communities, the occurrence and abundance of the species mirror a part of the complex of bonds between the community and the environment. Their use will be appropriate if the bonds are species-specific. The picture that will be shown illustrates the occurrence of two characteristic species in the area of Greater Prague. Skylark *Alauda arvensis* (Fig. 5a) is a typical species of open landscape biotopes. Swift *Apus apus* (Fig. 5b) is a typical species of urban biotopes. Its occurrence reflects the arrangement of urban residential areas. The abundance usually carries considerably more information than the occurrence does. The abundance of the swift (Fig. 6) mirrors not only urban areas but also their concentrations within the entire conurbation.

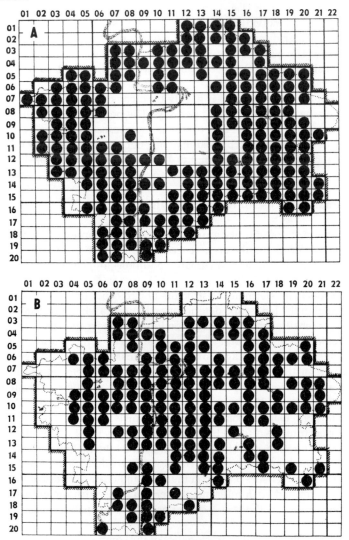

Figure 5. The occurrence of the two characteristic bird species on the territory of the Greater Prague (a) Skylark (*Alauda arvensis*), (b) Swift (*Apus apus*).

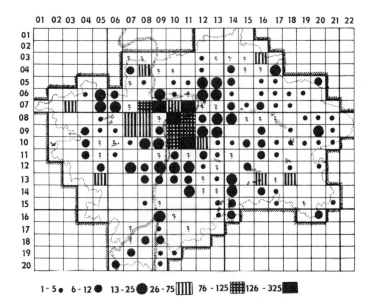

Figure 6. The number of breeding pairs of the Swift (*Apus apus*) in the Greater Prague area.

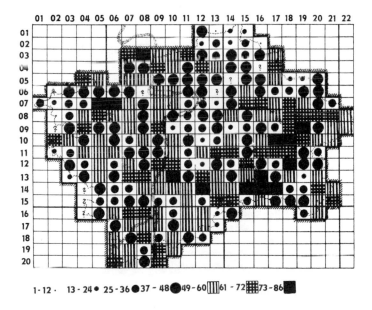

Figure 7. The number of the bird species in the Greater Prague area.

The basic synecological characteristics of the communities, number of species and total density, mirror a whole complex of bonds between the community and the environment. Their applicability thus is rather limited because the impact of individual factors cannot be deduced from them. The number of species in the community mirrors primarily the topographic heterogeneity of the environment. In the Zdíkov model area in the Sumava region (Fig. 8a) the highest values are typical of ecotones of forest biotopes and open landscape biotopes with rich fringe communities and dispersed vegetation, or ecotones of urban biotopes incorporating extensive parks and gardens. In the Greater Prague area (Fig. 7), on the other hand, the highest values characterize

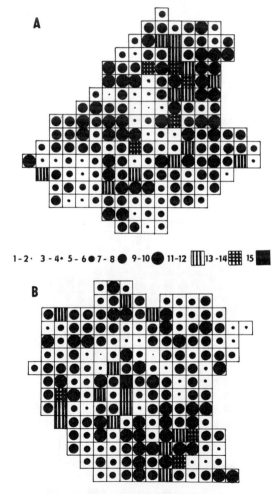

Figure 8. The number of the bird species in the two model areas in the Sumava mountains (a) foot-hills (the model area Zdíkov), (b) plateau (the model area Nové Hute).

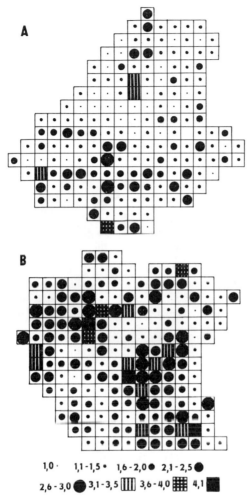

Figure 9. The faunal significance index in the two model areas in the Sumava mountains (a) foot-hills (the model area Zdíkov), (b) plateau (the model area Nové Hute).

areas which include extensive complexes of natural or seminatural forest ecosystems, usually combined with dispersed vegetation and waterlogged meadows. This difference is associated with different research approaches. In both cases, establishing the number of species enables the ecological stability of the landscape to be evaluated. The systems, however, are lower and higher order ones.

Characteristics of partial synusies are associated with bonds between selected species and selected environmental factors. Thus they have optimal properties to be employed with bioindicaton research. They can be evaluated either on their own, as autecological characteristics are, or in a complex manner, as synecological characteristics are.

Complex characteristics relying on life strategies of the species, or on their relictness of occurrence, cannot be applied to bird communities. This function, however, can be taken over by the faunal significance index, evaluating species with respect to their occurrence over the territory of Bohemia and Moravia. It mirrors the occurrence of natural biotopes or "relict" biotopes of anthropogenic origin. Figures 8 and 9 demonstrate the numbers of species and faunal significance index values for communities of the Zdíkov and Nové Hute model areas. While with respect to the former characteristics the two localities do not

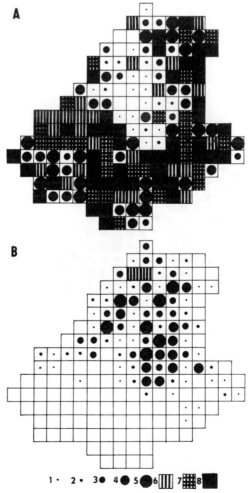

Figure 10. The number of bird species of the two basic topic synusies in the model area Zdíkov (a) the synusy of the woodland species, (b) the synusy of the openland species.

differ appreciably, the latter is markedly higher in the Nové Hute area. This difference is accounted for by the occurrence of rather rare species confined to biotopes of peat bogs or waterlogged meadows and mountain spruce forests. Thus, the faunal significance index can be used for evaluation of the ecological stability of the landscape at a qualitatively higher level; it can also serve as a measure of the overall anthropogenic effect.

In ornithocoenoses, the function of characteristics relying on the evaluation of life forms can be fulfilled by characteristics of partial synusies established based on topographic links of the species. They primarily mirror the abundances of the various biotopes in the landscape and also their internal structure. Figure 10 shows the absolute and relative abundances of the two main synusies established based on the links to the macrobiotope in the Zdíkov model area. Forest biotope species allow us to deduce which components of dispersed vegetation play the "role" of forests; open landscape species enable us to assess the "wholeness" of forest biotopes and extent of their disturbance by mining or emissions. The two synusies also mirror the horizontal and vertical structure of the vegetation: forest biotopes reflect particularly the age and species heterogeneity of the stands; open landscape biotopes, the occurrence of dispersed vegetation and also the intensity of agricultural exploitation of the soil.

The results of cluster analysis are related to the qualitative and quantitative similarity of the pictures of bird communities. They enable us to graphically evaluate the landscape structure at various similarity levels. Cluster analysis of communities of the Nové Hute model area (Fig. 11) form an objective basis for determining the degree of anthropogenic effect and the ecological importance of the partial areas.

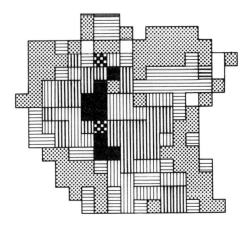

Figure 11. The cluster analysis of the bird communities at the similarity level 60% in the model area Nové Hute.

References

Bezel, E. and Rantfl, H. (1974) Vogelwelt und Landschaftsplannung. Eine Studie aus dem Werdenfelser Land (Bayern). *Tier und Umwelt,* **11/12,** 92.

Blondel, J., Ferry, C., Frochol, B. (1970) La méthode des indices ponctuels d'abondance (IPA) ou des reléves d'avifaune par "stations d'écoute". *Alauda,* **38,** 55-71.

Boháč, J. (1988) Die Ausnutzung von Raubkäfergemeinschaften (Coleoptera, Staphylinidae) für die Indikation der Umweltqualität. In J. Boháč and V. Ruzicka (eds.) *Proceedings of the Vth International Conference Bioindicatores Deteriorisationis Regionis, Instite of Landscape Ecology CAS, Ceske Budejovice.* 160-164.

Boháč, J. (1988a) *Staphylinid beetles - bioindicators of anthropogenous changes of the environment.* Thesis, Institute of Evolutionary Morphology and Ecology of Animals, Moscow, 403-411.

Buchar, J. (1983) Die Klassifikation der Spinnenarten Böhmens als ein Hilfsmittel für die Bioindikation der Umwelt. *Fauna bohemiae septentrionalis,* **8,** 119-135.

IBCC (1970) Recommendations for an International Standard for a Mapping Method in Bird Census Work. *Bulletins from the Ecological Research Committee,* Lund, **9,** 49-52.

Ruzicka, V. (1986) The structure of spider communities based upon the ecological strategy as the bioindicator of landscape deteriorisation. In J. Paukert, V. Ruzicka and J. Boháč (eds.) *Proceedings of the VIth International Conference Bioindicatores Deteriorisationis Regionis, Institute of Landscape Ecology, Ceské Budejovice.* 219-237.

Sharrock, J.T.R. (1974) The ornithological Atlas project in Britain and Ireland. Methods and preliminary results. *Acta ornithologica,* **14,** 269-285.

Biomonitoring of Environmental Change Using Plant Distribution Patterns

Erich Weinert

Martin-Luther-Universitat Halle-Wittenberg, Sektion Biowissen Schaften, DDR-4020 Halle, Neuwerk 21.

Key words: phytoindicator, distribution pattern, extinction, expansion, recession, pollution, environmental change, Central Germany

Abstract

Plant distribution depends upon autoecological requirements and behaviour of the species and on the environmental situation which is characterised by a complex interaction of biotic and abiotic factors. Biomonitoring of environmental change may be accomplished by recording changes in the regional and local distribution of indicator plants, as individuals, populations and in communities.

Distribution patterns of sensitive indicator plants are used to illustrate the range and intensity of human impact (air, soil and water pollution) which has led to environmental change in central European landscapes.

Introduction

Environmental change occurs on our planet at global, regional and local scales and may be monitored by biological, chemical and physical methods which have been developed for various factors in aquatic and terrestrial ecosystems (Schubert 1985; Dässler 1986; Salánki 1986; Arndt *et al.* 1987).

The growth of human population, technical development and increasing industrialization, traffic and transport systems result in various changes to our environment which are reflected in the behaviour and changing distribution of living organisms around us.

Climate change and increased nutrient deposition from the atmosphere will affect soils, plant productivity, biogeochemical cycles, vegetation structure, and species composition. Temporal and spatial patterns for temperature, precipitation, and the occurrence of extreme weather events, influence not only natural ecosystems, but also impose regional constraints on agriculture and forestry (IGBP 1990).

Bioindicators and Environmental Management
ISBN 0-12-382590-3

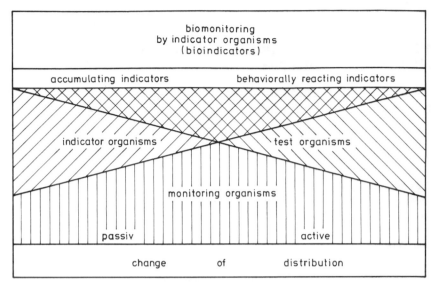

Figure 1. Relations between bioindicator types.

Bioindicators at different levels of biological organization, are means for monitoring these changes of life conditions (as macromolecules, cell, organ, organism, population, biocenosis) by considering
- biochemical and physiological reactions
- deviation from the norm in anatomic, morphologic, biorhythmic and behavioural terms
- change of floristic and faunistic populations (composition and distribution)
- changes in biocenoses including their distribution
- change in the structure and function of ecosystems including their distribution
- change in landscape characteristics.

In addition to active monitoring, which deals with test organisms exposed under standardized and field conditions, passive biomonitoring by phytoindicators provides by analysis of either visible or invisible damage or any deviation from a normal state. Thus by registering the change of the distribution pattern in nature it is possible to illustrate environmental change (Fig. 1.; Schubert 1985; Weinert 1986; 1990; Arndt *et al.* 1987).

Plant distribution and biomonitoring

Plant distribution patterns are, in general, highly dependent upon the combined effect of all climatical and other environmental factors including the effect of human impact on the various natural habitats. The amplitude and rate of change are especially critical for a change of the plant distribution and consequently for the build-up of a plant community. Not all changes in plant distribution are due to lethal threshold effects, some are due to non-lethal threshold effects and others may have been competitive as a result of variations in the ecological range. Each stage in the development of a plant constitutes a link in a chain of survival, any of which can dominate control of distribution. Early stages may be more sensitive to stress than adult ones (Woodward 1987). Extreme changes of environmental conditions are indicated in the long-term by changes in the population dynamics and by partial or complete alteration to the distribution pattern.

Biomonitoring by plant species, the phyto-indicators, appears to be characterized by a fluctuation of the individuals or of the species diversity within plant communities. Often more sensitive wild plants vanish, to be substituted by aggressive neophytes conquering the changed habitat. This can easily be recorded by mapping the species, the population or the plant community.

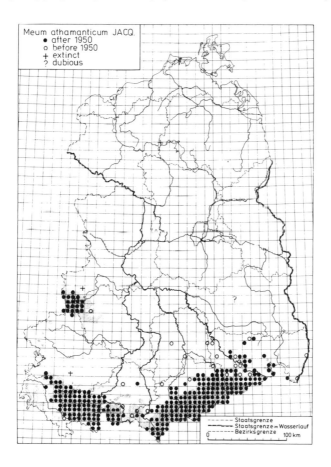

Figure 2. The distribution of *Meum athamanticum* JACQ.

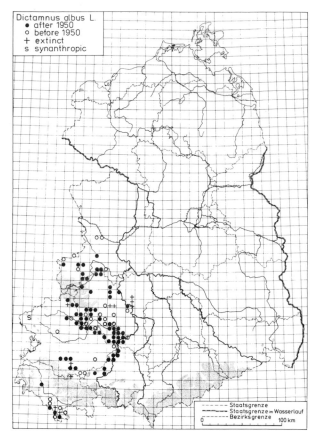

Figure 3. The distribution of *Dictamnus albus* L.

    The monitoring of long-term and large scale environmental trends in landscape development may be achieved by a mapping program of lichens, mosses, macrofungi, and vascular plants, as well as plant communities and sensitive ecosystems. Similar results can be expected from mapping resident animals. The mapping of the total distribution of particular vascular plants and cryptogams already provides useful basic information of the general ecological situation in parts of the continents and oceans, providing the ecological behaviour and requirements of the bioindicators are well known (Hultén 1964; 1971; Meusel *et al.* 1965; 1978; Jalas and Suominen 1972-1989; Weinert 1990). Actual results indicating the real environmental change can be recorded by mapping the flora on a regional or local grid system base. The scale and basic quadrat can be chosen due to the intended result of information. For the East German mapping program we used a basic quadrat of about 5.8 x 5.6 km², the so-called "quadrant" (Weinert 1984). A few examples from this mapping program illustrate biomonitoring of environmental changes by comparing distribution patterns of vascular plants.

The mid-mountain zone of the Hercynian Mts. in East Germany is covered by mixed montane forests of deciduous trees such as *Fagus sylvatica* and conifers such as *Abies alba* and *Picea abies*. These are associated with meadows characterized by *Meum athamanticum* (Apiaceae) (Fig. 2), which is a significant indicator plant for human affected, but less disturbed and less polluted, mountain grassland on neutral and mainly acidic soils. The distribution of this species delimits, more or less, the humid winter cold and summer cool mountain forest region.

In contrast to this pattern *Dictamnus albus*, a sub-continental perennial herb of the dry oak-hornbeam forests, is confined to the drier winter cold but summer warm hillside areas around the mountains, where fine sandy and calcareous soils predominate. This species is sensitive to human impact and it disappears in industrial and urban centres because of physical disturbance and chemical stress of pollutants (Fig. 3).

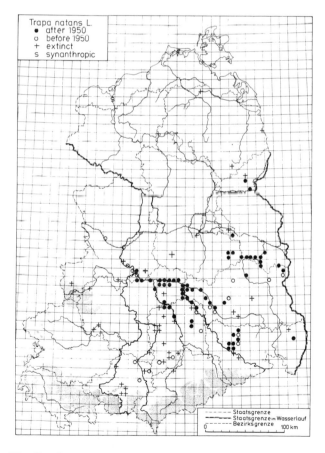

Figure 4. The distribution of *Trapa natans* L.

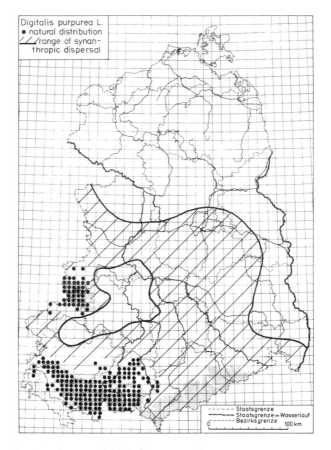

Figure 5. The distribution of *Digitalis purpurea* L.

Certain aquatic plants, such as the annual *Trapa natans*, are most sensitive to pollution and environmental stress (Fig. 4). This plant is threatened by extinction in the near future because of the heavy pollution along the rivers Mulde, Elbe and their tributaries by the industrialized centres around Bitterfeld, Halle and Leipzig. These are black spot areas in central Europe, comparable to the upper Silesia region in Poland (Schubert 1985; Weinert 1990; Jankowski 1990). Environmental change is simply indicated by the dying-back of the *Trapa* individuals with inhibition of germination of the plants. Another change in the East German landscapes can be monitored by mapping of the dispersal and expansion of the distribution of *Digitalis purpurea*, a native biennial herb of the spruce and beech forests of the western Hercynian Mts. zone which has not occupied an additional, secondary - synanthropic - area outside its natural habitats (Fig. 5).

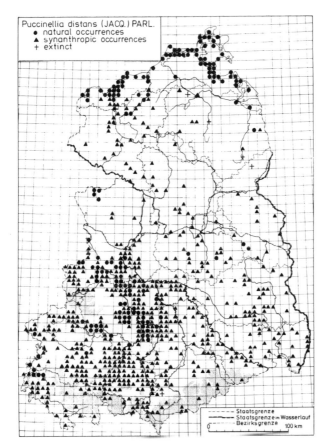

Figure 6. The distribution of *Puccinellia distans* (JACQ.) PARL.

The extreme eutrophication caused by liquid manures and application of extraordinarily high amounts of fertilizers onto agricultural fields and forests have drastically altered soil conditions and favoured the dispersal and growth of this nitrophilous *Digitalis* species in the forest areas. A further environmental change which currently happens during winter time is connected with the application of de-icing salts and salt solutions, which contain sodium and magnesium chlorides, onto the roads. The mapping of *Puccinellia distans*, a native halophyte on saline habitats along the coasts of the Baltic Sea and around inland salt springs, enables the monitoring of long-term salinization of the embankments and fringes of highways and streets (Fig. 6). Nowadays there is an increasing frequency of occurrence of *Puccinellia distans* populations, particularly in high traffic zones of central Germany.

In addition to the mapping of single plant species, or species combinations, current environmental change can also be illustrated by mapping the changing distribution patterns of plant communities. Maps of halophytic communities,

have been prepared comparing 1963 and 1986 for an inland salt spring area near
Hecklingen, close to Stassfurt in East Germany (Fig. 7). These indicate
significant local change in the soil and water conditions due to an altered
arrangement of the salt ditches. The most saline soils are free of vegetation. A
gradient from higher to lower salinity is responsible for the zonation of the plant
communities from Salicornietum to the Astero-Puccinellietum and other salt
tolerant communities.

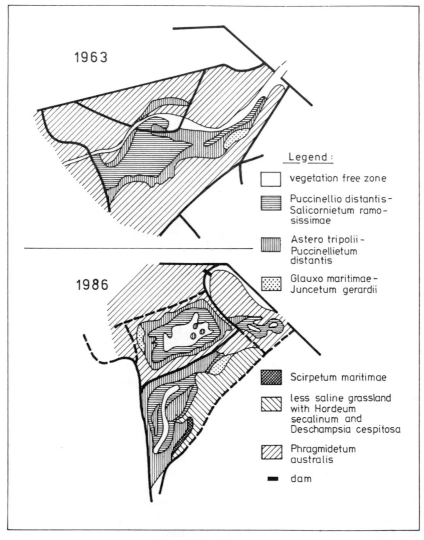

Figure 7. The distribution patterns of halophytic communities in
a salt-spring area near Hecklingen (East Germany) on the same saline plain in
1963 and 1986. After Schlag (unpubl. 1963) and Bank (unpubl. 1986) modified.

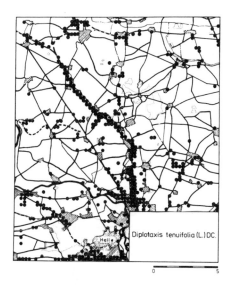

Figure 8. The distribution of *Diplotaxis tenuifolia* (L.) DC. in the countryside north of Halle (Saale) in East Germany. Map prepared by E. Grosse.

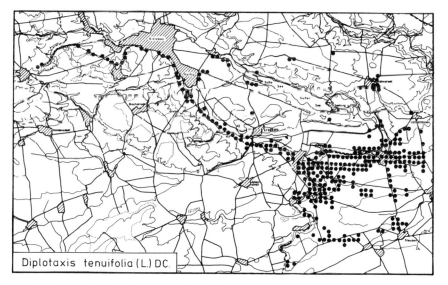

Figure 9. The distribution of *Diplotaxis tenuifolia* (L.) DC. in   the Mansfieldian lowland near Eisleben in East Germany characterized by salt tectonics, salt lakes and former copper mining industry. The occurrences mark the industrial centres (coalmines) and the railway. Map prepared by H. Volkmann.

Patterns of the total, regional and local distribution, including the patterns of the distribution of plant communities with a certain indicator value, may be interpreted as a function of the factorial effect of the habitat mosaic. The more anthropogenic factors overlap with natural factors, the more homophilous plant species and communities are represented within a countryside or urban area.

Most polluted habitats, for instance, in the vicinity of industrial centres and towns, as around Halle (Fig. 8), Eisleben (Fig. 9) and Bitterfeld (Fig. 10) which were subjected to a long-term and even large scale dust pollution, are now inhabited by many weeds. They are growing here often as invaders (*Atriplex tatarica, Solidago canadensis, Salsola kali, Diplotaxis tenuifolia*) (Figs. 8, 9, 10), particularly along railways, roads and motorways, in large numbers, as constituents of their own ruderal communities. *Diplotaxis tenuifolia* seems to be one of the most significant plants of extremely dust-polluted industrial centres.

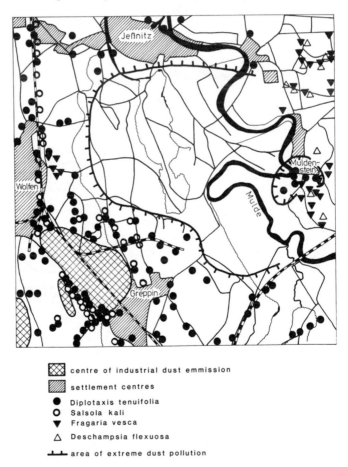

Figure 10. The range of industrial dust pollution in the industrial center of Bitterfeld (East Germany) indicated by plant distribution patterns.

Dot maps of such indicator plants, which represent the real occurrence in the field by one dot on a map, provide a means for understanding and assessing the real impact and spatial extent of an anthropogenic factor such as dust, or air and soil pollution. Figure 10 demonstrates the coincidence of the distribution of *Salsola kali* and *Diplotaxis tenuifolia* in most polluted industrial and settlement centres. Along the Mulde river the power station, where brown coal is fired, appears to be marked by this crucifer. Forests east of this power station currently display an increasing density and range expansion of the population of *Fragaria vesca* on formerly acidic sandy soils. *Salsola kali* is also growing here, increasing in number during the last few years along the railways. It occurs in dense carpets covering the ballast between the rails, a picture apparent in several other industrialized centres in East Germany.

Conclusion

Environmental change is signalled, in general, either by an expansion or a recession of plant populations and communities. Long-term and large scale change of the environmental conditions can be recognized by:-
- short-term and long-term reactions as changes in growth and distribution patterns of plants (extinction, recession, expansion);
- short-term and long-term changes of the species combination in plant communities (decline of diversity, arrival of new and/or invasive neophytes) and changes in their distributions;
- long-term evolutionary response in morphology, gene frequency and adaptability of plant populations controlling competition and distribution.

Although the mapping of single plant and animal species seems to be more time-consuming and expensive than the readily repeated mapping of the vegetation by remote sensing, both procedures are necessary in order to receive exact data and comprehensive information about the ongoing change in nature.

A large amount of literature is available about the relation between remote sensing techniques and vegetation (Merchant 1983; Küchler and Zonneveld 1988).

Airborn imaging spectrometers, which permit even more precise interpretations of the remote sensed data, will not solve the problem of less detectable single plant species distribution at this moment.

References

Arndt, U., Nobel, W. and Schweizer, B. (1987) *Bioindikatoren. Möglichkeiten, Grenzen und neue Erkenntnisse.* Stuttgart: Verlag Eugen Ulmer.

Däßler, H.-G. (ed.) (1986) *Einfluß von Luftverunreinigungen auf die Vegetation. Ursachen-Wirkungen-Gegenmaßnahmen.* 3rd ed. Jena: VEB Gustav Fischer Verlag.

Hultén, E. (1964) The circumpolar plants I. Vascular cryptogams, conifers, monocotyledons. *Kungl. Svenska Vetenskapsakad. Handlingar*, **8**, 1-275.

Hultén, E. (1971) The circumpolar plants II. Dicotyledons. *Kungl. Seveska Vetenskapsakad. Handlingar*, **13**, 1-463.

IGBP 1990. *The International Geosphere-Biosphere Programme: A Study of Global Change.* IGBP The Initial Core Projects. IGBP Global Change Report 12, Stockholm.

Jalas, J. and Suominen, J. (eds.) (1972) *Atlas Florae Europaeae 1* Pteridophyta (Psilotaceae to Azollaceae) Helsinki: Suomalaisen Kirjallisuuden Kirjapaino Oy (continued to vol. 8 Nymphaeaeceae to Ranunculaceae 1989).

Jankowski, A.T. (1990) The upper Silesia region as an area of ecological calamity. Seminar papers and IGBP WG 2 report. Inst. Geogr. Spatial Org. Polish Acad. Sci. Conf. papers 6: 118-132. Warszawa.

Küchler, A.W. and Zonneveld, I.S. (1988) *Vegetation mapping. Handbook of Vegetation Science*, **10**, 1-635. Dordrecht-Boston-London: Kluwer Academic Publishers.

Merchant, J.W. (1983) *Utilizing Landsat MSS Data in forest and range management, a guide to selected literature.* Lawrence, KS. University of Kansas, Space Technology Center.

Meusel, H., Jäger, E. and Weinert, E. (1965) *Vergleichende Chorologie der zentraleuropäischen Flora.* Jena: VEB Gustav Fischer Verlag.

Salánki, J. (ed.) 1986. *Biological Monitoring of the State of the Environment: Bioindicators. An overview of the IUBS Programme on Bioindicators 1985.* Oxford: IRL Press Limited.

Schubert, R. (ed.) (1985) *Bioindikation in Terrestrischen Ökosystemen.* Jena: VEB Gustav Fischer Verlag.

Weinert, E. (1984) Mapping the flora of the German Democratic Republic. *Norrlinia*, **2**, 69-74.

Weinert, E. (1986) Selected methods of bioindication in terrestrial ecosystems. In Salánki, J. (ed.). *Biological Monitoring of the State of the Environment: Bioindicators.* IUBS Monograph Series No. 1, 67-76. Oxford: IRL Press Limited.

Weinert, E. (1990) Vegetation of industrial centres under prospective climate change. Seminar papers and IGBP WG 2 report. Inst. Geogr. Spatial Org. Polish Academy of Sciences Conference paper 6, 108-117. Warszawa.

Woodward, F.I. (1987) *Climate and Plant Distribution. Cambridge Studies in Ecology.* Cambridge University Press, Cambridge.

Numerical Estimation of Climate from Indicator Plants

Rubén Retuerto[1] and Alejo Carballeira

Area de Ecología, Universidad de Santiago de Compostela, 15071 Santiago de Compostela, Spain.

[1]Present address: Department of Botany, University of Cambridge, Downing St., Cambridge CB2 3EA, U.K.

Key words: climatic factors, bioindicators, multiple regression, predictive models, northwest Spain.

Abstract

In many ecological studies it is crucial to know the values of certain climatic parameters. However, such values are often not available and their nature makes it impossible for them to be recorded inmediately. The strong spatial variability of the climate makes it inadvisable to interpolate from more or less distant areas. However, it is known that the presence, abundance and relative size of plants reflect the nature of the environment, so that they can be used as indicators.

This work seeks to estimate the climate of a locality from the indicator plants determined by Retuerto (1989). Various procedures used by other authors are tested and a new method, a modification of one of these, is proposed.

Predictions are made for 25 localities in northwestern Spain, and for 3 climatic parameters. These parameters, Baudiere's QE index, the mean minimum temperature in the coldest month and the mean temperature range in the coldest month, were chosen because, according to previous studies, they are thought to play a major role in regulating the distribution of the species considered in the area of study and similar areas.

All the methods tested enable us to establish the central values with acceptable accuracy. The numerical method proposed gives the best results, producing the smallest standard error estimates and is the most efficient predictor of the extreme values, for which the other methods produce the greatest residuals.

Bioindicators and Environmental Management
ISBN 0-12-382590-3

Introduction

In ecological studies it is crucial to know the values of certain environmental parameters, biotic or abiotic. Such values are often not available and their nature makes it impossible for them to be recorded immediately. This problem may be overcome using some type of indicators (Ellenberg 1979) or by means of predictive equations from other variables already known or measurable in a fast and economic way and in direct or indirect relation with the unknown parameters (Wikum and Wali 1974; Smith and Connors 1986; Carballeira *et al.* 1988; Hansen 1988; Uchijima and Seino 1988). The climatic parameters are considered to be first order variables to explain distribution and abundance of organisms (Woodward 1987). For these parameters to be considered representative of a locality, repeated measurements are required over many years and in standard conditions. Because of this such values are often not available. The strong spatial variability of the climate makes it inadvisable to interpolate from more or less distant areas. These factors have motivated studies that undertake the prediction of climatic parameters from other variables, commonly geotopographic ones (White 1977; 1979; Carballeira *et al.* 1981; Molina *et al.* 1983; Retuerto *et al.* 1983). On the other hand, it is known that the presence, abundance and relative size of plants reflect the characteristics of the environment, therefore, these aspects may be used as indicators. Vegetation is commonly used to estimate the productivity of forest ecosystems (Daubenmire 1976; Spurr and Barnes 1982; Carballeira *et al.* 1988); the quality of air (Ten Outen 1983; Posthumus 1983; de Wit 1983); water (Ringelberg 1983); soil (Brouwer 1983; Eijsackers 1983; Michener 1983) or environmental changes (de Boer 1983). Likewise, the knowledge of the close relationship between climate and vegetation (Rubel 1930; Holdridge 1947; Woodward 1987) has justified the employment of climatic characters to predict plant formation (Holdridge 1947; Sowell 1985); ecophysiognomic life forms (Box 1981a,b); and vegetation types (Looman 1983) or conversely, to estimate climatic parameters from the flora (Brisse and Grandjouan 1978; Ruffray 1978; Grandjouan 1982; Carballeira *et al.* 1984). Methods of estimation require a previous characterization of the species with regard to the parameters to be predicted and should preferably use groups rather than individual species. This is because the presence of a species in a locality is an imprecise testimony of the climate of this locality, since the species may be located on a wider climatic range than the range there represented. The species that are key indicators may or may not be present in a locality due to chance, historical reasons, or current competitive relationships; it is therefore more reliable to employ groups of indicator species (Dieterich 1970; Zonneveld 1983). Furthermore, the coexistence of different species in a locality provides complementary indications which correct one another and permit a more rigorous estimation (Grandjouan 1982).

The purpose of the present work is to test various procedures to estimate the climate of a locality from the indicator plants present there, taking as a basis an earlier study which characterized the climatic behaviour of a series of plants. A new method, a modification of one of these, is proposed.

Material and methods

Origin of data
A list of indicator taxa was obtained from Retuerto (1989) who characterized a series of species with regard to 92 climatic variables, using two frequential analysis methods (Brisse and Grandjouan 1978; Daget and Godron 1982). For sufficiently frequent species, several abundance levels were distinguished and climatically characterized. The entire range above each threshold defining what, following Brisse and Grandjouan (1978), he termed "plante". A "plante" was deemed indicator for a climatic variable when the indicator value that measures the degree of linkage between the "plante" and the climatic variable (Indicator Value of Brisse and Grandjouan 1978), was greater or equal to 50%, i.e. when its frequency profile was at least twice as concentrated as that of the whole set of "plantes" considered.

In order to test the different methods, predictions were made for 25 Galician localities (meteorological stations) in the northwest of Spain, and for 3 climatic parameters. These parameters, Baudiere's QE index, the mean minimum temperature in the coldest month and the mean temperature range in the coldest month, were chosen because, according to previous studies (Retuerto and Carballeira, in press), they are thought to play a major role in regulating the distribution of the species considered in the area of study and similar areas. These meteorological stations adequately represent the climatic diversity of an area on the frontier between the Eurosiberian and Mediterranean biogeographical regions (Izco 1982).

Floristic data were obtained according to a stratified sampling plan. In the surroundings of each meteorological station (radius 5 km) six stands located in sites under climatic control (i.e. eu-climatopes) were selected in order to obtain floristic inventories in places where mesoclimate was measured The choice of stands was made in order to maximize environmental heterogeneity under an identical climate. Within each of the selected stands four 10 m x 10 m plots were chosen. The cover of each species in each plot was evaluated in six 1 x 10 m belt transects located at random, 3 perpendicular to the other 3. We used the abundance-dominance scale of Braun-Blanquet (1932) as modified by Brisse and Grandjouan (1978). The value assigned to a species in a meteorological station was the maximum value of abundance found for that species in any of the plots associated with that meteorological station, since this value best expresses the species potentialities under a certain climate. Only woody formations were sampled, since these are relatively stable and hence more susceptible to the selective influence of climate than is the vegetation of a more ephemeral nature.

Methods

Limit plants (Ruffray 1978; Grandjouan 1982)
Climatically, a locality will be placed on the climatic range compatible with all the species present on the same. The value of the climatic variable on that locality

falls between the limits corresponding to the highest of the inferior extremes and the lowest of the superior extremes of all ranges of the indicator "plantes" found in that locality. The value assigned to the locality is the average of the two extremes.

Method of the absolute maximum and minimum (Ruffray 1978; Grandjouan 1982)
The predicted value for a locality is the average of two absolute extremes, a minimum and a maximum. These are obtained, respectively, from the first and third quartiles of the distributions showed by the indicator "plantes" present in the locality and are defined as follows:
Minimum: the third quartile of the distribution of first quartiles.
Maximum: the first quartile of the distribution of third quartiles.

Multiple regression
Stepwise regressions (P2R, RSWAP option, BMDP, Dixon 1983) were performed using climatic parameters as dependent variables and indicator species as independent variables.

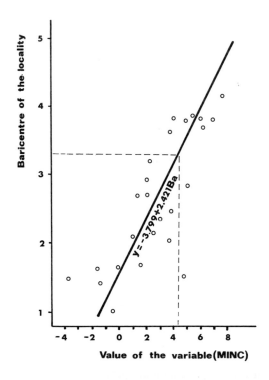

Figure 1. Numeric method proposed for the prediction of climatic factors from indicator plants.

Grandjouan's (1982) numeric method

It is based on the fact that the fidelities of the species of a locality to the classes of a climatic variable express a certain probability that the locality is in certain classes of that variable. The localities are ordered according to their value for the variable. This observed arrangement is compared, by means of regression, with the probable arrangement estimated from the "plantes" present in the localities. In the appendix an example shows how to estimate the probable position of a locality from the "plantes" present in it. The value of the regression coefficient indicates the magnitude of the deviations between the estimate positions and the observed positions, an aspect that we think could more appropriately be measured by means of the Spearman rank correlation coefficient. The regression equation, by means of a simple interpolation, allows us to know the absolute value of the variable corresponding to a certain estimate position, since the correspondences between the observed positions of the localities and the absolute values of the variables are known.

Proposed numeric method

We propose to compute the regression between the observed values of the variable in the localities and the correspondent baricentre of the locality calculated from the profile of frequency of the indicator "plantes" present in each locality (see appendix). The regression equation, through a simple interpolation (Figure 1), allows us to obtain the estimated value of the variable corresponding to the baricentre of a locality.

Results and discussion

Baudiere's QE index estimate

All methods showed the same tendencies in the predictions, as is revealed by the analysis of the residuals (Table 1, Figure 2). As a general rule all the methods generate negative residuals for the inferior values of the variable and positive residuals for the superior values. This may indicate that an experimental or logistic function would give a better fit. This homogeneity of the predicted data with respect to the observed data has been detected in similar studies (Ruffray 1978; White 1979; Carballeira *et al.* 1981; Molina *et al.* 1983; Retuerto *et al.* 1983).

Irrespective of the method used, the best predictions (lower absolute residuals) correspond to middle and lower values of the variable. This seems to be related to the fact that the indicator "plantes" for this variable are, predominantly, indicators of middle and lower positions (24 out of 32 indicator "plantes"). The localities placed on these ranges of the variable would have a higher number of indicator "plantes" and consequently the estimates would be more precise.

The best predictions, with lower standard errors of the estimate (4.6, only 8% of the range of variation observed in the real data) and lower dispersion of the residuals ($S = 4.5$), were given by the numeric method proposed. This method was also the best in estimating the higher values of the variable for which the other methods provided the greatest residuals. Only one locality presented a residual greater than 2 $ The Spearman rank correlation coefficient was very strong for all the methods ($r^2 > 0.93$), being higher in those corresponding to limit plants ($r^2 > 0.98$) and the proposed numeric method ($r^2 > 0.96$). Table 2 shows the predictive equations obtained for the variable considered. It may be seen that in all the cases $r^2 > 0.84$ and significance ($P < 0.01$).

In view of the high explicative power of the multiple regression method equations, we decided to select those equations which consider a small number of independent variables. In this way the equations would be easier and more extensively used. For this method the analysis of residuals was carried out from those residuals generated from the equations, taking into account five independent variables. The consideration of a greater number of variables would bring about a better fit for the method, but would lose significance.

Table 1. Results from the analysis of the residuals originated for the methods tested for the climatic factors considered.

	Limit Method	Quartiles Method	Numeric (Grandjouan)	Proposed Method	Multiple Regression
QE					
Mean	-0.20	4.56	0.48	0.00	0.00
Standard deviation of the residuals	4.91	7.32	4.96	4.50	5.81
Standard error of the estimate	5.01	8.86	5.09	4.61	6.51
Spearman rank correlation coeff.	0.98	0.95	0.95	0.96	0.93
MINC					
Mean	0.40	0.69	-0.09	0.00	0.08
Standard deviation of the residuals	1.48	1.63	1.55	1.57	1.10
Standard error of the estimate	1.56	1.82	1.59	1.59	1.32
Spearman rank correlation coeff.	0.90	0.89	0.87	0.87	0.87
CTR					
Mean	-0.36	-0.07	0.01	0.00	0.00
Standard deviation of the residuals	1.18	1.44	0.98	0.95	0.96
Standard error of the estimate	1.27	1.47	1.00	0.96	1.09
Spearman rank correlation coeff.	0.78	0.68	0.85	0.85	0.83

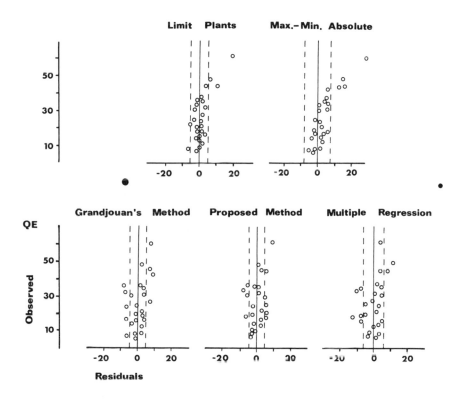

Figure 2. Plotting of residuals as contrasted with observed values for the Baudiere's OE index (— : standard deviation of residuals).

Mean minimum temperature for the coldest month (MINC)

The analysis of residuals (Table 1 and Figure 3) in all the methods shows the tendency mentioned above: homogeneity of predicted data with respect to the observed data. This tendency is manifest in the sign of the residuals, negative for the colder localities and positive for the warmer ones. No method generated residuals greater than 3 ⚲ Four of the tested methods showed only one locality with residuals greater than 2 ⚲ Apart from this general tendency the different methods showed distinct behaviours in the predictions. The limit plants and quartile methods gave good estimates of middle values but inaccurately predicted the extreme values, especially the higher ones, whose residuals exceed 1 ⚲ These values correspond to localities with a small number of indicator

Table 2. Predictive equations for the Baudiere's QE index.

Numeric method (Grandjouan 1982)
> Observed rank = 0.6703 + 0.9484 * Estimate rank
> $r = 0.95$; $r^2 = 0.90$; $F = 205.17$; $P < 0.0001$
> Standard error of the estimate = 5.09.

Proposed numeric method
> QE = 7.6208 + 0.6669 * e Baricentre
> $r = 0.95$; $r^2 = 0.90$; $F = 209.88$; $P < 0.0001$
> Standard error of the estimate = 4.61.

Multiple regression method

Variable	Coefficient
Castanea sativa	-3.3985
Erica cinerea	-6.4244
Ilex aquifolium	3.7917
Lithodora diffusa	7.4256
Quercus suber	-12.9708
The Y intercept =	31.1285

$r = 0.91$
$r^2 = 0.84$
$F = 19.51$
$P < 0.01$
Standard error of the estimate = 6.51

"plantes" for the variable (15 out of 18 indicator "plantes" are indicators of lower or middle positions). The lack of precision in the estimates provided by the other methods are of less magnitude and are not concentrated in a particular range of the variable. The proposed numeric method is the best predictor for the higher values of the variable; however, the regression multiple method is the best predictor for the lower values. This last method also provides the best global predictions, showing the smallest standard error of estimation (1.32, which represents 11% of the range shown for the observed values of the variable) and the smallest deviation in their residuals ($s = 1.1$).

The Spearman rank correlation coefficients are high for all the methods ($r^2 > 0.87$), especially for the limit plant method ($r^2 > 0.90$).

The equations originated for the prediction of these variables are shown in Table 3. In all the cases the r^2 are > 0.70 and significative ($P < 0.01$). For the multiple regression method the analysis of residuals was made from the values provided by the equations, taking into account five independent variables ($r^2 = 0.83$). These variables are, with the exception of *Ruscus aculeatus*, indicators of

Table 3. Predictive equations for the Mean minimum temperature for the coldest month (MINC).

Numeric method (Grandjouan 1982)

      Observed rank = 1.7634 + 0.8644 * estimate rank

      $r = 0.86$; $r^2 = 0.75$; $F = 67.43$; $P < 0.0001$

      Standard error of the estimate = 1.59

Proposed numeric method

      MINC = -3.7990 + 2.4209 * Baricentre

      $r = 0.83$; $r^2 = 0.70$; $F = 52.34$; $P < 0.0001$

      Standard error of the estimate = 1.59

Multiple regression method

Variable	Coefficient
Cytisus multiflorus	-0.6214
Ilex aquifolium	-0.2531
Quercus pyrenaica	-0.6129
Ruscus aculeatus	0.3711
Sorbus aucuparia	-0.7186
The Y intercept =	4.1040

$r = 0.91$

$r^2 = 0.83$

$F = 18.22$

$P < 0.01$

Standard error of the estimate = 1.32

---

lower positions, and both the sign and magnitude of their coefficients are in correspondence with the sign and magnitude of their climatic positions. The predictor power obtained with these five variables is high. In a study performed in the same territory, Molina *et al.* (1983) needed eight geotopographical variables to account for the same percentage of variation in the variable considered.

All the methods tested produced less accurate estimates for this variable than those generated for the Baudiere's QE index. This is shown by the lower regression and Spearman rank correlation coefficients and by the larger standard error of the estimate, representing a larger percentage in relation to the observed range shown by the variable. The explanation for this could be in the small number of indicator taxa (predictor variables) for this variable in comparison with Baudiere's QE index since the activity of the cold is of lesser importance than drought as a limiting factor for the vegetation, at least for the vegetation and territory studied (Diaz-Fierros *et al.* 1983; Retuerto and Carballeira, in press).

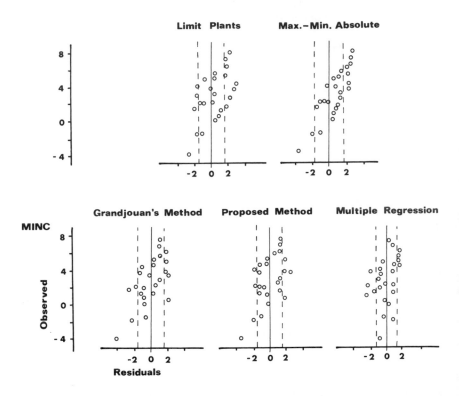

Figure 3. Plotting of residuals as contrasted with observed values for the mean minimum temperature for the coldest month (--- : standard deviation of residuals).

Mean thermal range for the coldest month

For this variable the residuals generated by the estimation methods showed the same tendency previously mentioned: overestimation of the low values and undervaluation of the high values (Table 1; Figure 4). The quartile and limit "plantes" methods are the least accurate in their predictions, especially for the lower values of the variable, for which the residuals generated exceed 1 $ The other methods tested provided smaller residuals and their predictions are not much more accurate for any segment in the range of the variable. No method generated residuals exceeding 3 $ with three of the methods showing only one locality surpassing 2 $ The numeric method proposed again achieved the lower standard error of the estimate, 0.96, which represents 14% of the observed variation range shown by the variable. This method also generated the best predictions for the extreme values and showed the smallest residual dispersions ($= 0.95$).

The Spearman rank correlation coefficients range from 0.68, for the quartile method, to 0.85, for the numeric method proposed. These values are clearly lower than those obtained for the two other variables.

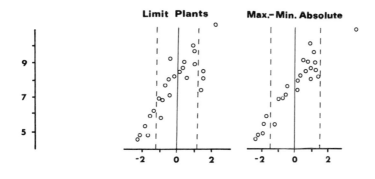

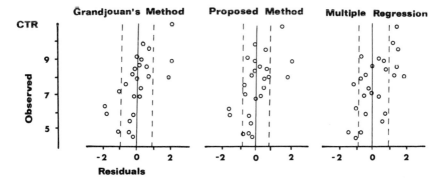

Figure 4. Plotting of residuals as constrasted with observed values for the mean thermal range for the coldest month (--- : standard deviation of residuals).

Table 4. Predictive equations for the Mean thermal range for the coldest month (CTR).

Numeric method (Grandjouan 1982)
    Observed rank = 1.9881 + 0.8471 * Estimate rank
    $r = 0.85$; $r^2 = 0.72$; $F = 58.11$; $P < 0.0001$
    Standard error of the estimate = 1.00

Proposed numeric method
    CTR = 0.3390 + 6.3894 * Baricentre
    $r = 0.84$; $r^2 = 0.71$; $F = 56.27$; $P < 0.0001$
    Standard error of the estimate = 0.96

Multiple regression method

Variable	Coefficient
Arbutus unedo	2.1986
Betula celtiberica	0.3820
Erica scoparia	-1.1604
Quercus suber	1.3847
Rubia peregrina	-0.7826
The Y intercept =	6.4872

$r =$ 0.80
$r^2 =$ 0.64
$F =$ 6.81
$P <$ 0.01
Standard error of the estimate = 1.18

The regression equations obtained for this variable are shown in Table 4. In any case the regressions are significant ($P < 0.01$) and the $r^2 > 0.70$. For the multiple regression method the analysis of the residuals was carried out from the equation with five variables. If we consider ten independent variables the method accounts for most of the variation ($r^2 = 0.78$) without the significance falling below 1%.

For this variable all the methods tested gave more inaccurate estimates than those obtained for the other two variables. This is indicated both by the lower correlation and regression coefficients obtained and by the higher standard error of the estimates (expressed as a percentage of the observed range for the variable). As a possible explanation for this result we suggest that the predictability of the climatic variables, at least from floristic variables, is related to the activity of this variable on the vegetation. According to Retuerto and Carballeira (in press) the activity of the mean thermal range for the coldest month (CTR) on the vegetation studied is less than that of OCEA and MINC, as is confirmed by the lower number of indicator "plantes" showing responsiveness to this variable.

Conclusions

All the methods tested allowed us to predict the middle values with acceptable accuracy, but they were more inaccurate for estimating the extreme values. In all the cases, the quartile method gave the worst predictions, showing higher standard errors for the estimates and larger dispersion in the residuals. The proposed numeric method, for the variables QE and CTR, and the multiple regression method, for the MINC, gave the best predictions, especially for the extreme values, those for which the other methods gave the larger residuals. The regression multiple method, apart from giving less accurate results in comparison with the numeric method proposed, has the disadvantage of being sensitive to the absence of the species when this has not by itself ecological significance. Certain "plantes" (independent variables in the regression equations) may not be in a locality due to extraclimatic factors (competition, anthropogenic influences, flora history, soil composition, etc) or may be present but have not been found in the sampling. This method is therefore not recommended for prediction purposes but may be very useful for increasing our knowledge of the relationship between plant and climate.

The predictive power showed by the floristic variables for the climatic factors seems to be higher than that of the geotopographic variables confirming the intensity of the relationship between climate and plant. The predictive power increases with the number of variables and with the intensity of indication that the "plantes" show for the climatic factor. The errors made in the prediction were smaller for the range of the variable presenting larger number of indicator "plantes".

Two implications have to be considered when the estimation methods are used. First, the predictor variables may not have significance beyond their ability to diagnose existing conditions. Secondly, to obtain reliable results, the estimation methods should not be applied in predicting conditions located out of the range of environmental conditions considered in the environmental characterization of the indicator elements.

Acknowledgements

We am grateful to R. Obeso for his critical comments on the manuscript and to D.Tricker for the stylistic revision.

References

Boer, TH. A., De. (1983) Vegetation as indicator of environmental changes. *Environmental Monitoring and Assessment*, 3, 375-380.

Box, E.O. (1981a) Predicting physiognomic vegetation types with climate variables. *Vegetatio*, 45, 127-139.

Box, E.O. (1981b) *Macroclimate and plant forms. An introduction to predictive modelling in phytogeography*. Junk. The Hague.

Braun-Blanquet, J. (1932) *Plant Sociology; the study of plant communities.* (Translated by G.D. Fuller and H.S. Conard), University of Chicago, III. (USA).

Brisse, H. and Grandjouan, G. (1978) Eléments de phytoclimatologie numérique. Principles de la classification climátique des "plantes"? Actes 3éme Colloque de l'Association Informatique et Biosphere, 55-184. París.

Brouwer, R. (1983) Soil properties indicative for quality. *Environmental monitoring and Assessment,* 3, 283-287.

Carballeira, A., Juste, J., Molina, A.,Retuerto, R. and Ucieda, F. (1981) Predicción de parámetros climáticos de interés ecológico en Galicia. I. Relación clima-topografía. *Anales de Edafologica y Agrobiologia,* XL, 2121-2137.

Carballeira, A., Retuerto, R. and Ucieda, F. (1984) Estima numérica de la productividad climática potencial a partir de la flora en Galicia. *Cuadernos del Area de Ciencias Agrarias,* 5, 41-53.

Carballeira, A., Reigosa, M.J. and Carral, E. (1988) Modelo predictivo de la calidad de sitio de *Pinus pinaster* en Galicia, construido con variables geotopográficas y su traducción climática. In A. Blanco (ed.), *Avances Sobre Investigación en Bioclimatología,* 229-241. C.S.I.C. Madrid.

Daget, P. and Godron, M. (1982) Analyse de l'écologie des espéces dans les communautés. Masson. Collection d'écologie. Paris.

Daubenmire, R.F. (1976) The use of vegetation in assessing the productivity of forest lands. *Botanical Review,* 42, 115-143.

De Wit, T. (1983). Lichens as indicators for air quality. *Environmental Monitoring and Assessment,* 3, 273-282.

Diaz-Fierros, V.F., Ucieda, F., Retuerto, R. and Carballeira, E. (1983) Productividad climática potencial de cultivos y bosques de Galicia. *Cuadernos del Area de Ciencias Agrarias,* 4, 27-40.

Dieterich, H. (1970) Die Bedeutung der Vegetationskunde für forstliche Standostskunde. *Der Biologieunterricht,* 6, 48-60.

Dixon, W.J. (1983) *BMDP Statistical Software.* University of California Press, Berkeley.

Eijsackers, H. (1983) Soil fauna and soil microflora as possible indicators of soil pollution. *Environmental monitoring and assessment,* 3, 307-316.

Ellenberg, H. (1979) *Zeigerwerte der Gefässpflanzen Mitteleuropas.* Scripta Geobotanica 9, 2nd ed. Göttingen.

Grandjouan, G. (1982) *Une methode de comparaison statistique entre les repartitions des "plantes" et des climats.* Thése Universitaire Louis Pasteur. Strasbourg.

Hansen, P.A. (1988) Prediction of macrofungal occurrence in Swedish beech forest from soil and litter variable models. *Vegetatio,* 78, 31-44.

Holdridge, L.R. (1947) Determination of world plant formation from simple climatic data. *Science,* 105, 367-368.

Izco, J. (1982) Problémes spatiaux et altitudinaux posés par la limite entre les ecosystémes Méditerranéens et Atlantiques. *Ecología Mediterránea,* VIII, 289-299.

Looman, J. (1983) Distribution of plant species and vegetation types in relation to climate. *Vegetatio,* 54, 17-25.

Michener, M.C. (1983) Wetland site index for summarizing botanical studies. *Wetlands*, **3**, 180-191.

Molina, A.M., Juste, J., Ucieda, F., Retuerto, R. and Carballeira A. (1983) Predicción de parámetros climáticos de interés ecológico en Galicia: II. Pluviometría y Termometría. *Anales de Edafologica y Agrobiologia*, **XLII**(1/2).

Posthumus, A.C. (1983) Higher plants as indicators and accumulators of gaseous air pollution. *Environmental Monitoring and Assessment*, **3**, 263-272.

Retuerto, R. (1989) *Fitoclimas de Galicia. Estudio numérico de los efectos del clima sobre la distribución y abundancia de las plantas.* Thesis. University Santiago de Compostela.

Retuerto, R. and Carballeira, A. (1990) (in press) Phytoecological importance, mutual redundancy and phytological threshold values of certain climatic factors. *Vegetatio*, (MS no 103).

Retuerto, R., Ucieda, F. and Carballeira, A. (1983). Predicción de parámetros climáticos de interés ecológico en Galicia: y III. Evapotranspiración y Balance hídrico. *Anales de Edafologica y Agrobiologia*, **XLII** (11/12).

Ringelberg, J. (1983) General remarks with regard to biological indicators used in water pollution studies. *Environmental Monitoring and Assessment*, **3**, 317-319.

Rubel, E.F. (1930). *Pflanzengesellschaften der Erde.* Hans Huber Verlag, Bern.

Ruffray, P. (1978) *Etalonnage climatique des "plantes" en Alsace et dans regions limitrophes.* Thése, Universitaire Louis Pasteur. Strasbourg.

Smith, K.G. and Connors, P.G. (1986) Building predictive models of species occurrence from total-count transect data habitat measurements. In *Wildlife 2000 : Modelling habitat relationships of terrestrial vertebrate.* University of Wisconsin Press, Madison.

Sowell, J. (1985) A predictive model relating North American plant formation and climate. *Vegetatio*, **60**, 103-111.

Spurr, S.H. and Barnes, B.V. (1982) *Ecología Forestal.* AGT Editor, S.A.México

Ten Houten, J.G. (1983) Biological indicators of air pollution. *Environmental Monitoring and Assessment*, **3**, 257-261.

Uchijima, Z. and Seino, H. (1988) An agroclimatic method of estimating Net Primary Productivity of Natural vegetation. *Japanese Agricultural Research Quarterly* , **21**, 244-250.

White, E.J. (1977) Computer programs for the estimation of selected climatic variables and of values of principal components expressing variation in climate, for any site in Great Britain. Merlewood Research and Development Paper no.70.

White, E.J. (1979) The prediction and selection of climatological data for ecological purposes in Great Britain. *Journal of Applied Ecology*, **16**, 141-160.

Wikum, D.A. and Wali, M.K. (1974) Analysis of a North Dakota Gallery Forest: vegetation in relation to topographic and soil gradients. *Ecological Monographs*, **44**, 441-464.

Woodward, F.I. (1987) *Climate and plant distribution.* Cambridge University Press. Cambridge.

Zonneveld, I.S. (1983) Principles of Bio-Indication. *Environmental monitoring and assessment*, **3**, 207-217.

Appendix

Calculation of the probable position of a locality from indicator "plantes":

Consider three localities A,B and C, in which the values of the variable are, respectively 2, 5 and 7. These values determine a rank of the localities:

Locality	Value of the variable	Observed position
A	2	1
B	5	2
C	7	3

Consider now the frequency profile corresponding to the indicator "plantes" present in each of the localities:

Locality A		Locality B		Locality C	
Plant	Profile	Plant	Profile	Plant	Profile
a	5 3 0 0 0	f	0 0 4 4 2	k	0 0 4 4 2
b	5 3 2 0 0	g	0 0 3 1 0	l	0 2 4 4 1
c	1 2 0 0 0	h	0 1 1 1 0	m	0 0 0 2 1
d	4 0 0 0 0	i	0 1 4 1 0	n	0 1 5 5 4
e	4 0 2 0 0	j	0 0 2 3 2	ñ	0 2 3 4 1
				o	0 1 3 5 3

Profile of the locality
    19 10 4 0 0   0 2 14 10 4   0 6 19 24 12

Climatic position (Brisse & Grandjouan, 1978) of the locality:
    -58   21   28
Baricentre (Daget & Godron, 1982) of the locality:
    1.54   3.53   3.69

From the climatic positions or baricentres a new ranking of the localities is made.

This ranking is compared with the observed one:

Locality	Climatic position	Baricentre	Estimated position
A	-58	1.54	1
B	21	3.53	2
C	28	3.69	3

In the proposed example, both the observed and estimated ranking coincide.

Land Use and Ecological Change in Areas of Outstanding Natural Beauty

Alan Cooper, Ronald Murray and Thomas McCann

Department of Environmental Studies, University of Ulster, Coleraine, Co. Londonderry BT52 1SA, Northern Ireland.

Key words: landscape ecology, land use change, land classification, monitoring, Northern Ireland.

Abstract

Redesignation of Areas of Outstanding Natural Beauty and provision for Environmentally Sensitive Areas in Northern Ireland has stimulated a programme of environmental baseline and monitoring research to facilitate ecological and landscape management.

The main aims of the research have been to (i) quantify the distribution of land use, ecological resources and landscape attributes; (ii) describe the structure, composition and management of resources; (iii) identify environmental problems and suggest management options; and (iv) prepare a database and map archive for monitoring environmental change.

Some of the applications of the database, relating to monitoring peatland exploitation in the Sperrins AONB, the effects of agriculture on species-rich wetland in County Fermanagh and the management of field boundaries in Northern Ireland are described.

Introduction

Major rural land use issues in Ireland include the intensification of agriculture and peatland exploitation (Cruickshank and Wilcock 1982; Aalen 1985). Associated with both are detrimental effects on the quality of ecological resources and landscape attributes. There is little quantitative information, however, on the distribution and current status of these environmental resources in rural areas of Northern Ireland (Cooper 1988). The integrated data needed to produce management guidelines for land users is not available. This is due mainly to the sectoral nature of work carried out by government agencies and the fact that monitoring the effects of rural land use has not been a statutory duty.

Recently, implementation of the Nature Conservation and Amenity Lands

(Northern Ireland) Order (1985) has resulted in the redesignation of Areas of Outstanding Natural Beauty (AONB) in Northern Ireland (Fig. 1). As part of the process, the Countryside and Wildlife Branch (Department of the Environment for Northern Ireland) have funded environmental baseline and monitoring research within the Mourne, Antrim Coast and Glens, Causeway Coast, Sperrins and North Derry Areas of Outstanding Natural Beauty (Cooper and Murray 1987a; 1987b; Cooper *et al.* 1988a) and in Fermanagh District. This has led to the re-evaluation of AONB management policies (DoE NI 1988; 1989). The Department of Agriculture for Northern Ireland (DANI) have also commissioned research (Cooper *et al.* 1988b; Cooper *et al.* 1989) to assess the effects of designating the Mournes and Glens of Antrim as Environmentally Sensitive Areas (ESA).

The main aims of these studies, carried out at the University of Ulster have been to:-

(a) Quantify the distribution of land use, ecological resources and landscape attributes.

(b) Describe the structure, composition and management of resources.

(c) Identify environmental problems and suggest management options.

(d) Prepare a database and map archive for monitoring long term environmental change.

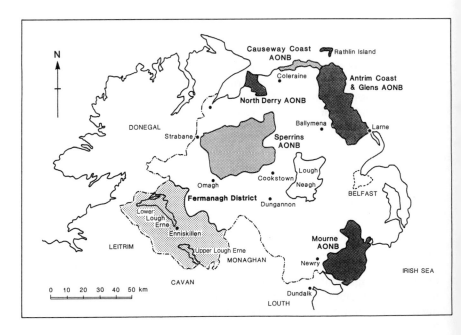

Figure 1. The distribution of Areas of Outstanding Natural Beauty.

This paper describes some of the applications of the database relating to monitoring peatland exploitation in the Sperrins, the effects of agriculture on species-rich wetland vegetation in Fermanagh and the management of field boundaries in Northern Ireland.

Methods

Our approach is based on the Institute of Terrestrial Ecology (Merlewood) method of multivariate land classification (Bunce *et al.* 1983). The technique classifies kilometre grid squares of the Irish Grid using map attributes related to climate, geology, soils, topography, hydrology, land use and rural settlement (Cooper 1986). The land classes are subsequently used as a sample stratification for ecological and landscape field survey. Samples are also stratified by geographical units of land. These are created by combining groups of land classes using criteria related to local land class discontinuities and distribution patterns.

Within each land class, 25 ha grid squares are selected, at a sampling intensity of approximately 2.5% of land class area. Resources are mapped onto computer coded field data sheets. Five main categories of resource are recorded, namely woodland, semi-natural vegetation, field boundaries, agricultural land use and landscape attributes. Ancillary data on management, structure and species composition are also recorded, together with target notes for descriptors not on the data sheets.

Land parcel areas and linear features are digitized directly from the field data sheets using a digitizing tablet linked to an IBM PS-2 microcomputer. Following this, information is stored as disk files and analyzed with the database software package dBASE III PLUS (Ashton-Tate 1985). Information from the database can be retrieved for any combination of resource type or management characteristic and can be presented for Northern Ireland as a whole, separate Areas of Outstanding Natural Beauty or specified localities.

Land classification maps are available for areas that have been surveyed. Resource distribution maps based on land classification have been produced by the geographic information system IDRISI (Eastman 1988) which has been linked to dBASE III PLUS by programming the latter.

Peatland exploitation in the Sperrins AONB

Peatlands, in the form of blanket bog, cover large areas of the uplands in the Sperrins AONB. They are an integral part of the landscape. The aim of our research was to contribute to the peatland exploitation debate by examining the types and distribution of peatland vegetation types and describing, in quantitative terms, the ways that they are currently being used.

Four major groups of land classes were selected for field survey within the study area as follows:-

a) Lower elevation, largely enclosed farmland including valley bottoms of the main glens.
b) Transitional hill land and farmed upper glens.
c) High elevation uplands with more unenclosed land, mountain sides and higher drumlins.
d) Mountain and upland plateau terrain.

A stratified random sample of 96, 25 ha squares was selected for field survey, representing 2.2% of the AONB. Information on the type, structure, management and main species present in semi-natural vegetation.

Five simplified categories of peatland vegetation were recorded:-
a) Wet bog: with mainly bog moss (*Sphagnum* spp.), cotton grasses (*Eriophorum* spp.) and ling heather (*Calluna vulgaris*).
b) Dry bog: consisting of less wet peat with species such as bog moss, cotton-grass (*Eriophorum vaginatum*), *Polytrichum*, heath rush (*Juncus squarrosus*) and soft rush (*Juncus effusus*). Grasses such as purple moor-grass (*Molinia caerulea*), mat-grass (*Nardus stricta*) and sweet vernal-grass (*Anthoxanthum odoratum*) were common.
c) Wet heath: with a dominant cover of ling heather in association with wetland species such as cross-leaved heath (*Erica tetralix*), cotton-grass, heath rush, deer-grass (*Scirpus caespitosus*), wavy hair-grass (*Deschampsia caespitosa*) and mosses.
d) Wet heath mosaic: mainly ling heather forming mosaics with wetland species and mat-grass.
e) Poor fen: dominated by either sharp-flowered rush (*Juncus acutiflorus/articulatus*) or soft rush, with mosses other than *Sphagnum*. Grasses are not predominant.

Table 1. Percentage cover of peatland vegetation types and forestry in the Sperrins AONB land class groups. Agricultural land and other land uses are not included.

| | Land Class Group | | | | |
	1	2	3	4	AONB
Wet bog	8.8	6.6	7.5	9.6	8.1
Dry Bog	1.2	5.7	14.1	22.3	12.6
Heath	0.1	4.8	16.2	22.2	12.6
Heath Mosaic	1.8	7.4	10.7	15.5	10.2
Poor fen	0.6	2.6	4.3	2.8	2.8
Forestry	<0.1	4.0	3.2	7.7	4.6

Resource distribution and land use

Wet heath and its mosaics, together with dry bog, were the most widespread vegetation types, particularly in the higher elevation land classes (Table 1). Wet bog of both raised and blanket types was less common but more uniformly distributed throughout the land classes. Conifer plantation covered less than 5% of the AONB, largely over peat in the hills.

The composition and structure of the peatlands, were particularly influenced by management factors (Tables 2 and 3). Hand peat cutting, largely abandoned, was proportionately more extensive at higher elevations where peat was a more abundant resource. Bog with abandoned hand cut peat often has a higher habitat and species diversity than intact peat, for example *Sphagnum* spp. are often found colonizing pools that have formed in old drainage ditches.

Machine peat cutting occurred throughout the AONB, largely for domestic purposes in plots of about 0.5-5.0 ha. Both types of cutting were more frequent on the less deep, more accessible peats, for example on heath mosaics. Unlike hand peat cutting, which is localized and removes peat from a vertical bank, recently introduced tractor-mounted auger machines remove a thin layer at any one time, from below the surface. The effect is to damage surface vegetation and compact the peat, destroying microtopographic variation and impeding drainage. This reduces species diversity in the short term with an immediate loss of *Sphagnum*, a reduction in heather cover (*Calluna vulgaris*) and an increase in cotton grass (*Eriophorum vaginatum*) and deer grass (*Scirpus caespitosa*). As the vegetation recovers, following cessation of cutting, increased surface wetness introduces a degree of habitat diversity not present in the drier peatland vegetation types.

Table 2. Peatland management and structure in the Sperrins land class groups. Results are expressed as a percentage cover of each land class.

	Land Class Group				
	1	2	3	4	AONB
Hand Cut	3	11	15	22	15
Machine Cut	0	4	3	3	3
Open Drains	0	8	12	33	17
Eroded	<1	2	6	24	10
Burnt	6	6	5	10	7

Table 3. Peatland vegetation management and structure in the Sperrins AONB. Results are expressed as a percentage cover of each vegetation type.

	Hand cut	Machine cut	Open drains	Eroded	Burnt
Wet bog	27	4	28	10	22
Dry bog	31	8	35	19	4
Heath	23	4	41	22	27
Heath Mosaic	44	12	23	27	12
Poor fen	5	0	19	2	0

Open drains dug with a single furrow plough are used to improve the grazing potential of many peatland types or as preparation for peat cutting. The effect is to decrease the wetness of the peat and reduce the *Sphagnum* cover.

Overgrazing by sheep has modified most of the peatland vegetation types over the years. Fence line observations have shown that the heather component has been lost or much reduced and that cotton grass (*Eriophorum vaginatum*) and purple moor-grass (*Molinia caerulea*) dominated vegetation, has replaced it.

Peat erosion is severe in the mountains, particularly on the summits. It is exacerbated by grazing, peat cutting, draining and burning. Burning is quite common throughout the land classes, most frequently on wet heath and wet bog. The calculated time interval between burns is between about 5-12 years (Cooper *et al.* 1988a). Regeneration of heather is good in spite of these short cycles.

Management issues

Peatland management problems are clearly widespread. They are also closely linked with socio-economic factors since farming, peat extraction, forestry and wildlife/landscape conservation interact. Upland farming is marginal in terms of economics and environmental constraints and the farming community has close cultural ties with the land. Most of the farms are small and owner-occupied and people do not want to give up their land. Farmers need to extract peat for fuel to supplement their income, which is an essential part of the farm economy. New access roads on hill farms, grant-aided to improve farm structure, have compounded the problems of peatland conservation and management by making extraction by machine at new and existing sites feasible.

Recent changes in Common Agricultural Policy, for example restrictions on milk production, have had the effect of increasing the numbers of sheep on the hills. Subsequent effects on peatland through erosion and overgrazing will occur in the long term. Whilst heather regeneration is reasonably good under current burning regimes, this might not be the case with heavier grazing pressures.

Government policy in Northern Ireland is to encourage private forestry. Recent examples of this in the Sperrins compare unfavourably, even with Forest Service monoculture plantations of Sitka spruce. There is little or no provision for wildlife or landscape conservation. The total area of afforestation, its distribution and allocation of land for forestry, all need to be considered if the ecological and landscape interest of the area is to be maintained.

The relevance of our work to this issue can be exemplified by considering the predicted distribution of coniferous forest plantation within the AONB (Fig. 2). The example is simplified but appropriately illustrative. Land class group 3 has a low percentage cover (3.2%) of coniferous forest (Table 1) but the greatest amount of poor fen and rush-infested grassland (12.0%). This vegetation type has little agricultural or nature conservation value in the area, but it does have a considerable potential for tree growth. If the socio-economic factors currently constraining land use for afforestation were to be altered, land class group 3 would be likely to change. This would have landscape effects, particularly in the north of the AONB.

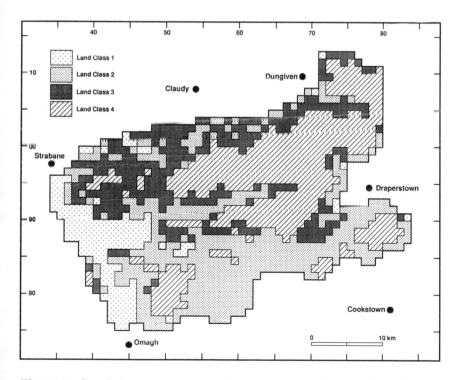

Figure 2. Land class groups in the Sperrins Area of Outstanding Natural Beauty.

Wetland management in Upper Lough Erne

The Upper Lough Erne basin (Fig. 1) is one of the most botanically rich wetland areas in the British Isles and also one of the most important areas for breeding waders (Partridge 1987). The aim of the ecological survey was to provide a quantitative description of the distribution and current status of environmental resources in the area as part of a land use assessment being prepared for Fermanagh District. One function of the work has been to provide background information for the possible designation of "Environmentally Sensitive Areas" in the District, involving the Department of Agriculture for Northern Ireland and the Countryside and Wildlife Branch (DoE NI).

Designation of ESA status is a site based statutory wildlife conservation measure currently operated by DANI and being considered for certain parts of Fermanagh. The aim of designation is to maintain and enhance the wildlife interest of particular areas of farmland by providing financial incentives designed to encourage farming practices that are sympathetic to the environment (DANI 1987). Farmers may join the scheme on a voluntary basis for a 5 year renewable period. This enables them to receive payment in return for following an agreed farm management plan.

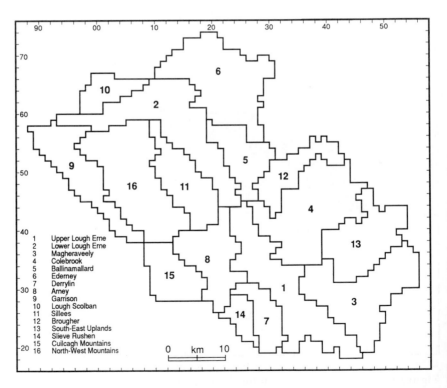

1   Upper Lough Erne
2   Lower Lough Erne
3   Magheraveely
4   Colebrook
5   Ballinamallard
6   Ederney
7   Derrylin
8   Arney
9   Garrison
10 Lough Scolban
11 Sillees
12 Brougher
13 South-East Uplands
14 Slieve Rushen
15 Cuilcagh Mountains
16 North-West Mountains

Figure 3. Landscape units in Fermanagh District.

In order to stratify the field sampling programme, local patterns of land class distribution and land class clusters were used to divide the district into 16 geographically distinct units (Fig. 3). The Upper Lough Erne unit (1) was defined as a farmed lakeland area consisting of kilometre squares encompassing the shoreline, islands and lake margins of the lough, lough channels and adjacent water bodies. The land consists of many small drumlins and inter-drumlin hollows, mainly at an elevation of between 40 m and 70 m. Soils are mostly heavy gleys associated with calp limestone, shale and some lower limestone.

Sample stratification was by landscape unit and land class. A random sample of 165, 25 ha sample squares was selected for field survey. The 20 squares lying within unit 1 represented a sampling intensity of 2.7%. A full inventory of semi-natural vegetation types and agricultural land use was recorded, together with structural and management descriptors.

Resource distribution and land use

The main ecological interest in the unit was wetland vegetation associated with the lakeside (Table 4). It was commonly only a narrow band but was often particularly species-rich and contained rarities such as marsh pea (*Lathyrus palustris*) and water hemlock (*Cicuta virosa*). Much of this fen vegetation was grazed (66%), often down to the water margin in a dry summer and 9% was disturbed in some other way, for example, poached by cattle. Fen carr was often contiguous with fen and was invasive under low grazing pressure. Species-rich ditch vegetation with a fen type flora was a common feature of field parcels near the lake margin

Species-rich wet grassland, occupying 9.7% of the unit, was dispersed throughout, on farmland. Large parcels were associated with inter-drumlin hollows and periodically inundated or low lying areas adjacent to the lakeside. Only in a minority of cases did species-rich vegetation, usually in the form of wet grassland, extend further inland onto farmed agricultural land.

Away from the water margin, agricultural grassland (65.1% cover) was the main resource type. Much occurred over reclaimed peat. It was largely unimproved (47% of total grasslands), being composed largely of *Holcus lanatus*, *Agrostis* spp., *Festuca pratensis* and *Anthoxanthum odoratum*. This grassland was species poor, with over 70% rush infested by either *Juncus effusus* or *Juncus articulatus/acutiflorus*. Improved ryegrass swards (17.9% of total grassland) were also relatively abundant.

Approximately 50% of all grassland was conserved for hay or silage with the most species-rich parcels being cut for hay in late June or early July. Suckler cattle and beef production were common forms of agriculture on the smaller and less intensive farms, whilst dairying was associated with the more intensive enterprises.

Table 4. The distribution and management of a) semi-natural vegetation and
b) agricultural grassland in the Upper Lough Erne area.

a) Vegetation type	Percentage mean area of land unit	Percentage mean area of vegetation Grazed	Disturbed
Freshwater vegetation	1.2	0	0
Reedbeds	2.4	10	1
Swamp	2.2	10	1
Fen	2.1	66	9
Water inundation vegetation	0.9	92	10
Rush pasture	1.3	86	10
Species-rich wet grassland	9.7	39	7
Ditch fen	0.3	10	0

b)	Percentage mean area of land unit	Conserved	Rush infested	Broadleaf weed infested	Rush and broadleaf weed infested
Italian ryegrass	0.2	74	0	45	0
Perennial	17.9	62	14	4	6
Mixed species grassland	11.1	12	89	52	48
Other agric. grassland	35.9	54	68	21	12
Total agric. grassland	65.1	49	57	33	16

Management issues

The present distribution and composition of the semi-natural vegetation of the
farmed landscape of Upper Lough Erne is strongly influenced by agricultural
practices. Whilst traditional methods of grassland management have largely
created and maintained parcels of ecologically interesting grassland and semi-
natural vegetation, current methods of intensive farming are reducing this
interest and have destroyed it over large areas.

Activities that are ecologically damaging include drainage, the application
of high rates of inorganic nitrogen, silage conservation, heavy grazing pressure
and grassland neglect, all of which decrease species diversity and benefit coarse
grasses, rushes and ruderal weeds.

There are three broad categories of vegetation around which prescription
can be modelled:-

a) Semi-natural vegetation: such as fen (fringing water bodies and in drainage ditches) and woodland. In terms of farm productivity, the contribution of these vegetation types is small but they are easily damaged by the trampling and grazing activities of farm stock. Prescription involves restricting access in species-rich locations, either to certain areas or to certain times of the year, by fencing. This is an operation that farmers are used to. It is unobtrusive and is not costly in relation to the benefits to wildlife. There are advantages to the farmer in that there is a danger of losing animals in soft ground at the waters edge if it is not fenced. Restricting grazing, however, could lead to the encroachment particularly of alder (*Alnus glutinosa*), an increase in species dominance, for example sedges, and the loss of regeneration niches.

b) Species-rich wet grassland and hay meadows: that are integral to the farm economy but on which there are severe environmental constraints on productivity, associated mainly with flooding, poor drainage and infertility. Prescription involves identifying species-rich grasslands and imposing restrictions on the degree and timing of grazing and grassland maintenance operations which ideally should be tailored to specific grassland parcels. At its simplest, however, prescription advocates more hay cutting and less intensive management on grassland parcels subject to severe constraints on farming. Such management under an ESA scheme may be welcomed by farmers as the sensible one on difficult land.

c) Agricultural grassland: ranging from productive swards to rush infested grassland in various stages of neglect. These are botanically uninteresting but provide habitats for birds such as the corncrake *Crex crex*, breeding waders (specifically snipe *Gallinago gallinago*, curlew *Numenius arquata*, lapwing *Vanellus vanellus* and redshank *Tringa totanus*) and wintering wildfowl, for example whooper swan *Cygnus cygnus*. Prescription for these grasslands involves considering of the timing of grassland management in relation to the breeding and feeding activities of the birds. Rush cutting on neglected grasslands is an example. It is carried out to maintain grass but results in the loss of cover necessary for nesting breeding waders and would be detrimental if implemented at the wrong time.

One of the main requirements for a successful ESA management prescription is that it should be practical for farmers to carry it out in a way that does not significantly upset the farm regime. In Fermanagh there is clearly much potential for integrating agriculture and wildlife conservation.

Field boundary structure and function

A characteristic feature of the Irish landscape is its field boundaries. They have an agricultural function but are also valuable wildlife habitats as well as contributing to the character of the landscape. Webb (1985), for example, has calculated that hedgerows occupy an area three times that of deciduous woodland in Ireland.

Throughout Northern Ireland, we have been recording the composition and structure of field boundaries across a wide range of landscape types and are able to use these attributes to describe their agricultural function and ecological status and to assess changes that are likely to take place if current agricultural practice continues.

Each boundary present within the 25 ha sample squares was mapped and assigned to a type based on structural and management attributes. The main species of shrubs, small trees (2-5 m) and large trees (>5 m) were also recorded.

Regional variation

The main types of field boundary present in the lower elevation farmed land classes present in the Mourne, Antrim Coast and Glens and Sperrins AONB's and Fermanagh District are shown in Table 5. The land classes present in each area differed in terms of their specific characteristics but were broadly comparable in consisting of the better types of farmed land, predominantly grassland.

Boundaries consisting entirely of post and wire fencing were ubiquitous (Table 5), constituting between 10-20% of the total. This type of fencing is much more extensive, however, since it is also added to many hedges and walls to make them stockproof. Earth banks were characteristic of the Sperrins and Antrim Coast and Glens AONB's, where there is sandy glacial till and coastal exposure respectively.

Dry stone walls were particularly abundant in the Mourne AONB, reflecting its geology, but in all regions, the proportion of ruined to maintained dry stone walls was high. Few hedgebanks were stockproof, this being defined as a hedgebank with <10% gaps. The majority were gappy or had degenerated into lines of overgrown scattered shrubs and trees. In Fermanagh, overgrown hedges, usually with a high proportion of trees, were a feature. Factors contributing to this are the mild oceanic climate, the bank and ditch structure of most field boundaries and the predominance of cattle rather than sheep farming.

Hedgebanks, dry stone walls and earth banks were originally constructed and maintained to control stock and for land division. The maintenance of traditional field boundaries, however, does not constitute a significant element of modern farming practice with its emphasis on mechanization. Since stock control is now almost entirely with wire fencing, traditional field boundaries are threatened either by dereliction or with removal. Throughout the study area boundary removal, calculated from the most recent series of Ordnance Survey maps, has been about 0.5% annually.

Variation with elevation

Major factors affecting the composition, structure and management of field boundaries within each of the AONB's were those associated with changes in topography and elevation from lowland to transitional hill land and then to the

Table 5. Field boundary type, structure and management in the lower elevation farmed land classes in the Mourne, Antrim Coast and Glens and Sperrins Areas of Outstanding Natural Beauty and Fermanagh District.

	Mourne	Antrim	Sperrins	Fermanagh
Wood post and wire fence	11.2	21.4	15.1	11.5
Earth bank	-	21.0	34.4	4.4
Stockproof dry stone wall and dry stone wall with gaps	32.5	4.2	2.6	-
Ruined dry stone wall	13.5	7.8	3.8	0.4
Stockproof hedgebank	5.9	4.2	14.4	8.7
Hedgebank with gaps	12.7	5.8	3.8	7.5
Hedgebank with scattered or overgrown shrubs	12.5	27.3	16.5	45.4
Other boundary	11.7	8.3	8.5	22.0

uplands. Field boundary characteristics in the Mourne AONB land classes are shown in Table 6. The land classes 1-4 represent a trend of increasing elevation from coastal areas (land class 1) and other low elevation farmland (land class 2), to higher elevation largely enclosed, rolling farmland (land class 3). Land class 4 represents transitional hill land with its greater cover of semi-natural vegetation, whereas land classes 5 and 6 correspond to the steeper mountain slopes and uplands.

Dry stone walls were particularly frequent in land class 2, where soils are often derived from sandy glacial deposits in which granite boulders are common. A considerable proportion of these walls were maintained as stockproof boundaries and they make a significant contribution to the character of the landscape. At higher elevation, dry stone walls were also predominant but with increasing elevation a much greater proportion were derelict.

There were relatively few stockproof hedgebanks, the majority consisting of lines of scattered, often overgrown hawthorn (*Crataegus monogyna*). The contribution of hedgebanks in the higher elevation land classes was less and in the mountains, post and wire fencing was common. The use of post and wire fencing in the mountains is a recent trend reflecting changing land holding patterns that are moving from a system of common grazing to land ownership. It detracts from the wilderness quality of the uplands.

Table 6. Field boundary type structure and management in land classes of the Mourne Area of Outstanding Natural Beauty.

	Land Class					
	1	2	3	4	5	6
Wood post and wire fence	15.7	11.2	11.2	7.9	14.0	69.9
Stockproof dry stone wall and dry stone wall with gaps	10.2	32.5	18.7	14.8	10.6	-
Ruined dry stone wall	4.2	13.5	11.9	34.9	37.2	30.1
Stockproof hedgebank	7.1	5.9	3.2	3.8	0.3	-
Gappy hedgebank	13.2	12.7	6.6	1.1	0.8	-
Hedgebank with scattered or overgrown shrubs	36.5	12.5	36.5	32.0	30.7	-
Other boundary	13.1	11.7	10.9	5.7	6.4	-

Estimates of field boundary dereliction emphasize the magnitude of any attempts at rehabilitation or management. There was a mean frequency of 151 field boundaries per square kilometre, equivalent to 6.0 km of dry stone wall and 6.3 km of hedgebank. About 50% of dry stone walls were neglected, amounting to 1 823 km in the AONB as a whole. Gappy hedges and lines of shrubs occupied 2 518 km. Field boundary removal from 1977 to 1986 was 4%.

Hedge management issues

Field boundaries are being lost or becoming derelict on an enormous scale throughout the most outstanding landscapes in Northern Ireland and in areas where ecological interest is correspondingly high. This raises a number of management issues. Firstly, should anything be done to maintain the landscape and wildlife value of field boundaries and how could this be carried out? Secondly, who will pay for and implement management? It is a major task that has been given some stimulus from recent changes in agricultural policy leading to the designation of the enclosed parts of the Mourne AONB as an ESA (DANI 1987).

Hedge management with a view to maintaining traditional structures and wildlife is labour intensive and skilled. It does not occur to any significant extent within the other Areas of Outstanding Natural Beauty, or in Fermanagh District. Certain individuals, however, within the farming community are motivated by aesthetics or an interest in wildlife and there is minor involvement of voluntary conservation organizations. On the more prosperous farms, hedges are kept tidy by trimming to a flat top, an operation that does not maintain their structure. Elsewhere hedges are simply cut back when they become overgrown. Scrub encroachment onto grassland and shading are the main reasons for doing this. Cutting overgrown hedges at ground level can lead to hedge loss, particularly if grazing by sheep follows.

Management strategies are needed that consider both regional and local variation. Field data stratified by land class and presented for particular landscape units can be used for this purpose. Proposals should accept that change is an integral part of economic agriculture and that not all field boundaries have a high ecological or landscape value.

Discussion

Multivariate land classification provides an explicit framework within which integrated resource management and monitoring studies can take place (Bunce and Smith 1978). Stratified field sampling programmes linked to land classification are efficient and cost effective. They can be extrapolated to obtain resource estimates and predictions of change for the whole of the study area (Bunce and Heal 1984).

Our work on monitoring ecological and landscape change based on land classification, is in its early stages. Most of the work so far has been on baseline studies that have been used to provide quantitative data for generating AONB and ESA management guidelines by inference (DoE NI 1988; 1989).

The types of change that are being monitored are largely those that result from abrupt land use change. In relation to long term ecological trends, these changes are highly significant but they are often related to and can point to other, more subtle ecological changes that are also of interest to land managers. These include the steady decline in quality of field boundaries as an ecological and landscape resource. It has been an integral part of our research to interact with government agencies, passing on this information and receiving feedback on its relevance to their research priorities. This is an aspect of environmental research that is often neglected by scientists but its importance cannot be over emphasized.

Our microcomputer database is currently being used regularly to extract information related to the abundance and nature of environmental resources. Peatland exploitation studies in the Sperrins AONB, for example, are currently being supplemented by a monitoring exercise to examine changes in the extent, rate and ecological effects of mechanical peat cutting. All sample grid squares within a specified group of upland land classes first surveyed in 1988 have been resurveyed between May and July 1990. This illustrates the cost-effectiveness of the technique. We consider this type of application of the database, i.e. rapid survey of a fairly common resource directed at a specific objective, to be one of the main strengths of our approach. The comprehensive nature of the original baseline means that monitored information can be interpreted in appropriate ecological context. This is also the case when interpreting other schemes for monitoring species distribution and population changes, for example Perring and Walters (1962).

The database summarizes the current status of environmental resources and provides a structured framework on which integrated land use and management policies can be developed. It is also a source of information from which guidelines on local management issues can be derived. In Fermanagh District, for example, the Upper Lough Erne basin has been identified as a potential ESA on the basis of comparative environmental resource studies. Management prescription guidelines have been derived from the landscape ecology database using indicator attributes based on both the composition and the structure of semi-natural vegetation. Similarly, in the Mourne ESA we have selected simple field boundary management criteria based on their structure and composition that can be used as guidelines for the preparation of farm conservation plans.

If ESA designation proceeds in Upper Lough Erne, the database will be used to monitor the success of the scheme by comparing areas that have taken part in the scheme with areas in the same land class, that have not. The first opportunity to monitor the success of ESA designation may be for the Mourne ESA in 1992 when the first set of five year farm plans implemented under the scheme are due for renewal.

The emphasis of our research until now has concentrated on field survey and the preparation of database files for the more scenic and species-rich areas. Further work will involve extending the field survey of sample squares to the wider countryside and the completion of a sample set for the whole of Northern Ireland. Subsequent analysis will involve comprehensive land classification and landscape unit description for the province and the creation of an integrated data management, monitoring and mapping system.

Acknowledgements

The research was funded by the Countryside and Wildlife Branch of the Department of the Environment for Northern Ireland. Andrew Stott provided helpful discussion throughout. Joyce Forsythe and Debbie Rainey assisted with aspects of data preparation. Kilian McDade and Nigel McDowell did the reprographics. The authors recognize these contributions and thank everyone for their help.

References

Aalen, F.H.A. (1985) The rural landscape : change, conservation and planning. In F.H.A. Aalen (ed.) *The Future of the Irish Landscape*, 1-25. Trinity College, Dublin.

Ashon-Tate (1985) *dBASE III PLUS*. Torrance. Ashton-Tate

Bunce, R.G.H., Barr, C.J. and Whittaker, H. (1983) A stratification system for ecological sampling. In R.M. Fuller (ed.) *Ecological Mapping from Ground, Air and Space*, 39-46. Institute of Terrestrial Ecology Symposium Number 10. Institute of Terrestrial Ecology, Cambridge.

Bunce, R.G.H. and Heal, O.W. (1984) Landscape evaluation and the impact of changing land-use on the rural environment : the problem and an approach. In R.D. Roberts and T.M. Roberts (eds.) *Planning and Ecology*, 164-188. Chapman and Hall, London.

Bunce, R.G.H. and Smith, R. (1978) *An Ecological Survey of Cumbria*. Cumbria County Council and Lake District Special Planning Board, Kendal.

Cooper, A. (1986) *A System of Land Classification for the Mourne ANOB*. Report to the Countryside and Wildlife Branch, Department of the Environment for Northern Ireland. Department of Environmental Studies, University of Ulster, Jordanstown.

Cooper, A. (1988) Ecological survey, databases and environmental impact assessment. In W.I. Montgomery *et al.* (eds.) *The High Country : land use and land use change in Irish uplands*, 92-102. Institute of Biology and Geographical Society of Ireland, Belfast.

Cooper, A. and Murray, R. (1987a) *A Landscape Ecological Study of the Mourne AONB*. Report to the Countryside and Wildlife Branch, Department of the Environment for Northern Ireland. Department of Environmental Studies, University of Ulster, Jordanstown.

Cooper, A. and Murray, R. (1987b) *A Landscape Ecological Study of the Antrim Coast and Glens and Causeway Coast Areas of Outstanding Natural Beauty*. Report to the Countryside and Wildlife Branch, Department of Environment, Northern Ireland. Department of Environmental Studies, University of Ulster, Jordanstown.

Cooper, A., Murray, R., McCann, T. and Forsythe, J. (1988a) *A Landscape Ecological Study of the Sperrins and North Derry Areas of Outstanding Natural Beauty*. Report to the Countryside and Wildlife Branch, Department of Environment, Northern Ireland. Department of Environmental Studies, University of Ulster, Coleraine.

Cooper, A., Murray, R. and McCann, T. (1988b) *The Mourne Environmentally Sensitive Area Land Use Database*. Report to the Department of Agriculture for Northern Ireland. University of Ulster, Coleraine.

Cooper, A., Taylor, D. and Murray, R. (1989) *A Land Use and Ecological Database of the Glens of Antrim Environmentally Sensitive Area*. Report to the Department of Agriculture for Northern Ireland. University of Ulster, Coleraine.

Cruickshank, J.G. and Wilcock, D.N. (1983) *Northern Ireland: Environment and Natural Resources*. Queen's University Belfast and University of Ulster, Belfast.

Department of Agriculture for Northern Ireland (1987) *The Mourne Environmentally Sensitive Area*. DANI, Belfast.

Department of the Environment for Northern Ireland (1988) *Antrim Coast and*

Glens Area of Outstanding Natural Beauty : guide to designation. HMSO, Belfast.

Department of the Environment for Northern Ireland (1989). *Mourne Area of Outstanding Natural Beauty : policies and proposals*. HMSO, Belfast.

Eastman, J.R. (1988) *IDRISI : a grid-based geographical analysis system*. Clark University, Massachusetts.

Partridge, J.K. (1987) *Northern Ireland Breeding Wader Survey*. Report to the Department of the Environment, RSPB, Belfast.

Perring, F.H. and Walters, S.M. (1962) *Atlas of the British Flora*. Nelson, London.

Webb, R. (1985). Farming and the landscape. In F.H.A. Aalen (ed.) *The Future of the Irish Landscape*, 80-92. Trinity College, Dublin.

The Use of Radish as a Bioindicator in an International Programme for Evaluating the Effects of Air Pollution on Agricultural Crops

M.B. Jones[1], C.E. Booth[2] and E. Shanahan[1]

[1]Department of Botany, Trinity College, Dublin 2, Ireland
[2]Department of Physiology and Environmental Science, School of Agriculture, University of Nottingham, Sutton Bonington, Loughborough, U.K.

Key words: ozone, *Raphanus sativus*, radish, EDU, growth, biomass, biomonitor.

Abstract

Tropospheric ozone (O_3) concentrations in Europe and North America have possibly doubled during the last few decades. When climatic conditions are suitable, substantial amounts of O_3 are produced in the form of "ozone episodes", as a result of anthropogenic activity. The workplan for the United Nations Economic Commission for Europe (UNECE) Convention on Long-Range Transboundary Air Pollution includes the requirement for international cooperative programmes to evaluate the effects of air pollution in a number of impact areas including agricultural crops.

   In 1987 a programme on agricultural impacts was initiated with the objective of evaluating the effects of ambient levels of O_3 on crop production. The aim was to allow the participation of all UNECE countries using a common experimental protocol and a standard plant species as a biomonitor. The plant species selected was radish (*Raphanus sativus* L. cv. Cherry Belle) and the basic experiment involved the use of the antioxidant ethylenediurea (EDU), which is thought to protect plants from O_3 damage. Experiments carried out in a rural area of Ireland during 1989 and 1990 using the common protocol showed that episodes in July of each year were sufficient to affect the growth of radish plants. EDU provided some protection against O_3 damage but the amount of protection appears to be variable.

Introduction

There is evidence that tropospheric ozone concentrations in Europe and North America have increased substantially during the last few decades, possibly by a factor of two (Penkett 1988). The lifetime of ozone in the troposphere is short (a

Bioindicators and Environmental Management
ISBN 0-12-382590-3

few weeks), and the variable source and sink strengths of different regions of the world results in substantial spatial variability in ozone concentrations. This spatial variability makes it difficult to extrapolate local trends in concentration to global scales.

The sources of ozone (O_3) in the troposphere include transfer from the stratosphere and photochemical ozone production within the troposphere. The production of ozone in the troposphere results from the photolysis of nitrogen dioxide:

$$NO_2 \longrightarrow NO + O \text{ (for wavelengths} < 400nm)$$
$$O + M + O_2 \longrightarrow O_3 + M$$

Where M can be any molecule such as nitrogen or oxygen, which dissipates the energy released in the reaction and thereby prevents the ozone decomposing. The ozone produced may react with nitric oxide to form NO_2 and a photostationary state is established. However, when any nitric oxide is oxidized to nitrogen dioxide by other processes, net formation of ozone occurs. When this occurs it is primarily through the reaction of peroxy radicals, produced by the photochemical degradation of hydrocarbons, reacting with nitric oxide to produce nitrogen dioxide. This process, occurring during daylight, shifts the equilibrium in favour of ozone production. Consequently, when climatic conditions are suitable, substantial amounts of ozone are produced as a result of anthropogenic activity. These 'ozone episodes' have been shown to occur throughout Europe, particularly when anticyclonic weather conditions prevail in summer over northern Europe (Colbeck 1988). The ozone concentrations produced at these times in rural areas of Britain and Ireland exceed 60 ppb and have been shown to be high enough to damage plants (Ashmore et al. 1978; 1980).

Surveys of damage to plants by ozone have, until now, been limited in their time scale and geographical distribution. However, the workplan of the United Nations Economic Commission for Europe (UNECE) Convention on Long-Range Transboundary Air Pollution includes the requirement for international cooperative programmes to evaluate the effects of air pollution in a number of impact areas including agricultural crops. The agricultural impact programme aims to (i) evaluate the effects of ambient concentrations of air pollution on crop production, (ii) assess the importance of environmental conditions in affecting crop response to air pollutants, and (iii) determine the influence of different aspects of pollutant exposure (e.g. combinations of pollutants, concentration peaks and means) on crop response. The rationale for this approach is that although atmospheric pollution may be transported over long distances, its effects on plants may be quite variable across regions or continents as a result of different exposure conditions. Although some progress has been made in the development of exposure-response relationships for various crops and pollutants, the accuracy of predictive models developed under one set of conditions is uncertain under another set of conditions. For example, plants exposed to conditions producing drought stress which induces stomatal closure may be less sensitive to ozone damage than those not subject to such stress.

It is envisaged that the formation of both national and international pollution control strategies will be based on the results of the UNECE cooperative programmes which will be presented in terms of critical loads and critical levels of air pollutants. Although critical loads do not, as yet, appear to be very precisely defined they are clearly a more complex parameter than critical levels which have been used in the establishment of air quality standards quite widely already. Critical levels can be defined as the maximum concentration of a pollutant at which adverse effects will not occur on sensitive targets or receptors. However, the response of receptors is complex and depending on exposure conditions, the physiological state of the target and the co-occurrence of more than one pollutant, the critical levels for each region could be quite different. The UNECE Convention on Long-Range Transboundary Air Pollution has already established a Task Force on Mapping Critical Loads/Levels. The long-term critical levels for SO_2, NO_2 and O_3 which are currently being used in the mapping exercise are respectively 11 ppb (annual mean), 15 ppb (annual mean) and 25 ppb (9.00-16.00 hrs summer mean) (Ashmore *et al.* 1990). However, it is clear that these values will have to be adjusted on a regional basis in future to take account of new information concerning receptor sensitivity. It will also probably be necessary to classify areas into vegetation or land use types in order to be more precise about the response at any one location. Ultimately it should be possible to compare critical levels across Europe with actual pollutant emissions and identify areas where critical levels are exceeded.

As far as agricultural crops are concerned it will be necessary to take into account the differing sensitivity of the major crops to pollutants. Even though agriculture is largely dependent on a very small number of plant species, there is considerable variation across Europe in the crops grown and in the time at which they grow. Because of this, it is probably more realistic in the first instance to identify a single species which could be used widely as a convenient bioindicator of phytotoxic effects of pollutants. Bioindicators are biological systems which show a quantifiable response to a perturbation of their environment. With respect to air pollution, their growth and development is influenced in a predictable fashion by the presence of the pollutant, changes taking place in characteristics such as dry weight increase of shoots and roots, development of leaf area, foliar damage and physiological characteristics such as net photosynthesis, transpiration and stomatal conductance.

The selection of plant bioindicators for pollution work is dependent on characteristics of both the plant and the environment in which it will be grown. The aim is to use a species which is representative of the vegetation type for which critical levels of pollutants will be established. Plant species have long been known to show different physiological responses and different types of foliar injury in response to air pollutants. In general, plant species are often classified as sensitive or insensitive to air pollutants, but even within a species there are cultivars or individuals which differ in their sensitivity. The choice of plant bioindicators should therefore depend on their position in the spectrum of sensitivities and, on the whole, they should represent the more sensitive species. It is also important that, if a bioindicator is used in different locations for comparison of responses, then the same variety and seed source should be used. Other favourable characteristics of plant bioindicators are (i) a short growing cycle and the ability to harvest the plant after short time intervals, (ii) a structure which enables relatively easy separation into shoot and root components, and (iii) a leaf form which permits non-destructive physiological measurements such as photosynthesis and transpiration. A species with many of these characteristics which has been used quite widely in work on pollution effects on plants is radish (*Raphanus sativus* L.) (Tingey *et al.* 1971; Reinert *et al.* 1972). Consequently a sensitive variety of radish cv. 'Cherry Belle' (Gillespie and Winner 1989) was chosen as a suitable plant to use for the UNECE international cooperative programme to evaluate the effects of air pollutants on agricultural crops in Europe and North America.

We describe here the basic form of the cooperative experimental programme on agricultural impacts which was initiated in 1987 and continued in subsequent growing seasons. A minimum of two treatments were used; one group of plants was treated with an antioxidant which is though to protect plants from ozone damage, and the others were untreated. The same cultivar of radish (*Raphanus sativus* L. cv. 'Cherry Belle') was used at all sites and the effect of ambient ozone levels in producing visible injury and changes in growth and yield was investigated. Results obtained in Ireland are used to illustrate some of the principal conclusions which can be drawn from the work so far.

Methods

In order to assess the impact of ozone on crop yield the basic experiment involved the use of an antioxidant, ethylenediurea (EDU), which is thought to protect plants from ozone damage. The biochemical basis of this protection is not known but there is one report of increased levels of the oxyradical scavenger, superoxide dismutase, in EDU-treated bean (*Phaseolus vulgaris*) (Lee and Bennett 1982). EDU has been applied both as a foliar spray and soil drench. In a study with 40 plant species both methods provided equal protection (Cathey and Heggestad 1982). There has, however, been little or no standardization of EDU application methods or application rates. In most cases several applications have been made during the growth of plants, application intervals ranging from 7 to 21 days (Heagle 1989).

Although it was felt that EDU should provide a simple and effective way of protecting plants from ozone damage, other methods of providing protection from ozone were also used when available because the mechanism of EDU action and possible side effects are not well understood. The alternatives normally involved growing the plants in open-top-chambers (OTC's) where the air entering was filtered through activated charcoal which removed most of the ambient levels of ozone.

The experimental procedure was developed at the co-ordinating centre at the Department of Physiology and Environmental Science, University of Nottingham. All centres involved in the co-operative programme adopted the same protocol. Radish seeds were planted four per pot (10cm x 10cm) at a depth of 0.7 cm. After germination the seedlings were thinned to one per pot. The growing medium was a mixture of 3 parts vermiculite with 1 part pure clay (cat litter). To each m^3 of growing medium was added 3.4 kg of slow release fertilizer (14:14:14 NPK 'Osmocote') and 3.3 kg calcium sulphate. The experiment in Ireland was carried out in 1989 and 1990 at Oak Park Research Station, Carlow, using an open-top chamber facility described by Jones *et al.* (1989). After planting, ten pots were placed in each of four charcoal filtered chambers and four unfiltered chambers. EDU was applied as a soil drench to half the pots using 100 ml per pot. In 1989 a 200 ppm solution was applied for the first time when the first true leaves were 1-2 cm long and for the second time 14 days after. In 1990 a 150 ppm solution of EDU was used.

The growing medium water content was kept at field capacity by standing the pots overnight in trays of water at a depth of approximately 3 cm. The irrigation schedule was adjusted according to the weather conditions in order to avoid either water stress or water logging. Pots were not irrigated on the day of EDU treatment or the day after.

Visual observations of injury were made at weekly intervals by estimating separately the percentages of leaf area showing chlorosis and necrosis. After 28 or 35 days from planting the plants were harvested by removing the whole plant from the growing medium and washing off attached soil particles. The plants were divided into above ground (shoots) and below ground (hypocotyl and roots) components and the fresh weight determined. The leaf area was determined using a leaf area meter (Delta-T Devices, Cambridge), plant parts were dried at 80°C for 48 hours and then weighed.

Daily meteorological measurements were recorded at a weather station adjacent to the experimental site. Ambient levels of ozone, SO_2 and NO_x were determined using Monitor Labs analyses. Ozone concentrations were recorded as 7 hour (9.00-16.00) and daily means.

Results and discussion

Results obtained for the radish crops grown during the month of July in 1989 and 1990 are presented. In both years these were months when ozone episodes were observed at the experimental site (Fig. 1). In 1989 there were two distinct episodes. The first was at its peak at the time of seedling emergence (5-6 July) and the 7 hour mean reached 80 ppb while the maximum 24 hour mean was 55 ppb. The second episode occurred during the third week of growth of the radish plants. The peak concentrations were probably lower than in the first episode but two days of records around the time of the maximum ozone values were lost due to equipment failure. In 1990 two ozone episodes were also recorded, the first peaking at a 7 hour mean of 56 ppb on the 17th day after sowing and the second reaching 48 ppb on the 24th day after sowing. During these months the concentration of SO_2 and NO_x remained below 5 ppb, which is the limit of sensitivity of the analyzers used.

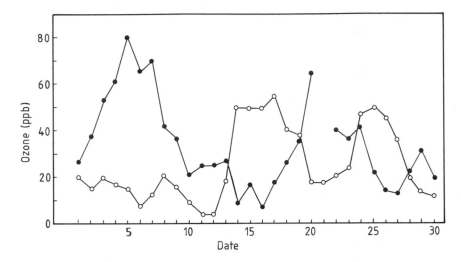

Figure 1. The daily 7-hour (9.00-16.00) means of ozone concentrations during the month of July in 1989 (closed circles) and 1990 (open circles) at Oak Park Research Station, Carlow, Ireland.

In 1989 the EDU treatment gave no protection against ozone damage, and actually caused a significant reduction in growth of the radish plants. Subsequently it was found that the EDU was not chemically pure (Dr N. Ainsworth, pers. comm.). Filtration of the air by activated charcoal in the open-top chambers removed most of the ozone and plants grown in these chambers produced significantly (p<0.01) more fresh weight than those in the non-filtered chambers (Fig. 2). No visual symptoms of ozone damage were observed. The reduction in fresh weight yield in the unfiltered treatment for the hypocotyl+roots was greater (-17.0%) than for shoots (-9.7%).

Reductions in dry weight in the unfiltered compared with the filtered treatment were observed for both hypocotyl+roots (-14.5%) and shoots (-7.5%) but these differences were not statistically significant. There was, however, a statistically significant (p<0.005) reduction in leaf area in the unfiltered treatment. The shoot to root ratio calculated on both a fresh weight and dry weight basis was reduced in filtered air but neither of these differences were statistically significant.

In 1990 the yield of the radish plants was also reduced in the unfiltered treatment (Fig. 3) but in this case the dry weight differences were statistically significantly (p<0.001). Leaf area was reduced by a similar amount in both years (-16.8% and -22.0% respectively). In 1990 the EDU had no effect on the yield of the plants in the filtered chambers (Fig. 3). In the unfiltered chambers the reduction in yield of the EDU-treated plants was significantly less (p<0.001) than

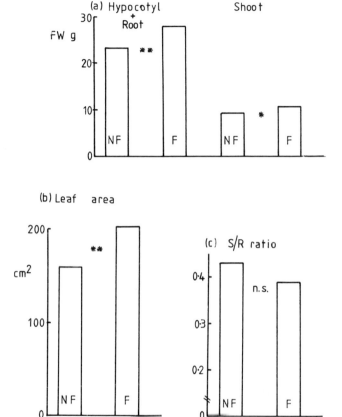

Figure 2. The effect on radish plants of removing ambient ozone in charcoal filtered (F) compared with non-filtered (NF) open-top chambers during July 1989. a) hypocotyl plus root and shoot fresh weight; b) leaf area; and c) shoot to root ratio. (*, p<0.05; **, p<0.005; n.s. not significant).

for the non-EDU treated plants (Fig. 3). The reduction in dry weight of the hypocotyl+roots in the EDU treated plants was -18.5% compared with -23% in the untreated plants. The EDU treatment therefore appears to confer some protection against ozone damage in the unfiltered air treatment.

The amount of protection provided by EDU in the present treatment is apparently far from complete. In previous work the variable effectiveness of EDU has been noted. For example Toivonen *et al.* (1982) found that the amount of protection of white bean ranged from 80% to <20%, and Heagle (1989) has pointed out that the amount of EDU protection from ozone injury has not so far been quantified for any crop. If EDU is to be used in assessing the impact of ozone on crop yield it should give a predictable level of protection from ozone damage. There are clearly a number of factors which can determine the

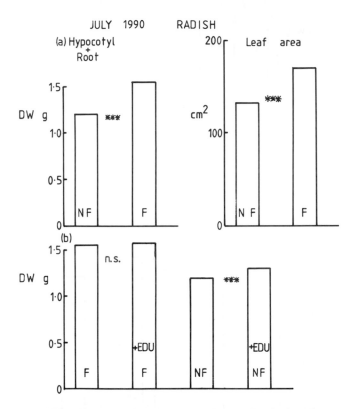

Figure 3. a) The effect of removing ambient ozone in charcoal filtered (F) compared with non-filtered (NF) open-top chambers on radish hypocotyl plus root dry weight and leaf area; b) the effect on hypocotyl plus root dry weight of treating radish plants with a soil drench of ethylenediurea (EDU) in charcoal filtered (F) compared with non-filtered (NF) open-top chambers. (*** p<0.01; n.s. not sifgnificant).

effectiveness of EDU used as a soil drench. One of the most important is likely to be the soil characteristics, but climate conditions and growth rate also probably have an influence. It is therefore important to establish dose response relationships for radish plants grown under controlled environment conditions in a uniform soil type. Under these conditions it should be possible to assess quantitatively the protection against ozone damage afforded by EDU.

The experiments carried out in Ireland in 1989 and 1990 have shown that in the month of July in both years the ozone concentrations appear to be high enough to reduce the yield of Cherry Belle radish plants. Several yield parameters were used but hypocotyl fresh weight and leaf area were the only two for which there were statistically significant differences between charcoal filtered treatments in both years. In subsequent months in both years, filtering the air to remove ozone had no significant effects on yield. During these months the 7 hour mean values of ozone were consistently lower than in July. On the basis of these observations it would appear that the critical level of ozone required to bring about a reduction in radish yields at this location corresponds to the occurrence of 7 hour means in excess of about 80 ppb during episodes which last at least 4 or 5 days. These results also suggest that Cherry Belle radish plants are sensitive enough biomonitors to be used in coordinated programmes to evaluate the effects of air pollution, particularly ozone, on crop plants. However, similar measurements at other centres have produced more ambiguous results (unpublished) and the reasons for this need to be investigated further.

The use of EDU to protect radish plants against ozone damage, and therefore to provide a control treatment, has been shown to be less than fully effective. However, even if EDU only provides partial protection it is still possible to use it to show that ozone has a damaging effect on plants, even though it is not possible to quantify this effect unless EDU provides complete protection. The use of Cherry Belle radish in conjuction with the ozone protectant EDU therefore provides the opportunity to carry out geographically wide-ranging surveys of the effects of ozone on plants.

Acknowledgements

We thank the U.K. Department of the Environment for funding the Nottingham co-ordination centre and providing supplies of EDU.

References

Ashmore, M.R., Bell, J.N.B. and Reilly, C.L. (1978) A survey of ozone levels in the British Isles using indicator plants. *Nature*, **276**, 813-815.

Ashmore, M.R., Bell, J.N.B. and Reilly, C.L. (1980) The distribution of phytotoxic ozone in the British Isles. *Environmental Pollution* (Series B), **1**, 195-216.

Ashmore, M.R., Bell, J.N.B. and Brown, I.J. (1990) *Air pollution and forest health in the European Community Air Pollution Series of Environmental Research Programme, Report 29.* CEC Brussels

Cathey, H.M. and Heggestad, H.E. (1982) Ozone sensitivity of herbaceous plants: Modification by ethylenediurea. *Journal of American Society of Horticultural Science*, **107**, 1035-1042.

Colbeck, I (1988) Photochemical ozone pollution in Britain. *Science Progress*, Oxford, **72**, 207-226.

Gillespie, C.T. and Winner, W.E. (1989) Development of lines of radish differing in resistance to O_3 and SO_2. *New Phytologist*, **112**, 353-361.

Heagle, A.S. (1989) Ozone and crop yield. *Annual Review of Phytopathology*, **27**, 397-423.

Jones, M.B., Jackson, N. and Richardson, D.H.S. (1989) The growth of agricultural crops in open-top chambers in rural Ireland. In Bonte, J. and Mathy, P. (eds.) *Air Pollution Series of Environmental Research Programme, Report 19*, 157-167. CEC Brussels.

Lee, E.H. and Bennett, J.H. (1982) Superoxide dismutase. A possible protective enzyme against ozone injury in snap beans (*Phaseolus vulgaris*). *Plant Physiology*, **69**, 1444-1449.

Penkett, S.A. (1988) Indications and causes of ozone increase in the troposphere. In F.S. Rowland and I.S.A. Isaksen (eds.) *The Changing Atmosphere*, 91-103. Wiley, Chichester.

Reinert, R.A., Tingey, D.T. and Carter, H.B. (1972) Ozone induced foliar injury in lettuce and radish cultivars. *Journal of the American Society for Horticultural Science*, **97**, 711-714.

Tingey, D.T., Heck, W.W. and Reinert, R.A. (1971) Effects of low concentrations of ozone and sulphur dioxide on foliage growth and yield of radish. *Journal of the American Society for Horticultural Science*, **96**, 369-371.

Toivonen, P.M.A., Hoftra, G. and Wukash, R.T. (1982) Assessment of yield losses in white bean due to ozone using the antioxidant EDU. *Canadian Journal of Plant Pathology*, **4**, 381-386.

Mussel as a Test Animal for Assessing Environmental Pollution and the Sub-lethals Effect of Pollutants

J. Salánki[1], T.M. Turpaev[2] and M. Nichaeva[2]

[1]Balaton Limnological Research Institute of the Hungarian Academy of Sciences, H-8237, Tihany, Hungary.
[2]Institute for Development Biology of the Academy of Sciences, USSR, Moscow.

Key words: mussel, filtering activity, heavy metals, toxicity test.

Abstract

Freshwater and marine mussels are proving to be versatile subjects as biomonitors of pollution. Mussels, especially the gills, accumulate heavy metals and other substances from the environment indicating both the occurrence and level of the given pollutant in water and sediment.

Recording the filtering activity of mussels (water flow through the siphon and/or opening and closing of the valves) gives good indication on the effect of pollutants in a concentration dependent manner.

By recording the water flow from the outflow siphon the effects of Cd, Cu, Hg, Pb and Zn ions were studied, and the usefulness of the method in testing the toxicity of anthropogenic pollutants in laboratory was demonstrated. The attenuation of rest periods, and/or the shortening of active periods, indicate harmful effects on the animals behaviour.

Introduction

Mussels (Pelecypoda) are widely distributed in aquatic ecosystems, and the very common structural and functional characteristics of different species living in the sea, in lakes and rivers, offer opportunities for their general application as bioindicators. On one hand they are comparatively tolerant of changes in the temperature, salinity, oxygen supply and chemicals by surviving adverse circumstances. On the other, however, they respond in a specific way to them, namely by blocking water flow through the siphon or by closing their shells for shorter or longer periods. Mussels are also good indicators of pollution because they accumulate substances from the water or food into their body, retaining

Bioindicators and Environmental Management
ISBN 0-12-382590-3

these for weeks and months. Mussels possess a long life span (10-15 years), they survive in laboratory conditions for months without special oxygen and food supply and being sessile do not need space for motility.

The behavioural reaction of mussels mentioned above has a very important ecological significance. When animals are active, they filter the water, take up oxygen and food whilst releasing excretory products, and simultaneously they clean the water. When, as a result of adverse conditions, they close their siphons or shells, not only feeding and respiration will be reduced or stopped, resulting in retardation of all metabolic processes, but also cleaning from the water of bacteria and detritus will decrease. Because of the high number of mussels in waters their filtering behaviour is important for the health of the aquatic ecosystems.

Use of mussels in detecting the presence of pollutants

It has been shown in a number of previous studies that all aquatic animals, including mussels, are capable of accumulating and storing organic and inorganic chemicals occurring in the water as anthropogenic pollutants. The so called "mussel watch" (Goldberg et al. 1978) is based on these observations. Our own measurements, concerning the occurrence of heavy metals in various animals of Lake Balaton, showed that the highest concentrations of Cd, Cu, Hg, Pb and Zn occurred in mussels (Salánki et al. 1982). Concentrations were especially high in their gills, surpassing several times the concentrations found in other organs of the same animal. Since a small part of the body is usually sufficient for measurement, in order to reduce error and facilitate comparison we recommend the use of the gills of mussels for detection of heavy metal pollution (Salánki and V.-Balogh 1989).

When using mussels, or other living organisms, for detecting the level of pollution of the environment one should take into consideration that uptake is not a simple, linear, one-directional process. The uptake of substances into the animal will depend on the substance, time of exposure, temperature and other circumstances. It will be partly counterbalanced by a depuration process. Differences in the measured concentrations in various animals and at different locations may reflect these circumstances and not the level of pollution. Our laboratory experiments with mussels showed that: (1) measurable accumulation occurs only after one day of exposure (shorter, episodic pollution will not be detected this way); (2) saturation does not occur under low heavy metal pollution within four to five weeks; (3) in metal free water, half-depuration time is longer than 2-3 weeks for most heavy metals in case of the gills (Salánki and V.-Balogh 1985).

These circumstances require care in the comparison and quantitative evaluation of the presence of pollutants in different areas. Nevertheless, measuring the concentration of anthropogenic chemicals in the gills of mussels is one of the best methods for monitoring the level of environmental pollution.

Use of mussels for the fast detection of sub-lethal toxic effect of pollutants

Recently a new method has been worked out for the fast testing of the adverse effect of toxic substances using adult mussels in the laboratory. The principle of the method is based on the observations that the inflow and outflow siphons are sensitive to certain chemicals and respond to them by closure (Lukacsovics and Salánki 1968). We constructed a rig in which the opening and closure, as well as the open or closed state of the outflow siphon, can be monitored by recording the water flow out of the siphon. Figure 1 shows the siphons in open (a) and closed (b) position. Figure 2 is the scheme of the recording system. When the siphons are open, because of ciliary and body movements, there is a flow of water in and out of the animal. To measure this flow, a small umbrella fixed to a fine lever is positioned 0.5 cm from the outflow. The water flow causes dislocation of the umbrella. The force of water flow is proportional to the opening of the siphon and can be recorded by using a transducer, amplifier and recorder system (Véró and Miller, 1979).

The equipment is quite sensitive, and even in the open state of the siphon small rhythmic waves can be recorded (Fig. 3a), corresponding to regular changes in the water flow. From time to time irregular, fast, brief closure of the siphon (Fig. 3b) occurs without any other visible reaction. At other times it is correlated with temporary closure of the shells (Fig. 3c), caused by fast contraction of the adductor muscles. In this case, a much stronger water flow is generated, causing a substantial dislocation of the umbrella. Periodically, but rarely, a shorter or longer closure of the siphon can be recorded (Fig. 3d) even in control conditions. We used *Anodonta* and *Unio* species only when the shells were open and the water flow from the siphon was not blocked by the shells.

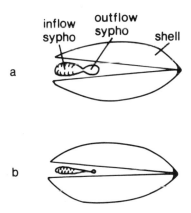

Figure 1. Schematic representation of the inflow and outflow siphons in open (a) and closed (b) position.

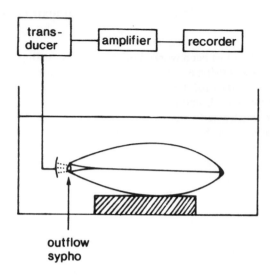

Figure 2. Set up recording the water flow coming from the exhalant siphon.

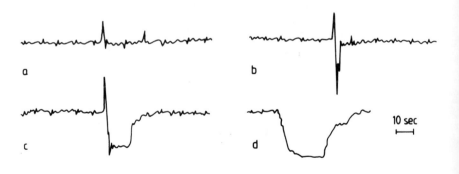

Figure 3. Types of water flow recordings in control conditions. Upward movement is increase of water flow, downward movement means decrease of water flow (corresponding to opening and closing of the siphon). a) the siphon is open and fine rhythmic changes in the force of water flow are present; b) at open state of the siphon a fast closing and opening preceded by increasing of the water flow occurs; c) 10 sec long fast closing of the siphon; d) 20 sec long slow closing and opening of the siphon.

The animal was placed in a 2 litre volume aquarium with a perfusion system. One valve of the animal was attached to a fixed stage lying on the bottom. Test chemicals were added to the water from stock solutions and the desired concentration was reached by fast mixing.

The perfusion system served to wash out the test substances. Consecutive testings were made at least at 30 min intervals, depending on the duration of the previous test. Before testing the effect of a chemical, a 5 minute control recording was made.

Effect of heavy metals on the siphon-activity

Varying concentrations of Cd, Cu, Hg, Pb and Zn-ions were tested, using stock solutions of $CdCl_2$, $CuCl_2$, $HgCl_2$, $PbCl_2$ and $ZnCl_2$. Generally, the type of reaction was substance and concentration dependent. From the reactions observed, both the threshold concentration and the effect of higher concentrations could be monitored.

Effect of Cd^{2+}: Under the influence of cadmium ions, there was a slight but definite change in the siphon activity. The control, short period waving of the water flow remained intact. However, at 0.2 mg l^{-1} concentration, fast outflow pulses occurred and the opening of the outflow siphon (force of water flow) tended to be smaller (Fig. 4b). At higher (1.2 mg l^{-1}) Cd-concentration the water flow remained permanent but at a weaker level, due to periodic short restriction of the opening of the outflow siphon (Fig. 4c).

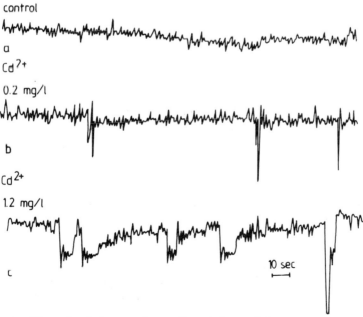

Figure 4. Effect of cadmium on the outflow siphon activity.

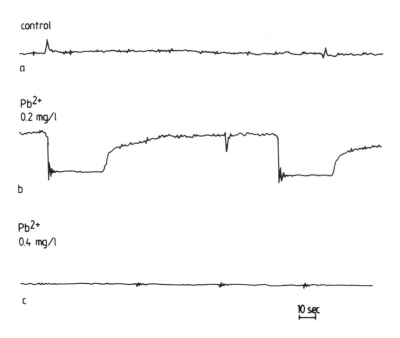

Figure 5. Effect of lead on the outflow siphon activity. At "c" the siphon is permanently closed as a result of the presence of 0.4 mg l^{-1} Pb^{2+}.

Effect of Pb^{2+} and Cu^{2+}: Under the influence of lead another type of reaction was observed. At 0.2 mg l^{-1} concentration it caused periodic, 50-80 sec-long full closure of the outflow siphon, with arrest of the water flow (Fig. 5b). The closures occurred at 100-150 sec intervals. At a higher lead concentration (0.4 mg l^{-1}) total closure of siphon was caused, interrupted periodically by rare, very weak openings (Fig. 5c). A similar effect to that of lead was caused by copper at 0.1-0.2 mg l^{-1} concentration.

Effect of Zn^{2+}: The concentration dependence of the siphon activity is well demonstrated for zinc by Figure 6. Low concentration (0.78 mg l^{-1}) caused periodic, very short closures of the siphon (Fig. 6b). A concentration of 12.2 mg l^{-1} evoked 30-50 sec-long closures, with regular intervals (Fig. 6c). A higher concentration (50 mg l^{-1}) resulted in full closure of the siphon, interrupted periodically by short, weak openings (Fig. 5d).

Effect of Hg^{2+}: At a concentration of 0.0125 mg l^{-1}, mercury caused periodic alternation of open and closed states of the outflow siphon at 20-30 sec intervals (Fig. 7). This response was checked for 80 minutes, and during this time its character did not change. Nevertheless, after 20-30 min the water flow was weaker during activity, relating to the reduced siphon opening.

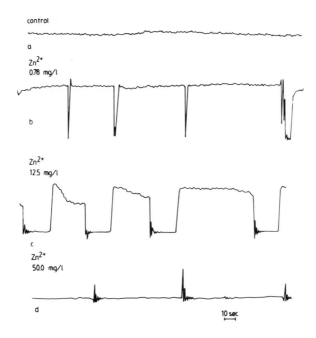

Figure 6. Effect of zinc on the outflow siphon activity. At "d" the siphon is closed as a result of the presence of 50 mg l^{-1} Zn^{2+}.

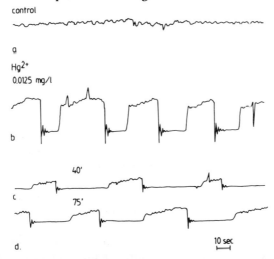

Figure 7. Effect of mercury on the outflow siphon activity. 0.0125 mg Hg^{2+} caused periodic alteration of open and closed position of the outflow siphon for more than 80 min. With time the force of water flow during active periods became weaker (c and d).

Use of mussels in long term detection of sub-lethal toxic effects of pollutants

This method is based on monitoring the pumping movements performed by the contraction and relaxation of the adductor muscles (Barnes 1955). The opening and closing of the shells are recorded (Salánki and Balla 1964) when one of the shells is fixed while the other is connected to a lever.

Using a slow speed recorder, several days or weeks activity can be continuously registered and analysed. With appropriate technical equipment the method can also be used in the natural habitat as well as in the laboratory (Véró and Salánki 1969; Kramer *et al.* 1989).

By recording the opening and closing of the shells of both freshwater and marine mussels, an alteration of active and rest periods can be detected, each lasting for several (10-30) hours (Salánki 1966; Salánki *et al.* 1970). As a rule the time of active periods is much longer, being 3-5 times the duration of rest periods.

It has been shown that in the presence of various pollutants in the water, the periodic activity changes in a special way: the duration of active periods becomes shorter, while as a rule the duration of rest periods increases. As a result the proportion of rest during a given period (a day or week) increases considerably. This type of response has been described under the influence of heavy metals as well as some pesticides (Salánki 1960; 1979; Salánki and Varanka 1976; 1978). The effect can be easily analysed and can be used as a good indicator of pollution.

Discussion

Mussels are amongst the most common aquatic animals. They can be easily collected and survive well in the laboratory. Most species are similar in basic anatomical structure and function. Filter feeding behaviour leads to bioaccumulation and the use of gill tissue as a biomonitor of pollutant concentration.

The siphon and pumping activity is a physiological characteristic of all mussels, and can be used to monitor sub-lethal pollutant concentrations rapidly. Long term responses may be monitored by recording shell activity. In using different mussel species one should take into consideration some anatomical differences in the location of the siphons as well as the fact that some mussels (*Pecten, Cardium* spp.) have no "catch" portion of the adductor muscle and so they are unable to keep a long closure of the shells. These latter species are filtering permanently and no periodic alteration of activity and rest can be observed. The reduction of water flow through the siphon and of the filtering (pumping) activity of the whole animal is the sign of "discomfort" and they are clear responses to undesirable influences. The recording of these behavioural reactions is used for biological indication of the presence of harmful substances, assuming that chemicals which do not reduce the animal's activity are not harmful to them.

Besides using accumulation capacity and behavioural reactions of mussels other approaches can also be considered with these animals. Feeding, respiration and growth rate are also indicative of pollution (Widdows and Donkin 1989) and also neurochemical (Salánki and Hiripi 1990) as well as immunological correlates (Stefano 1989) can be used in monitoring and testing the effect of environmental pollutants and toxicants.

References

Barnes, G.E. (1955) The behaviour of *Anodonta cygnea* L., and its neurophysiological basis. *Journal of Experimental Biology*, **32**, 158-174.

Goldberg, E.D., Bowen, V.T., Farrington, J.W., Harvey, G., Martin, J.H., Parker, P.L., Risebrough, R.W., Robertson, W., Schneider, E. and Gamble, E. (1978) The mussel watch. *Environmental Conservation*, **5C**, 101-125.

Kramer, K.J.M., Jenner, H.A. and de Zwart, D. (1989) The valve movement response of mussels: a tool in biological monitoring. *Hydrobiologia*, **188/189**, 433-443.

Lukacsovics, F. and Salánki, J. (1968) Data to the chemical sensitivity of freshwater mussel (*Anodonta cygnea* L.). *Annals of Biology (Tihany)*, **35**, 75-81.

Salánki, J. (1960) On the dependence of the slow rhythm of periodical activity of *Anodonta cygnea* on the condition of sulfhydryl groups in protein bodies. *Journal of General Biology*, **21**, 229-232. (in Russian).

Salánki, J. (1966) Comparative studies on the regulation of the periodic activity in marine lamellibranches. *Comparative Biochemistry and Physiology*, **18**, 829-843.

Salánki, J. (1979) Behavioural studies on mussels under changing environmental conditions. In J. Salánki and P. Biró (eds.) *Human Impacts on Life in Freshwaters. Symposia Biologica Hungarica*, **19**, 169-176. Akadémiai Kiadó, Budapest.

Salánki, J. and Balla, I. (1964) Ink-lever equipment for continuous recording of activity in mussels (musselactograph). *Annals of Biology (Tihany)*, **31**, 117-121.

Salánki, J., Glaizner, B. and Lábos, E. (1970) On temporal organization of periodic and rhythmic activity in freshwater mussel (*Anodonta cygnea* L.). *Journal of Interdisciplinary Cycle Research*, **1**, 123-134.

Salánki, J. and Hiripi, L. (1990) Effect of heavy metals on the serotonin and dopamine systems in the central nervous system of the freshwater mussel (*Anodonta cygnea* L.). *Comparative Biochemistry and Physiology*, **95C**, 301-305.

Salánki, J. and Varanka, I. (1976) Effect of copper and lead compounds on the activity of the freshwater mussel. *Annals of Biology (Tihany)*, **43**, 21-27.

Salánki, J. and Varanka, I. (1978) Effect of some insecticides on the periodic activity of the freshwater mussel (*Anodonta cygnea* L.). *Acta Biologica Academiae Scientiarum Hungaricae*, **29**, 173-180.

Salánki, J. and V.-Balogh, K. (1985) Uptake and release of mercury and cadmium in various organs of mussels (*Anodonta cygnea* L.). *Symposia Biologica Hungarica*, **29**, 325-342.

Salánki, J. and V.-Balogh, K. (1989) Physiological background for using freshwater mussels in monitoring copper and lead pollution. *Hydrobiologia*, **188/189**, 445-454.

Salánki, J., V.-Balogh, K. and Berta, E. (1982) Heavy metals in animals of Lake Balaton. *Water Research*, **16**, 1147-1152.

Stefano, G.B. (1989) (personal communication).

Véró, M. and Miller, T. (1979) Sensitive tension and force transducer. *Medical and Biological Engineering and Computing*, **17**, 662-666.

Véró, M. and Salánki, J. (1969) Inductive attenuator for continuous registration of rhythmic and periodic activity of mussels in their natural environment. *Medical Biological Engineering*, **7**, 235-237.

Widdows, J. and Donkin, P. (1989) The application of combined tissue residue chemistry and physiological measurements of mussels (*Mytilus edulis*) for the assessment of environmental pollution. *Hydrobiologia*, **188/189**, 455-461.

Behavioural Responses to Pollutants - Application in Freshwater Bioassays

D. Pascoe, D.E. Gower, C.P. McCahon, M.J. Poulton, A.J. Whiles and J. Wulfhorst

School of Pure and Applied Biology, University of Wales College of Cardiff, P.O. Box 915, Cardiff CF1 3TL, U.K.

Key words: bioassays, toxicity tests, behavioural toxicity, sub-lethal responses, freshwater invertebrates.

Abstract

Changes in animal behaviour resulting from exposure to toxicants can be highly variable and difficult to quantify. However, in some cases, they can provide a rapid and sensitive bioindication of poor environmental quality. Several behavioural responses observed in freshwater macroinvertebrates are described and their possible role as bioassays of water quality discussed

Behavioural toxicity studies

Since it is reasonable to assume that all behavioural responses are subject to toxicant-induced changes, the scope for research is virtually limitless. In practice, however, relatively few responses have been investigated in depth by aquatic toxicologists. These include responses of individuals, such as feeding (Hara and Thompson 1978) and learning behaviour (Rand and Barthalmus 1980), and inter-individual responses such as reproductive (Bloom et al. 1978), social (Henry and Atchison 1979) and predator-prey (Finger et al. 1985) interactions. However, the field has clearly been dominated by studies of locomotor responses, such as swimming activity and preference-avoidance investigations, particularly with fish (Cherry and Cairns 1982; Little et al. 1990). A general review of the literature concerning aquatic behavioural toxicity is provided by Little et al. (1985), and Rand (1985) recommends techniques for experimental investigations.

Although changes of behaviour brought about by toxic chemicals are of intrinsic biological interest, they may also be of value in a more applied sense in bioassays for detecting and/or monitoring the presence of pollutants in water. For this purpose responses occurring rapidly, at low toxicant levels and in a concentration-related manner will be of most value. This paper describes several individual and inter-individual behavioural toxicity studies which have been carried out in the author's laboratory with freshwater macroinvertebrates and discusses their possible role as bioassays of water quality. These studies can be classified into investigations concerning (1) construction skills (2) feeding activity and (3) reproductive behaviour.

(1) Construction skills

(a) Net-spinning activity of the trichopteran *Hydropsyche angustipennis* (Curtis).

Larvae of the caddis fly *H. angustipennis* occur on stony substrates in lowland streams and rivers and are relatively tolerant of organic pollution (Badcock 1975; 1976). The animals spin silken secretions to form nets comprising a tunnel-shaped dense web of fine silk within which the larvae lives (the retreat) and an additional area, with regular rectangular openings which trap detritus and small animals, and acts as a feeding net.

Studies on net spinning were carried out in the laboratory using an experimental channel in which the temperature and velocity of circulating water could be controlled. Individual fifth instar larvae (identified by head capsule width) were placed in open chambers suspended in the channels for 24 hours. They were then removed from the chambers and any nets spun were mounted on glass slides for microscopic examination, and measurements were taken using a Quantimet image analyser.

In a study of 146 fifth instar larvae, net spinning activity was found to be both velocity and temperature dependent with most nets constructed at the maximum tested water velocity and temperature (i.e. 40 cm s^{-1} and 20°C). When copper was introduced into the test channels net spinning was reduced in proportion to the exposure concentration (Table 1). In addition, at copper concentrations greater than 64 µg l^{-1} there were differences in the fine structure of the feeding nets compared with the controls. In particular there was a significant ($p > 0.001$) increase in the size of the mesh and a decrease in the number of meshes ($p > 0.001$). Clearly such anomalies could affect the filtering efficiency of the net with possible consequences for larval energetics and growth. For example, Petersen and Petersen (1983, 1984) noted anomalies in the nets of *H. angustipennis* exposed to kraft pulp mill effluent and speculated on the link between these and the delayed emergence of animals from effluent treatments.

(b) Case building of the trichopteran *Agapetus fuscipes* Curtis.

Larvae of the caddis fly *A. fuscipes* are widely distributed in British rivers preferring stony substrata in fast flowing waters (Hickin 1967). As they grow the larvae leave their old cases and construct new ones using small stones and sharp edged sand grains (Hanna 1960; 1961), unlike most cased trichoptera which increase the case size by addition of material as they mature. Clearly the larvae are vulnerable to predation during this building stage, a situation which could be compounded if case building speed and/or efficiency were reduced by the presence of pollutants.

Table 1. Effect of copper on the percentage of *Hydropsyche angustipennis* larvae spinning nets.

Nominal Copper Concentration ($\mu g\ l^{-1}$)	% larvae spinning nets	% larvae spinning nets with feeding section
Control	90	46
13	96	36
64	56	26
130	46	28
640	36	0

To examine this possibility, acute toxicity studies were conducted in the laboratory and these confirmed previous observations (Brown and Pascoe 1988) that trichoptera are relatively tolerant of cadmium. It was also demonstrated (McCahon *et al.* 1989a) that for each of the instars tested (1st, 3rd and 4th) uncased larvae were more sensitive than cased animals (Table 2). In order to study case building efficiency under toxicant stress 30 larvae of 1st, 3rd and 4th instar stages were removed from their cases and exposed to a range of cadmium concentrations for 24 hours. They were then provided with disrupted case material or fine sand grains and their attempts to rebuild cases observed. It is evident (Table 3) that their ability to reconstruct decreased as the concentration of cadmium to which the animals were exposed increased. The structure and strength of the constructions also deteriorated, cases becoming little more than a loose collection of stones, at the highest concentration.

Table 2. Median Lethal Concentrations (LC50) in mg Cd l^{-1} for 3 larval instars of *Agapetus fuscipes* both cased (c) and uncased (u) (data from McCahon, Whiles and Pascoe 1989).

Instar Number	24h LC50		96h LC50	
	c	u	c	u
1	860	295	210	50
3	940	300	250	<90
4	>1000	450	320	<90

D. Pascoe et al.

Table 3. Percentage of the first, third and fourth larval instar of *Agapetus fuscipes* constructing cases using their normal material (a) and fine sand (b), in the presence of cadmium over a twenty four hour period. (Data from McCahon, Whiles and Pascoe 1989).

Instar	Time from start (hrs)	Mean recorded cadmium concentration (mg Cd l^{-1})							
		0.002 ± 0.001 control		46 ± 2.4		97 ± 1.9		186 ± 3.6	
		a	b	a	b	a	b*	a	b
	3	60	67	7	13	13	-	0	7
1	9	80	80	20	20	13	-	0	7
	24	80	80	33	27	13	-	0	7
	3	73	80	40	27	13	-	13	20
3	9	87	80	40	27	27	-	13	20
	24	87	87	40	40	27	-	20	20
	3	73	0	60	0	47	-	27	13
4	9	80	0	67	13	60	-	40	20
	24	80	0	67	27	73	-	47	33

* Case building activity in fine sand not measured at this concentration.

This study clearly implies that the unusual behaviour of *A. fuscipes*, i.e. leaving its case to construct a new one, could render it susceptible to pollutant discharges, particularly those of an episodic nature. It is also evident that a pollution incident occurring during this sensitive stage could impair the construction of new cases, and since these are thought to be important anti-predator devices (Otto and Svensson 1980) the larvae may be at risk from both pollutant and predator (McCahon *et al.* 1989a).

(2) Feeding activity

This laboratory started studies on feeding several years ago when it was noted with a variety of animals such as *Lymnaea peregra* (Müller), *Asellus aquaticus* (L.) and *Gammarus pulex* (L.) that feeding rates were depressed when the animals were exposed to toxicants. Attempts were therefore made to develop a system which would quantify this reduction in feeding rate using the freshwater amphipod *G. pulex* which is known to be sensitive to many pollutants (Williams *et al.* 1984). The method eventually adopted is based upon the change in weight of 1 cm diameter discs of horse chestnut (*Aesculus hippocastanum* L.) leaves which have been conditioned in organically enriched water to permit the growth of surface fungi and bacteria. Five *G. pulex* are allowed to feed on a single leaf disc for 24 hours and the weight of leaf consumed (or the area of disc ingested as measured by image analysis) is related to the weight of animal present or to the number of animal hours for which the leaf disc was exposed.

Table 4. The mean (± standard deviation) wet weight (mg) of leaf disc consumed by *Gammarus pulex* per animal hour exposed ($\times 10^{-3}$) for uninfected animals and those infected with the acanthocephalan *Pomphorhynchus laevis* (data from McCahon, Brown and Pascoe).

Nominal cadmium concentration (μg l⁻¹)	uninfected *G. pulex*	parasitized *G.pulex*
Control	34.55 ± 12.6	15.75 ± 9.3
30	34.11 ± 14.3	12.27 ± 1.4
300	12.40 ± 3.6	2.39 ± 1.7

This feeding test was employed in a study carried out to compare the sensitivity to cadmium of *G. pulex* infected with the acanthocephalan parasite *Pomphorhynchus laevis* (Müller) to the sensitivity of those which were uninfected. The results (Table 4) show that the feeding of parasitized animals was greatly depressed in comparison to uninfected gammarids and that for both groups, exposure to the higher cadmium concentration led to much reduced feeding. Such a sub-lethal toxic effect could rapidly be reflected in impaired growth and development of the affected organisms and eventually in changes of population structure. It could also be useful as a rapid indication of poor water quality.

3. Reproduction behaviour

(a) Ovipositing by *Chironomus riparius* Meigen
It is vital that animals with an aquatic egg stage should oviposit into water of a suitable quality for egg survival and development. Female mosquitoes are reported to be able to select water of this sort (Hudson 1955) but there is very limited information on the ability of insects to identify and avoid toxicant - contaminated water.

To examine the extent to which *C. riparius* adults can detect and avoid cadmium, several hundred newly emerged adults were introduced daily into a chamber containing randomly allocated dishes of cadmium at a range of concentrations. To avoid any selection related to the position of the dishes, the pattern of distribution was changed daily. The number of egg ropes oviposited in each cadmium solution was recorded after successive 24 hour test periods (Williams *et al.* 1987). The production of egg ropes over a ten day period is shown in Figure 1 and it is evident that the higher cadmium concentrations have been avoided by females. The mechanism by which the animals discriminate between the available solutions is not clear, although Sutcliffe and Carrick (1973) have suggested that chemoreceptors on insect antennae could be involved in testing water quality.

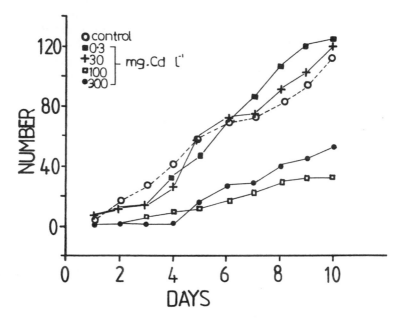

Figure 1. The cumulative number of egg ropes by *Chironomus riparius* in various concentrations of cadmium and in control water.

(b) Precopulatory behaviour of *G. pulex* (L.)

A characteristic phase of reproduction in *G. pulex* is precopulatory guarding where male and female swim together for up to ten days before the female moults and mating takes place. This behavioural strategy has probably evolved as a result of competition between males for productive females and several studies have shown that it can be disrupted by environmental stressors (Davis 1978; Linden 1976).

Table 5. Median Induced Separation Times (IST50-secs) in 2-phenoxyethanol with 95% confidence intervals for *Gammarus pulex* precopulatory pairs pre-exposed to cadmium for 24 hours (data from Poulton and Pascoe 1990).

Recorded Cd Concentration ($\mu g\ l^{-1}$)	<0.1 control	8.3	10.4	22.0	27.6	36.0	46.7
IST50	776	798	680	663	518	456	294
95% C.I.	665–881	697–915	580–796	535–820	422–637	374–555	246–350

A method was devised to demonstrate the effect of pollutants on this important phase of the reproductive process. Precopula pairs (57-87 replicates) were exposed to cadmium concentrations (0-45 µg l^{-1}) for 24 hours and then transferred to 2-phenoxyethanol, an anaesthetic known to induce separation. It was found (Table 5) that the Median Induced Separation Time (IST50) decreased as the concentration of cadmium to which the animals were pre-exposed increased (Poulton and Pascoe 1990). These authors also discuss the possible causes and consequences of precopulatory separation e.g. it could be advantageous to desert a mate if this increases the chance of avoiding the pollutant.

As well as indicating an important behavioural toxic effect this work suggested that the separation phenomenon could be useful as a bioassay of water quality.

Role of behavioural toxicity tests as bioassays of water quality

These changes in behaviour described above, occurring as a result of pollutant presence could all have a significant effect upon the biology of the species involved and upon their ability to survive within the community.

Since the behavioural changes may also be detected relatively quickly and at toxicant concentrations below those causing mortality they could provide a more sensitive indication of poor water quality than the more conventional acute lethal toxicity test. However, for practical purposes, it is preferable, though not essential, that such bioassays involve simple and easily detected concentration-related responses which can be measured rapidly with a minimum of sophisticated equipment, if necessary in the field. It is apparent that several of the behavioural responses discussed earlier do not satisfy these criteria. For example, the ovipositing behaviour of *C. riparius* was modified only at concentrations of cadmium which are, in fact, lethal to first instar larvae and so this behaviour would not form the basis of a sensitive bioassay. Similarly, the net spinning activity of *H. angustipennis* and the case building of *A. fuscipes*, although relatively sensitive to the toxicants tested, are difficult to demonstrate and measure experimentally. However, the results obtained for the feeding activity and precopulatory behaviour of *G. pulex* suggest that these responses could form the basis of useful bioassays.

The detritivorous amphipod *G. pulex* occurs widely in freshwaters and is sensitive to a range of common pollutants (Williams *et al.* 1984). It can be cultured easily in the laboratory (McCahon and Pascoe 1988), an important attribute of any bioassay organism, and has been used extensively in toxicity studies (Wright 1980; Brown and Pascoe 1988).

(a) *G. pulex* feeding activity
The laboratory study described above demonstrated that *G. pulex* stressed by a toxicant consumed measurably less food than non-exposed animals. The

application and possible role of this effect as a bioassay has been evaluated by deploying the feeding test system during two experiments in which pollution incidents were simulated in natural streams. In both studies, (i) an acid-aluminium pollution episode (McCahon *et al.* 1989b) and (ii) a farm waste pollution episode (McCahon *et al.* unpublished data), the feeding activity of *G. pulex* was significantly depressed during the pollution episode in comparison to that of untreated animals.

This test system therefore appears to have some potential for evaluating water quality although further development is probably required to increase the sensitivity and discriminating power of the bioassay.

(b) *G. pulex* precopulatory behaviour

The laboratory study demonstrated that the response time for *G. pulex* precopulatory pairs induced to separate in anaesthetic was related to previous toxicant exposure. A similar study (unpublished data) revealed that separation could also occur as a result of direct exposure to a toxicant and that this separation time was also correlated with the exposure concentration.

Consequently two bioassays, one based upon 'induced separation' and the other on 'direct separation' have been devised (Poulton and Pascoe 1990) and evaluated in the field during simulated acid-aluminium and farm waste pollution incidents. The tests have proved to be more rapid and sensitive than lethal bioassays and the feeding bioassay described above. The 'direct separation' method is particularly useful because, not being dependent upon use of an anaesthetic, it permits animals to re-establish pairing after the pollution event and is therefore very relevant to the evaluation of episodic pollution such as that resulting from the periodic release of farm waste.

Conclusions

As the quest for more sensitive, rapid and reliable bioassays of water quality continues it is inevitable, and desirable, that behavioural responses will play an increasingly significant role. However, behavioural changes are frequently highly variable, difficult to record and measure and one must be selective in identifying those which merit further investigation and eventual incorporation into laboratory and field bioassay programmes.

Acknowledgements

We are grateful to the German Academic Exchange, NERC, SERC, WRc and the CEC for financial support and to Christine Gould for typing the paper.

References

Badcock, R.M. (1975) The hydropsychidae (Trichoptera) in Staffordshire. *North Staffordshire Journal of Field Studies*, **15**, 10-18.

Badcock, R.M. (1976) The distribution of hydropsychidae in Great Britain. *Proceedings of the First International Symposium on Trichoptera*, Lunz am See (Austria), 49-57.

Bloom, H.D., Perlmutter, A. and Seeley, R.J. (1978) Effect of a sub-lethal concentration of zinc on an aggregating pheromone system in the zebrafish *Brachydanio rerio*. *Environmental Pollution*, **17**, 127-131.

Brown, A.F. and Pascoe, D. (1988) Studies on the acute toxicity of pollutants to freshwater macroinvertebrates: The acute toxicity of cadmium to twelve species of predatory macroinvertebrates. *Archiv für Hydrobiologie*, **114**, 2, 311-319.

Cherry, D.S. and Cairns, J. Jr. (1982) Biological Monitoring. Part V. Preference and avoidance studies. *Water Research*, **16**, 263-301.

Davis, J.C. (1978) Disruption of precopulatory behaviour in the amphipod *Anisogammarus pugettensis* upon exposure to bleached pulpmill effluent. *Water Research*, **2**, 273-275.

Finger, S.E., Little, E.F., Henry, M.G., Fairchild, J.F. and Boyle, T.P. (1985) Comparison of laboratory and field assessment of fluorine, Part 1: Effects of fluorine on the survival, growth, reproduction, and behaviour of aquatic organisms in laboratory tests. In T.P. Boyle (ed.) *Validation and Predictability of Laboratory Methods for Assessing the Fate and Effects of Contaminants in Aquatic Ecosystems*, 120-133. ASTM STP 865, Philadelphia.

Hanna, H.M. (1960) Methods of case-building and repair by larvae of caddis flies. *Proceedings of the Royal Entomological Society of London*, (A), **35**, 97-106.

Hanna, H.M. (1961) Selection of materials for case-building by larvae of caddis flies (Trichoptera). *Proceedings of the Royal Entomological Society of London*, (A) **36**, 37-47.

Hara, T.J. and Thompson, B.E. (1978) The reaction of whitefish, *Coregonus clupeaformis*, to the anionic detergent sodium lauryl sulphate and its effects on their olfactory responses. *Water Research*, **12**, 893-897.

Henry, M.G. and Atchison, G.J. (1979) Influence of social rank on the behaviour of bluegill *Lepomis macrochirus* Rafinesque exposed to sublethal concentrations of cadmium and zinc. *Journal of Fish Biology*, **15**, 309-315.

Hickin, N.E. (1967) Caddis Larvae. Larvae of the British Trichoptera. Hutchinson, London.

Hudson, B.N.A. (1955) The behaviour of the female mosquito in selecting water for oviposition. *Journal of Experimental Biology*, **33**, 478-492.

Linden, O (1976) Effect of oil on the reproduction of the amphipod *Gammarus oceanicus*. *Ambio*, **5**, 36-37.

Little, E.E., Flerov, B.A. and Ruzhinskaya, N.N. (1985) Behavioral approaches in aquatic toxicity investigations: a review. In P.M. Mehrle, R.H. Gray and R.L. Kendall (eds.) *Toxic Substances in the Aquatic Environment: An International Aspect*, 72-98. American Fisheries Society Maryland.

Little, E.E., Archeski, R.D., Flerov, B.A., and Kozlovskaya, V.I. (1990) Behavioral indicators of sublethal toxicity in rainbow trout. *Archives of Environmental Contamination and Toxicology*, **19**, 380-385.

McCahon, C.P. and Pascoe, D. (1988). Culture Techniques for three freshwater macroinvertebrate species and their use in toxicity tests. *Chemosphere*, **17**, 2471-2480.

McCahon, C.P., Whiles, A.J. and Pascoe, D. (1989a). The toxicity of cadmium to different larval instars of the trichopteran larvae *Agapetus fuscipes* Curtis and the importance of life cycle information to the design of toxicity tests. *Hydrobiologia*, **185**, 153-162.

McMahon, C.P., Brown, A.F., Poulton, M.J. and Pascoe, D. (1989b) The effect of liming on the response of fish and invertebrates to a simulated acid episode in a Welsh stream. *Water Air and Soil Pollution*, **45**, 345-359.

Otto, C. and Svensson, B.S. (1980) The significance of case material selection for the survival of caddis larvae. *Journal of Animal Ecology*, **49**, 855-865.

Petersen, L.B-M. and Petersen, R.C. (1983) Anomalies in hydropsychid capture nets from polluted streams. *Freshwater Biology*, **13**, 185-191.

Petersen, L.B-M. and Petersen, R.C. (1984) Effect of kraft pulp mill effluent and 4,5,6 trichloroguaiacol on the net spinning behaviour of *Hydropsyche angustipennis* (Trichoptera). *Ecological Bulletin*, **36**, 68-74.

Poulton, M. and Pascoe, D. (1990) Disruption of precopula in *Gammarus pulex* (L.) - development of a behavioural bioassay for evaluating pollutant and parasite induced stress. *Chemosphere*, **20**, 403-415.

Rand, G.M. (1985) Behaviour. In G.M. Rand and S.M. Petrocelli (eds.) *Fundamentals of Aquatic Toxicology*, 221-263. Hemisphere, Washington.

Rand, G.M. and Barthalmus (1980) Use of an unsignalled avoidance technique to evaluate the effects of the herbicide 2,4 dichlorophenoxyacetic acid on goldfish. In I.G. Eaton, P.R. Parrish and A.C. Hendricks (eds.) *Aquatic Toxicology*, 341-353. ASTM STP 707, Philadelphia.

Sutcliffe, D.W. and Carrick, T.R. (1973) Studies on mountain streams in the English Lake District. I. pH, calcium and the distribution of invertebrates in the River Duddon. *Freshwater Biology*, **3**, 437-462.

Williams, K.A., Green, D.W.J. and Pascoe, D.D. (1984) Toxicity testing with freshwater macroinvertebrates: methods and application in environmental management. In D. Pascoe and R.W. Edwards (eds.) *Freshwater Biological Monitoring*, 81-93. Pergamon Press, London.

Williams, K.A., Green, D.W., Pascoe, D. and Gower, D. (1987) The effects of cadmium on oviposition and egg viability in *Chironomus riparius* (Diptera - Chironomidae). *Bulletin of Environmental Contamination and Toxicology*, **38**, 86-90.

Wright, D.A. (1980). Cadmium and calcium interactions in the freshwater amphipod *Gammarus pulex*. *Freshwater Biology*, **10**, 123-133.

Detection of Antialgal Compounds of Water Hyacinth

Yu Shu-wen, Sun Wen-hao and Yu Zi-wen

Institute of Plant Physiology, Academia Sinica, Shanghai 200032, China.

Key words: allelopathy, antialgal compound, bioassay, Water hyacinth, *Chlamydomonas reinhardtii.*

Abstract

An allelopathic effect of water hyacinth *Eichhornia crassipes* Solms against algae has been discovered. In order to isolate and identify the allelopathic compounds it is necessary to establish a simple, rapid, reliable and sensitive bioassay. Hence an agar-filter paper disc method was developed to detect the antialgal activity of allelopathic compounds. Filter paper discs were placed on agar blocks in petri dishes, and a certain amount of algal suspension was added to each disc and then cultured in the light for 2-4 days. Growth was measured by the determination of acetone extracted chlorophyll. If the extract of allelopathic compounds was added to the filter paper discs before the addition of the algal suspension, the growth of algae would be inhibited. IC_{50} may be obtained by using a series of concentrations of allelopathic compounds extract. This method can also be used to monitor all pollutants which are toxic to algae. *Chlamydomonas reinhardtii* was used as an indicator species in the bioassay since it is sensitive to the allelopathic compounds and its chlorophyll is convenient to extract.

Introduction

Water hyacinth is an aquatic weed widespread in tropical countries. In eutrophic rivers where water hyacinth plants are cultivated, the water becomes clear owing to the elimination of green algae.

Eliminating the competition for light and mineral nutrients between water hyacinth and algae in experimental conditions, it was demonstrated that the inhibitory effect of water hyacinth on algae was due to allelopathy (Sun *et al.* 1988; 1989; 1990). This meant that water hyacinth could secrete allelochemicals which were toxic to algae. In order to isolate, purify and identify these antialgal compounds, a simple, rapid, reliable and sensitive bioassay was needed. This paper presents such a method using *Chlamydomonas reinhardtii* as an indicator species.

Bioindicators and Environmental Management
ISBN 0-12-382590-3

Materials and methods

1. Material

Water hyacinth *Eichhornia crassipes* was cultured in 0.1N modified Hoaglands solution in plastic tanks measuring 415 x 285 x 200 mm in a green house. Temperature was $25 \pm 3°C$, relative humidity about 70%, and light natural. The plant-free culture water was used in the bioassay.

   Chlamydomonas reinhardtii was cultured in BBM solution (Bold 1949) in conical flasks. Culture temperature was 24-26 °C. Fluorescent lamps were used and the light intensity was 80 μmol $m^{-2} s^{-1}$. The light period was 12 hours. The culture solution was ventilated and stirred by passing through filtered air. The algae grown at the logarithmic phase were used in the bioassay.

2. Extraction and separation of the allolepathic compounds

XAD-2 resin (20-50 mesh, Ruhm and Hass Company) was used. Pretreatment of XAD-2 and packing were performed according to the method of Junk (Junk *et al.* 1979). The packing tube was 1.5 cm in diameter and packing was 7.5 cm in height. 400 ml of filtered culture water was passed through the column at a rate of 20-30 ml min^{-1}, washing with 100 ml redistilled water and then eluting with 50 ml redistilled methanol. The eluate was evaporated in a vacuum rotary evaporator at 40°C. The concentrated solution was made to 10 ml volume (pH 5-6) by adding redistilled water, then extracted by an equal volume of purified ether twice. The two portions of ether extract were combined and dried over anhydrous $CaCl_2$, and then evaporated in a vacuum rotary evaporator once again until a small amount of concentrated extract remained. It was transferred to a 5 ml graded tube, vacuum evaporated to dryness, and finally was made to 2 ml volume by adding anhydrous ether. This crude extract was used for bioassay or for further separation.

   Silica gel dry column chromatograph was chosen for separation (Shriner *et al.* 1980). Nylon tube was used for the column tube and petroleum ether:acetone (3:1 v/v) as developer. After development the column was cut into six separate parts. All of them were extracted separately with methanol for bioassay. They were denoted as EC-W-A1-6 separately.

3. Preparation of agar medium and filter paper discs

Petri dish (60 x 15 mm) and agar medium BBM were autoclaved at 120°C for 20 min. The hot medium was poured into petri dishes, each for 8 ml. After cooling the agar plate was about 2.5 mm in thickness. The plate in each petri dish was punched into 5 blocks with diameter of 12 mm. The remaining parts of the agar plate were removed. Filter paper (Sin-hua No 1) was punched into discs with diameter of 11 mm.

4. Addition of allelopathic compound and inoculation of algae

Filter paper discs were placed on a rack made from nylon gauze. A given amount of extract was added to each disc with a micropipette. There were five addition amounts generally, i.e. 0 (control), 15, 30, 45 and 75 μl $disc^{-1}$. After the solvent evaporated, these five discs were placed on five agar blocks separately in one dish. Then 15 μl algae suspension (OD_{662} is 0.12, cell density 1.3 x 106 ml^{-1}) was added evenly on each paper disc. They were incubated with the same temperature and light regime described above.

5. Measurement of chlorophyll-a content of *C. reinhardtii*
The filter paper discs were taken out after 48 hours inoculation and put into 10 ml test tubes. To each tube was added 3.5 ml of 90% acetone. After sealing, the tubes were shaken 10 times and then placed in the dark for 1 hour at room temperature. Chlorophyll-a content was measured by fluorescence method (APHA *et al.* 1980). It may be also measured by a spectrophotometer, in which case the incubation time would be prolonged to 96 hours.

6. Dose-response relationship
In each treatment the growth inhibition percentage could be calculated by comparing with control. According to the inhibition percentage and the corresponding dose logarithm of extract of allelopathic compounds, the dose-response curve and the half inhibition dose (IC_{50}) could be obtained by the linear regression method. IC_{50} could be also obtained approximately by the nomographic method (Yang 1983 APHA *et al.* 1980)

Experimental

1. Blank control problem
Although the XAD-2 had been pretreated and methanol and ether had been purified, it was possible that a small amount of impurities might remain. As the solvent evaporated, the remaining impurities might interfere with the growth of *C. reinhardtii*. Therefore a resin-solvent control was established, i.e. an equal volume of redistilled water instead of culture water was passed through the column, eluted by methanol and dissolved in ether etc. in order to compare with blank control. The result is shown in Table 1. It indicates that there was no significant difference among the four resin-solvent controls and also no significant difference between the resin-solvent control and blank control ($F=2.1489$, $P>0.05$). Therefore the resin-solvent control may be substituted by the blank control.

2. The selection of algal species.
The inhibition effects of extracts of water hyacinth exudates on four algal species were compared. They were *Chlorella pyrencidosa, Scenedesmus obliquns, C. reinhardtii* and *Anabaena azollae,* all being common to eutrophicated waters. Experimental results indicated that *C. reinhardtii* was very sensitive to the allelopathic compounds. Its IC_{50} was 8.7 μl disc^{-1}, lower than that of *C. pyrencidosa* (38.0 μl disc^{-1}) and *S. obliquus* (13.2 μl disc^{-1}), but close to that of *A. azollae* (7.9 μl disc^{-1}). Furthermore, *C. reinhardtii* had two flagella which rendered them hardly precipitable. Its growth was faster than the other three.

Table 1. Comparison between XAD resin-solvent control and blank control (chlorophyll-a content, μg disk^{-1}, of *Chlamydomonas reinhardtii*)*

Repeat no.	Blank control	XAD resin-solvent control			
	0	15	30	45	75
1	12.47	13.38	11.56	10.14	11.46
2	12.17	13.18	12.78	12.78	11.97
3	11.76	12.47	11.36	11.26	11.05
4	12.13	13.01	11.90	11.39	11.49

Amount of Addition (μl disk^{-1})

* Incubated for 96h, Chl-a was measured by spectrophotometry

Thus we selected *C. reinhardtii* as the indicator species in the bioassay of allelopathic compounds from water hyacinth.

3. Problem of chlorophyll extraction.

Another advantage of *C. reinhardtii* was the convenience in extracting its chlorophyll, because its cell wall was thin and its chloroplast was close to the cell wall. It was unnecessary to grind the algal cells as they could be extracted by 90% acetone directly. Extracting for 1 hour at room temperature, the chlorophyll was almost all transferred to the solution. As the turbidity was slight, it could be measured by the spectrophotometer. Table 2 shows that there was no significant difference between the chlorophyll-a content measured by direct extraction and that by ground routine method. Therefore the substitution of the ground method by a 90% acetone direct extraction is feasible for *C. reinhardtii*. When we used this method to extract the chlorophyll of the other three species, the efficiency was not sufficently high.

Table 2. Comparison between ground method and extraction method to determine the chlorophyll-a amount* of *Chlamydomonas reinhardtii*.

	Chl-a	Phe-a	Chl-a content	Significant difference (t-test)
Ground method	2.74	0.34	3.08	/
Extraction method	2.84	0.36	3.20	p 0.05

* Chl-a content (mg l^{-1}) was measured by spectrophotometry

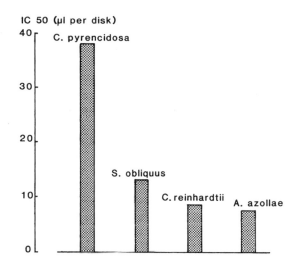

Figure 1. Inhibitory effect (IC$_{50}$) of root exudates from water hyacinth against four species of algae.

4. The culture time of *C. reinhardtii* in bioassay

Using EC-W-A4 as the inhibiting substance in bioassay, and sampling at 24 hour intervals, we measured the chlorophyll amount of *C. reinhardtii* by fluorescent method, and calculated the IC$_{50}$. The results (Fig. 2) showed that when cultured for 24 hours, IC$_{50}$ was the highest, and then decreased to a minimum at 48 hours, thereafter it rose gradually. So 48 hours was considered a suitable culture time with a higher sensitivity of detection.

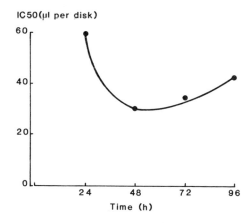

Figure 2. Relationship between the culture time of *Chlamydomonas reinhardtii* and IC$_{50}$ in bioassay. Allelopathic compounds --EC-W-A4 (1 μg μl^{-1}).

Discussion

A lot of investigation has been conducted on the utilization of algae for the detection of the biological effects of chemicals and environmental pollutants. In the book *Algae as Ecological Indicators*, Whitton (1984), Pipe and Shubert (1984), Eluabarawy and Welter (1984) and others presented detailed reviews. Algal assays may be broadly divided to two types - water solution culture methods and solid culture methods. The former, termed the "bottle test", was assigned in 1974 as the standard bioassay for water pollutants (APHA *et al.* 1980). In 1975 Klotz developed an algal assay SAAP characterized by the circulation culture (see Trainor 1984 p.11). The volume of culture solution was 10 ml and the culture time 5 days. Recently Wren and McCarroll (1990) reported a simple, sensitive algal test. The volume of culture solution reduced to 0.5 ml and the culture time 3 days. The latter might be called "plate method". Wright (1975), Chan *et al.* (1980) and Van Aller *et al.* (1985) proposed that a small piece of filter paper be dipped with solution of a toxic chemical, and then placed on the centre of an agar plate which had been inoculated with algae. The inhibitory effect could be determined by the diameter of the zone of inhibition of growth. Kruglov and Kwat-Kowskaya (see Pipe and Shubert 1984, p.225) established another type of plate method. At first algae were impregnated to Seitz filter pads, then transferred to the surface of soil which had been treated with a toxic substance such as a herbicide. Growth was measured by determination of ethanol extracted pigments. The common disadvantage of these methods was that a large quantity of tested substances was required. Since the allelopathic substances secreted by plants were in trace quantities, the extracts reduced even more after several fractionation and purification procedures.

Therefore, it was necessary to use apparatus of a smaller size, decrease the amount of substance applied, and increase the sensitivity of detection. In the agar-filter paper disc method presented, the allelochemicals were initially applied on the paper discs, the discs were transferred onto agar blocks, and then inoculated and cultured. Thus *C. reinhardtii* could absorb abundant nutrients from the agar medium and grow fast. On the other hand, the allelochemicals could react with the algae directly. Furthermore, agar blocks were used instead of plates to limit the diffusion of allelochemicals transversely, so the efficiency of utility of allelochemicals was raised, and the amount of allelochemicals used could be decreased.

A marked characteristic of this method was that the apparatus used was smaller, occupying less space. There were five agar-filter paper discs in one petri dish, including one blank control and four concentration treatments. Experimental results showed that there was no significant difference between the blank control put in the same dish with the concentration treatments and the blank control alone put in a separate dish. This demonstrated that there was no interference among the various concentration treatments in the one dish.

Because *C. reinhardtii* grew on filter paper discs, rarely on and/or into the agar blocks, only the filter paper discs needed to be taken out, and almost all algal cells could be retrieved. The method omitted the centrifugation or filtration steps, which were necessary in the elution culture method in order to collect the algal cells. It was very efficient to extract chlorophyll-a from *C. reinhardtii* by means of direct extraction with 90% acetone, omitting the addition of $MgSO_4$, grinding, filtration or centrifugation which were necessary in the ground method. It was not only simplifying the procedure, but also avoided the errors which might be induced by these processes.

The substances being detected in the solution culture method must be readily water soluble. The agar-filter paper disc method is free from the limitation of water solubility, and therefore has a wider range of application. It can be used to detect the biotoxicity of not only allelopathic compounds, but also a lot of environmental pollutants, insecticides, herbicides etc. The key to a successful detection is the selection of appropriate algae which are sensitive to the substances being tested and are convenient for culture and measurement. Different algal species respond differently to different toxic substances. Now it is known that there are many aquatic and terrestrial algae which are sensitive to various pollutants and agrochemicals and which can serve for our selection (Pipe and Shubert 1984, Van Aller *et al.* 1985, p.389).

References

APHA-AWWA-WPCF (1980) *Standard Methods for the Examination of Water and Wastewater* (M. Franson ed.), 15th edition.

Bold, H.C. (1949) The morphology of *Chlamydomonas chlamydogama*. November *Bulletin of the Torrey Club*, **76**, 101-103.

Chan, A.T., Anderson, R.J., LeBlanc, M.J. and Harrison, P.J. (1980) Algal plating as a tool for investigating allelopathy among marine microalgae. *Marine Biology* **59**, 7-13.

Elnabarawy, M.T. and Welter, A.N. (1984) Utilization of algal cultures and assays by industry. In L.H. Schubert (ed.) *Algae as Ecological Indicators*, 317-328. Academic Press, London.

Junk, G.A., Richard, J.J., Witiak, D., Witiak, J.L., Argvello, M.D., Vická, R., Svec, H.J., Fritz, J.S. and Calder, G.V. (1974) Use of macrorcticular resins in the analysis of water for trace organic contaminants. *Journal of Chromatography*, **99**, 745-762.

Pipe, A.E. and Shubert, L.E. (1984) The use of algae as indicators of soil fertility. In L.H. Shubert (ed.) *Algae as Ecological Indicators*, 213-233. Academic Press, London.

Shriner, R.L., Fuson, R.C., Curtin, D.Y., Morrill, T.C. (1980) (eds.). *The Systematic Identification of Organic Compounds*, 6th edition, John Wiley and Sons, New York, London.

Sun, W.H., Yu, Z.W. and Yu, S.W. (1988) Inhibitory effect of *Eichhornia crassipes* (Mart.) Solms on algae. *Acta Phytophysiologica Sinica*, **14**, 294-300 (in Chinese with English abstract).

Sun, W.H., Yu, Z.W. and Yu, S.W. (1989) The harness of an eutrophic water body by water-hyacinth. *Acta Scientiae Circumstantiae*, **9**, 188-195 (in Chinese with

English abstract).

Sun, W.H., Yu, Z.W., Tai, G.F. and Yu, S.W. (1990) Sterilized culture of water hyacinth and its application in the study of allelopathic effect on algae. *Acta Phytophysiologica Sinica*, **16**, 305-309 (in Chinese with English abstract).

Trainor, F.R. (1984) Indicator algal assays : laboratory and field approaches. In L.E. Shubert (ed.) *Algae as Ecological Indicators*, 11. Academic Press, London.

Van Aller, R.T., Pessoney, G., Rogers, V.A., Watkins, E.J. and Leggett, H.G. (1985) Oxygenated fatty acids: a class of allelochemicals from aquatic plants. In A.C. Thompson (ed.) *The Chemistry of Allelopathy*, 337-386. American Chemical Society, Washington D.C. Press.

Whitton, B.A. (1984) Algae as monitors of heavy metals in freshwaters. In L.E. Shubert (ed.) *Algae as Ecological Indicators*, 257-280. Academic Press, London.

Wren, M.J. and McCarroll, D. (1990) A simple and sensitive bioassay for the detection of toxic materials using a unicellular green alga. *Environmental Pollution*, **64**, 87-91.

Wright, S.J.L. (1975) A simple agar plate method, using micro-algae, for herbicide bioassay or detection. *Bulletin of Environmental Contamination and Toxicology*, **14**, 65-70.

Yang, J.K. (1983) (ed.) *Applied Biostatistics*, Science Press, Beijing, (in Chinese).

Lichens as Biological Indicators - Recent Developments

D.H.S. Richardson

Department of Botany, Trinity College, Dublin 2, Ireland.

Key words: lichens, bioindicators, pollution, sulphur dioxide, heavy metals, radionuclides, acid rain.

Abstract

In general, the public, legislators and politicians know only that lichens are very sensitive to air pollution, which is an oversimplification. It is true, however, that lichens have evolved a survival strategy which results in many species being very pollution-sensitive. Studies on the distribution pattern of a range of lichen species around urban or industrial areas can yield valuable data on air quality and this has lead to lichens being among the most widely accepted bioindicators of air quality. In addition, by analysing lichens collected at different distances from pollution sources, the size of the fallout zone and the nature of the emitted pollutants whether it be metal, non-metal, organic or radionuclide can be determined. Several thousand scientific papers on the use of lichens as bioindicators have appeared in international journals and are listed in a compendium entitled 'Literature on air pollution and lichens', which is published regularly in the *Lichenologist*. This paper summarizes some of the more recent approaches using lichens as bioindicators.

Introduction

Lichens are symbiotic plants in which a photosynthetic alga or cyano-bacterium is surrounded by fungal tissue. Lichens lack cuticles, stomata and functional roots. In consequence they have evolved both highly efficient nutrient uptake systems and an ability to absorb and lose water rapidly; being metabolically active when moist (Brown 1985; Laundon 1986). Where the absorbed water contains dissolved pollutants, the more sensitive species of lichens may be poisoned and this leads to a change in the community of lichen species growing on a particular substratum such as soil, rocks or tree trunks. This feature enables lichens to be used as bioindicators. If the occurrence and abundance of particular species is recorded, it is then possible to calculate,

Bioindicators and Environmental Management
ISBN 0-12-382590-3

for example, indices of atmospheric purity (Nash and Wirth 1988). Where the lichens are not actually killed, enhanced levels of atmospheric pollutants are found when samples are collected, in a distance-related way, from an urban or industrial emission source. By analysing samples collected at different distances, lichens provide valuable bioindicators of a wide range of atmospheric pollutants including metals, non-metals, radionuclides and organic substances such as chlorinated hydrocarbons (Richardson 1988; Henderson 1990).

Sulphur dioxide

The ways in which lichens are used as bioindicators is now being re-examined because the distribution of lichen species has, and is, being altered by changing pollution climates throughout the industrialized world. In the latter half of the last and first half of this century, high sulphur dioxide and smoke levels affected relatively limited areas around the larger and more industrialized cities. By examining the lichen flora around such cities, and relating the distribution of particular species to measured winter mean sulphur dioxide levels, Hawksworth and Rose (1970) were able to establish a qualitative scale for the estimation of sulphur dioxide air pollution, using lichens growing on tree bark. A series of 10 zones were delimited, with zone 1 having no lichens (a lichen desert) and sulphur dioxide levels in excess of 170 ug m^{-3}, to zone 10 with a wide range of lichens typical of ancient woodlands and less than 10 ug m^{-3}.

Since the publication of this scale, lichens have been used to assess air quality in many cities around the world. In general a single substratum, such as deciduous tree bark, is selected for study and the number and/or abundance of lichens recorded (Hawksworth and Rose 1976). The collected data can be analysed mathematically in various ways to produce, for example, Indices of Atmospheric Purity and a map of air quality (Nash and Wirth 1988; Zobel 1988). It is even possible, with training, to use school children to collect the lichen distribution data (Richardson 1987).

More recently, a reduction of ground-level sulphur dioxide concentrations has resulted from Clean-Air legislation banning the domestic burning of high sulphur coal, and from the introduction of high stack emission policies in industry. However, these moves, together with the increasing emissions from internal combustion engines, has lead to acid rain, ozone and PAN becoming pollutants of significant importance over large areas of the developed world. This change has resulted in the Hawksworth and Rose Scale being more difficult to interpret, because pollution tolerant lichens are re-invading cities where they were formerly absent. For example in northwest London, the mean winter sulphur dioxide levels fell from around 130 ug m^{-3} in 1980 to 29-55 ug m^{-3} in 1988. What was interesting was that an assemblage of zone 4-5 species on the scale, had failed to colonize the trees, although more sensitive species from zones 6 and 7 had. Some of the latter produce abundant reproductive propagules (soredia). This may mean that under conditions of very rapidly falling sulphur dioxide levels, recolonization does not follow an orderly sequence, where the more pollution tolerant species invade first (Hawksworth and McManus 1989). Another lichen which is currently showing a remarkable change in distribution is the beard lichen *Usnea* spp. In the British Isles, it was widespread until about 1800 but it then disappeared from an area covering some 70 000 km^2 as a result of increased pollution, especially sulphur dioxide (Seaward 1987; 1989). However, as sulphur dioxide levels have fallen, *Usnea* has re-established itself on trees such as *Fraxinus* and *Salix*. These trees have bark with a higher pH relative than most other trees (Farmer *et al.* 1990). As urban smoke and sulphur dioxide levels fall, a group of pollution-tolerant crustose lichens, including *Lecanora muralis* and *Xanthoria elegans*, have also progressively invaded roofs and stonework in what used to be lichen deserts (Henderson-Sellers and Seaward 1979). By monitoring such changes in distribution in a given area on a regular basis, it is possible to document the changing pollution climate with time.

Acid rain

The high stack emission policies have lead to transboundary pollution and an increasing awareness of acid rain. The very pollution sensitive species of ancient woodlands, especially those containing cyanobacteria as photobionts, appear to be damaged by acid rain. Sensitive genera such as the lungworts *Lobaria* spp. seem to be disappearing from many regions. This has been well-documented for *Lobaria scrobiculata* in southern Sweden where the species was formerly recorded from more than 300 localities. Recently, it could only be confirmed as still present at just two of 50 thoroughly investigated old sites (Hallinback 1989). It may be that even weak acid rain, from transboundary sources, progressively overcomes the buffering capacity of the tree bark to the point where the surface pH of the bark falls significantly (Nieboer *et al.* 1984). This may prevent the germination of *Lobaria* spores or the growth of the symbiotic cyanobacterium and thus prevent regeneration of thalli (Richardson 1988). By monitoring changes in the distribution of *Lobaria*, and recording whether it is changing its preferred substratum from all trees, to only those with more basic bark like ash, it is possible to use these plants as bioindicators and gain information as to the impact of transboundary acid rain. Another recent threat to lichens is ammonia emissions from intensive cattle rearing, which in parts of The Netherlands and Denmark, are injuring populations of *Cladonia portentosa*, *C. arbuscula* and *C. ciliata* on dunes and heathlands (Sochting 1987; De Bakker 1989).

Damage due to gaseous air pollutants

Our knowledge as to how gaseous air pollutants affect lichens is far from complete (Richardson 1988). The ability of lichens to accumulate anions (including dissolved gaseous pollutants such as sulphur dioxide) actively and rapidly, the lack of a protective cuticle, and poikilohydrous water relations, all contribute to the sensitivity of these plants. Of the various metabolic processes in lichens, laboratory exposure studies using sulphur dioxide have generally shown the following sequence of sensitivity: nitrogen fixation < photosynthesis, respiration < pigment status < potassium efflux. The damage mechanisms include: (1) enzyme perturbations; (2) interaction with reactive biomolecules; (3) formation of free radicals (Richardson and Nieboer 1983). Ultrastructural changes in the photobiont are among the earliest consequences of sulphur dioxide fumigation. As damage progresses, it is reflected by changes in the fluorescence of the algal cells which appear orange or white rather than red under the UV microscope. This feature provides another way in which to use lichens as bioindicators and assess air pollution in an area. Lichen thalli are transplanted into an industrial or urban area and the degree of pollution-induced damage assessed after a given time by retrieving the transplants. Sections are cut off retrieved lichen samples and the proportion of cells showing abnormal fluorescence recorded (Holopainen and Kauppi 1989). Recently it has also been found that fluorescence of the photobiont can be used to study the combined influence of temperature, lead and herbicide stress on lichens (Luhmann et al. 1989).

Recent studies in Switzerland indicate that the long-term effects of pollution on lichen metabolism may be different from that following briefer laboratory exposures or transplantation to urban sites. Thus investigations on thalli of *Parmelia sulcata* growing naturally in different urban pollution zones have shown that the overall protein content, dark respiration, gross and net photosynthesis of *Parmelia sulcata* did not change between lichens from urban and suburban habitats. However, lichens from the urban habitat exhibited up to 7 times less growth, and the amount of assimilate released by the algae was 15 times less than in lichens from the periphery of the city. The higher chlorophyll content recorded in lichen samples from the central city area was thought to be related to the stimulatory effects of nitrogen oxides from traffic emissions on chlorophyll synthesis (von Arb and Brunold 1989). In spite of all the research done to date, the reasons why different species of the same lichen genus, containing apparently closely related algal symbionts, vary in their pollution sensitivity has still to be revealed. However, the fact that different lichen species do exhibit a marked variation in pollution sensitivity has meant that studies on lichen distribution can yield useful information on air quality. As a result lichens are now widely accepted bioindicators of air quality (Hawksworth and Rose 1976; Henderson 1990; Nash and Wirth 1988).

Metallic pollutants

The ability of lichens to accumulate high levels of various metals and other elements has led to them being excellent bioindicators of atmospheric fallout around isolated pollution sources such as electricity generating stations, smelters, steel-making facilities, abandoned mining areas. Typically high elemental levels are found in lichens collected close to the emission source and these levels fall off rapidly at first and then more slowly (Bargagli *et al.* 1987; Nieboer and Richardson 1981; Henderson 1990). A similar phenomenon has been recorded in terms of fluoride accumulation near aluminium smelters and volcanoes (Perkins and Millar 1987a,b; Davies and Notcutt 1988). Lichens have also been used as bioindicators of changes in the pattern of elemental fallout over larger areas. Recently, in the case of urban areas, the advantage of combining data from lichen distribution studies with that derived from elemental analyses has been realised, for example, in Budapest and in various parts of Switzerland (Farkas *et al.* 1985; Herzig *et al.* 1989; Urech *et al.* 1990). In this way it is possible to come to conclusions about the composition and amount of important active pollutants. Deposition patterns for particular elements have also been measured over areas as large as Finland and Eastern Canada (Zakshek and Puckett 1986; Takala *et al.* 1990).

The efficiency of the metal uptake systems in lichens (see below) has resulted in samples of *Parmelia* being packed into porous PVC tubes and suspended in a tributary of the Calcasieu River in Louisiana for two weeks. The lichen proved a useful bioindicator for a wide variety of metal contaminants including Zn, Cu, Cr, Ni Fe, Mn and Hg (Beck and Ramelow 1990). Lichen material has also been successfully used to construct lichen-modified carbon paste electrodes to produce electrochemical biosensors (Connor *et al.* 1991).

Damage due to metallic pollutants

In general, lichens can accumulate high levels of metallic elements without apparent harm (Nash 1989). Lichens accumulate metallic elements by trapping insoluble particulates (usually oxides, sulphates and sulphides) (Milford and Davidson 1985; Tyler 1989), and taking up metal ions by extracellular ion exchange processes (Richardson 1988). The latter may be accompanied by a slower uptake into the cells (Brown 1985). It is the nature and the form of the accumulated element that dictates whether the lichen will die or show damage symptoms. Where metals are presented in soluble form, are present in excess, or are by their nature highly toxic (Nieboer and Richardson 1980), lichens are adversely affected (Nieboer *et al.* 1978, Richardson *et al.* 1978). The effects of zinc and particularly lead and copper ions are very damaging to lichens. The effects of zinc can be seen on tree trunks below the points at which barbed wire is attached (Seaward 1974) or on lichen-covered roofs in rural areas below galvanised supports of television aerials. The impact of copper telephone wires can also be seen in such situations. Gravestones are sometimes inlaid with lead to make the inscription more permanently visible and beneath the lead, leached ions prevent lichen growth.

Radionuclides

The efficiency with which lichens absorb radionuclides from the air and the importance of the lichen-reindeer-man foodchain has been known since the 1960s following the time when atmospheric nuclear bomb testing was widespread (Aberg and Huneal 1969). Studies revealed that Lapps had much higher Cs-137 levels than non Lapps due to the fact that Lapps consumed meat from the reindeer which had been feeding on lichen as a major winter food (Richardson 1975). However, it was the Chernobyl accident in April 1986, which lead to unacceptably high levels of radioactivity in the reindeer meat (over 10 000 becquerels of Cs-137 per Kg in many animals, with a legal limit for sale of only 300 becquerels per Kg) that has brought further focus to the subject (Mackenzie 1986; O'Clery 1986). Scientists around the world now conclude that lichens are excellent bioindicators of atmospheric aerosols owing to their high surface-to-mass ratio, slow growth rates and the fact that they derive their nutrients from atmospheric sources. These scavenging properties make them cost-effective monitoring devices for radionuclides (Schuepp 1984; Smith and Ellis 1990). The Chernobyl plume arrived in Canada 11 days after the accident and its movement across Canada provided a unique opportunity to validate a model describing radionuclide uptake by lichens. A deposition velocity of 1.1 cm sec^{-1} gave good agreement between the model and experimental results. Using this result, combined with atmospheric residence time data, it was possible to calculate that the height of the radioactive cloud as it passed over Canada was 10 000 m (Smith and Ellis 1990). Thalli of *Umbilicaria* in Southwest Poland exhibited higher levels of radioactivity when collected above 800 m. The lichen showed up to 165 times the radioactive levels recorded before the Chernobyl accident with maximum values in excess of 35 000 becquerels per kilogram (Seaward *et al.* 1988; pers. comm.). Similar results have been recorded for Austria (Turk 1988).

Chlorinated hydrocarbons

Reindeer lichen *Cladonia alpestris* was collected over a ten year period from 1960 to 1972 in Sweden. Analyses of the samples revealed a progressive increase in the levels in lichens. Results suggest that transport by air is one of the main routes in the dispersal of PCBs and DDTs and that the mean residence time in the atmosphere was in the order of 2-3 years (Villeneuve and Holm 1984).

Lichens have also been used successfully as bioindicators of chlorinated hydrocarbons in the Antarctic. a-HCH levels were an order of magnitude lower than samples from Sweden and Finland and up to two orders of magnitude lower than those from Italy. HCB levels were closely similar from all sites with Italian values being slightly higher. The predominance of L-HCH in the Antarctic samples which makes up more than 90% of the isomer in current products, may indicate a rather recent arrival of these components in the

Antarctic (Bacci *et al.* 1986). More recent data from lichens collected in southern France reveals rather similar levels to those from Italy. Generally, the concentration of insecticides and toxaphene in the lichens increased with altitude while individual PCBs were irregularly distributed. It is thought that in insecticides were carried by the wind from agricultural areas in the Rhone Valley. Concentration factors in lichens have been calculated for PCBs and are around 1×10^5 while that for Arochlor 1254 was three times this. Such values indicate that lichens are suitable bioindicators of organochloride compounds (Villeneuve *et al.* 1988).

Concluding remarks

It is clear from the above review that lichens are useful bioindicators. The value of these plants can be optimized by selection of the appropriate experimental designs, use of standard collection protocols, and the application of suitable computer-aided data analysis. A full discussion of these topics is provided by Nash and Wirth (1988). In studies on the elemental content of lichens, there is a wide range of available analytical techniques. Non destructive methods such as X-ray fluorescence analysis or neutron activation have distinct advantages in some situations; e.g. where a baseline study which is part of an environmental impact statement is required to be repeated at regular intervals after start-up of an industrial complex. With non-destructive techniques, samples can be archived and the original samples checked against later collected samples for elements not originally quantified. For destructive analyses, atomic absorbtion spectrometry, either flame or electrothermal has been the most popular technique to date, although inductively coupled plasma analysis with or without mass spectrometry is now becoming more common. Finally, electrochemical methods such as square wave voltammetry using modern microprocessor controlled polarographs provide rapid, very sensitive methods for a range of inorganic and organic pollutants. With the availability of these powerful analytical techniques, the use of lichens as bioindicators is not limited to establishing zones of air quality in urban areas or predicting levels of gaseous air pollutants. Lichens act as integraters and are therefore increasingly used to provide qualitative and sometimes quantitative information, as well as fallout-distance data on a wide range emissions from urban, industrial and agricultural sources.

References

Aberg, B. and Huneal, F.P. (1969) *Radioecological Concentration Processes.* Pergamon Press, Oxford.

Bacci, E., Calamari, D., Gaggi, C., Fanelli, R., Focardi, S. and Morosini, M. (1986) Chlorinated hydrocarbons in lichen and moss samples from the Antarctic Peninsula. *Chemosphere*, **15**, 747-754.

Bargagli, R., Iosco, F.P. and Barghigiani, C. (1987) Assessment of mercury disposal in an abandoned mining area by soil and lichen analysis. *Water, Air, and Soil Pollution*, **36,** 219-225.

Beck, J.N and Ramelow, G.J. (1990) Use of lichen biomass to monitor disolved metals in natural waters. *Bulletin Environmental Contamination and Toxicology*, **44,** 302-308.

Brown, D.H. (1985) *Recent Advances in Lichen Physiology.* Plenum Press, London.

Connor, M., Dempsey, E., Smyth, M.R. and Richardson, D.H.S. (1991) Determination of some metal ions using lichen-modified carbon paste electrodes. *Electroanalysis*, **3,** (in press).

Davies, F.B.M. and Notcutt, G. (1988) Accumulation of fluoride by lichens in the vicinity of Etna volcano. *Water, Air, and Soil Pollution*, **42,** 365-371.

De Bakker, A.J. (1989) Effects of ammonia emission on epiphytic lichen vegetation. *Acta Botanica Neerlandica*, **38,** 337-342.

Farkas, E., Lokos, L. and Verseghy, K. (1985) Lichens as indicators of air pollution in the Budapest agglomeration. *Acta Botanica Hungarica*, **31,** 45-68.

Farmer, A.M., Bates, J.W. and Bell, N.J.B. (1990) A comparison of methods for the measurement of bark pH. *Lichenologist*, **22,** 191-194.

Hallinback, T. (1989) Occurrence and ecology of the lichen *Lobaria scrobiculata* in Southern Sweden. *Lichenologist*, **21,** 331-342.

Hawksworth, D.L. and McManus, P.M. (1989) Lichen recolonization in London (UK) under conditions of rapidly falling sulphur dioxide levels, and the concept of zone skipping. *Botanical Journal of the Linnaean Society*, **100,** 99-110.

Hawksworth, D.L. and Rose, F. (1970) Qualitative scale for estimating sulphur dioxide air pollution in England and Wales using epiphytic lichens. *Nature* (London), **227,** 145-148.

Hawksworth, D.L. and Rose, F. (1976) *Lichens as Air Pollution Monitors.* Studies in Biology no 66. Edward Arnold, London.

Henderson, A. (1990) Literature on air pollution and lichens XXXI. *Lichenologist*, **22,** 173-182.

Henderson-Sellers, A. and Seaward, M.R.D. (1979) Monitoring lichen reinvasion of ameliorating environments. *Environmental Pollution*, **19,** 207-213.

Herzig, R., Urech, M., Liebendorfer, L., Ammann, K., Guecheva, M. and Landolt, W. (1989) Lichens as biological indicators of air pollution in Switzerland: passive biomonitoring as part of an integrated measuring system for monitoring air pollution. In H. Lieth and B. Markert (eds.) *Element Concentration Cadasters in Ecosystems.* VCH Verlagsgesellschaft, Osnabruck.

Holopainen, T. and Kauppi, M. (1989) A comparison of light, fluorescence and electron microscopic observations in assessing the SO_2 injury of lichens under different moisture conditions. *Lichenologist*, **21,** 119-134.

Laundon, J.R. (1986) *Lichens.* Shire Publications, Aylesbury.

Luhmann, H.J., Wietschorke, G. and Kreeb, K.H. (1989) Influences of combined temperature, lead and herbicide stresses on lichen fluorescence and their mathematical modelling. *Photosynthetica*, **23**, 71-76.

Mackenzie, D. (1986) The rad-dosed reindeer. *New Scientist*, **1539**, 37-40.

Milford, J.B. and Davidson, C.I. (1985) The size of particulate trace elements in the atmosphere - a review. *Air Pollution Control Association Journal*, **35**, 1249-1260.

Nash, T.H. (1989) Metal tolerance in lichens. In J. Shaw (ed.) *Heavy Metal Tolerance in Plants: Evolutionary Aspects*, 119-132. CRC Press, Boca Raton.

Nash, T.H. and Wirth, V. (1988) *Lichens, Bryophytes and Air Quality*. Cramer, Berlin-Stuttgart.

Nieboer, E., Richardson, D.H.S., Lavoie, P. and Padovan, D. (1978) The role of metal ion binding in modifying the toxic effects of sulphur dioxide on the lichen *Umbilicaria muhlenbergii*. I. Potassium efflux studies. *New Phytologist*, **82**, 621-632.

Nieboer, E., McFarlane, J.D. and Richardson, D.H.S. (1984) Modifications of plant cell buffering capacities by gaseous air pollutants. In M. Koziol and F.R. Whatley (eds.) *Gaseous Air Pollutants and Plant Metabolism*, 313-330. Butterworths, London.

Nieboer, E. and Richardson, D.H.S. (1980) The replacement of the nondescript term heavy metals by a biologically and chemically significant classification of metal ions. *Environmental Pollution*, Series B, **1**, 3-26.

Nieboer, E. and Richardson, D.H.S. (1981) Lichens as monitors of atmospheric deposition. In S.J. Eisenreich (ed.) *Atmospheric Pollutants in Natural Waters*, 339-388. Ann Arbor.

O'Clery, C. (1986) The Chernobyl fallout: How Lapland is paying the price. *Irish Times* (weekend supplement), September 13.

Perkins, D.F. and Millar, R.O. (1987a) Effects of airborne fluoride emissions near an aluminium works in Wales: Part 1 - Corticolous lichens growing in broadleaved trees. *Environmental Pollution*, **47**, 63-78.

Perkins, D.F. and Millar, R.O. (1987b) Effects of airborne fluoride emissions near an aluminium works in Wales: Part 2 - Saxicolous lichens growing on rocks and walls. *Environmental Pollution*, **48**, 185-196.

Richardson, D.H.S. (1975) *The Vanishing Lichens: Their History, Biology and Importance*. David and Charles, Newton Abbot.

Richardson, D.H.S. (1987) Lichens as pollution indicators in Ireland. In D.H.S. Richardson (ed.) *Biological Indicators of Pollution*, 155-168. Royal Irish Academy, Dublin.

Richardson, D.H.S. (1988) Understanding the pollution sensitivity of lichens. *Botanical Journal of the Linnean Society*, **96**, 31-43.

Richardson, D.H.S. and Nieboer, E. (1983) Ecophysiological responses of lichens to sulphur dioxide. *Journal of the Hattori Botanical Laboratory*, **54**, 331-351.

Richardson, D.H.S., Nieboer, E., Lavoie, P. and Padovan, D. (1978) The role of metal ion binding in modifying the toxic effects of sulphur dioxide on the lichen *Umbilicaria muhlenbergii* II. 14C-fixation studies. *New Phytologist*, **82**,

633-643.

Schuepp, P.H. (1984) Observations on the use of analytical and numerical models for the description of transfer to porous surface vegetation such as lichen. *Boundary-Layer Meteorology*, **29**, 59-73.

Seaward, M.R.D. (1974) Some observations on heavy metal toxicity and tolerance in lichens. *Lichenologist*, **6**, 158-164.

Seaward, M.R.D. (1987) Effects of quantitative and qualitative changes in air pollution on the ecological and geographical performance of lichens. In T. Hutchinson (ed.) *The Effects of Atmospheric Pollutants on Forests, Wetlands and Agricultural Ecosystems*, 439-450. Springer-Verlag, Berlin.

Seaward, M.R.D. (1989) Lichens as pollution monitors: adapting to modern problems. In M.A. Ozturk (ed.) *Plants and Pollutants in Developed and Developing Countries*, 307-319. University of Bornova, Izmir.

Seaward, M.R.D., Heslop, J.A., Green, D. and Bylinska, E.A. (1988) Recent levels of radionuclides in lichens from southwest Poland with particular reference to Cs-134 and Cs-137. *Journal of Environmental Radioactivity*, **7**, 123-129.

Smith, J.N. and Ellis, K.M. (1990) Time dependent transport of Chernobyl radioactivity between atmospheric and lichen phases in Eastern Canada. *Journal of Environmental Radioactivity*, **11**, 151-168.

Sochting, U. (1987) Injured reindeer lichens in Danish lichen heaths. *Graphis Scripta*, **1**, 103-106.

Takala, K., Olkkonen, H., Jaaskelainen, J. and Selkainaho, K. (1990) Total chlorine content of epiphytic and terricolous lichens and birch bark in Finland. *Annales Botanici Fennici*, **27**, 131-137.

Turk, R. (1988) Bioindikation von Luftverunreinigungen mittels Flechten. In D. Grill and H. Guttenberger (eds.) *Okophysiologische Probleme durch Luftverunreinigungen*, 13-27. Karl-Franzens-Universitat, Graz.

Tyler, G. (1989) Uptake, retention and toxicity of heavy metals in lichens: a brief review. *Water, Air, and Soil Pollution*, **47**, 321-333.

Urech, M., Peter, K., Liebendorfer, L. and Herzig, R. (1990) Bioindikation mit Flechten im Kanton Luzern. *Mitteilungen der Naturforschenden Gesellschaft Luzern*, **31**, 219-232.

Villeneuve, J.P. and Holm, E. (1984) Atmospheric background of chlorinated hydrocarbons studied in Swedish lichens. *Chemosphere*, **13**, 1133-1138.

Villeneuve, J.P., Fogelqvist, E. and Cattini, C. (1988) Lichens as bioindicators for atmospheric polluton by chlorinated hydrocarbons. *Chemosphere*, **17**, 399-403.

Von Arb, C. and Brunold, C. (1989) Lichen physiology and air pollution. I. Physiological responses of *in situ Parmelia sulcata* among air pollution zones within Biel, Switzerland. *Canadian Journal of Botany*, **68**, 35-42.

Zakshek, E.M. and Puckett, K.J. (1986) Lichen sulphur and lead levels in relation to deposition patterns in Eastern Canada. *Water, Air and Soil Pollution* **30**, 161-169.

Zobel, K. (1988) The indicator value of a set of lichen species assessed with the help of log-linear models. *Lichenologist*, **20**, 83-92.

Natural Bacterial Communities as Indicators of Pollutants in Aquatic Environments

J. Martinez[1], Y. Soto [2], J. Vives-Rego[1] and M. Bianchi [2(*)]

[1]Departament de Microbiologia, Universitat de Barcelona, Av Diagonal, 645, 08028-Barcelona, Spain.
[2]Microbiologie Marine, Facult des Sciences de Luminy, Case 907, Avenue de Luminy, 13288 Marseille Cedex 9, France.
(*) Corresponding author.

Key words: thymidine incorporation, bacterial assemblages, ecotoxicity, aquatic environments.

Abstract

The aquatic toxicity of compounds depends on environmental parameters as well as on the organisms being tested. Because of self-depuration mechanisms and detoxification processes, the effect of pollutants on marine bacteria is of great interest. The toxicity of Cu^{++}, Ni^{++} and linear alkylbenzene sulfonate (LAS) to the bacterioplankton in the Rhône River plume (Mediterranean Sea), has been assessed by ^{3}H-thymidine incorporation into trichloroacetic acid-insoluble material. The concentration of the test chemical causing 50% inhibition of the activity measured was EC_{50}. This method was useful to evaluate the toxic effect of the three common pollutants to the natural bacterial population in a system with permanent mixing between freshwater and seawater. The toxic effect that we measured in this work referred to bacteria uptaking thymidine, i.e. active or duplicating bacteria.

An important variability in the toxicity has been detected irrespective of bacterial densities or activities. According to the observed mean EC_{50}, the relative toxicity was in decreasing order: $Cu^{++} > Ni^{++} > LAS$. A high negative correlation was observed between Cu^{++} and particulate matter. There was an antagonistic interaction between Ni^{++} and particulate matter, salinity and nitrite-nitrate, as shown by the positive and high correlation between this pollutant and the corresponding parameters. In the case of LAS, the positive correlation with ammonia and phosphate clearly suggested an antagonistic interaction.

In conclusion, ^{3}H-thymidine incorporation is a sensitive methodological approach in the evaluation of pollutant impacts on bacterial populations in stressed aquatic environments, such as rivers.

Bioindicators and Environmental Management
ISBN 0-12-382590-3

Purpose of the procedure

Natural aquatic environments are subjected to increasing ecotoxicological problems because of the release of toxic contaminants due to industrialization. Because of their capabilities to respond rapidly to environmental changes, especially in aquatic systems, bacteria are good indicators of pollution (Bianchi and Colwell 1985). Usually, single-species tests are used to assess relative toxicity of the compounds (Bitton and Dutka 1986). In aquatic environments, however, (i) the toxicity of compounds can be modified by specific environmental parameters, i.e. temperature, salinity, organic matter, which can vary as a function of the season or the latitude; (ii) the response of *in situ* microbes will depend not only on environmental parameters, but also on specific biotic factors, i.e. eutrophication or oligotrophy of the ecosystem, numerical abundance, taxonomic composition and physiologic state of the individuals.

Furthermore, because of basic ecological processes such as self-depuration and detoxification mechanisms, it is of great interest to check the effect of pollutants on the natural microbial communities. For a particular site, such specific determinations lead to appropriate predictive measures.

Principle of method

In studies of aquatic microbial ecology, heterotrophic bacteria have been recognized as important components in the function of the ecosystems. Rates of bacterial production can be used to estimate the average growth rate of bacterial assemblages and, consequently, as an indicator of the response of bacteria to fluctuations in environmental conditions.

Determining the rate of bacterial DNA synthesis is conceptually an elegant approach for estimating bacterial production in aquatic environments. A widely used field procedure is the incorporation of (methyl-^{3}H)thymidine into DNA, because of its specificity for heterotrophic bacteria, and its experimental simplicity.

In the assay, environmental water samples are incubated with the radio-labeled thymidine, and the amount of radioactive label incorporated, is measured as cold-trichloroacetic acid (TCA)-insoluble material (Fuhrman and Azam 1982). The results are expressed as rates of incorporation. The inhibitory effect of the tested pollutant is demonstrated by comparing the rate of thymidine incorporation into TCA-insoluble fraction of bacterial assemblages, between samples receiving the toxicant with controls (Bauer and Capone 1985; Vives-Rego *et al.* 1985). The measured toxic effect can be referred to the active population when the experiment is complete as bacteria absorbing thymidine. On the other hand, with *in situ* samples, possible synergistic or antagonistic effects between environmental characteristics and toxicants can be established.

The (methyl-^{3}H)thymidine incorporation method, as a criterion of toxicity and to assay the toxic effect of pollutants in aquatic environments, has already been used for punctual samples (Bauer and Capone 1985; Jonas *et al.* 1984; Martinez and Vives-Rego 1988; Riemann and Lindgaard-Jorgensen 1990).

Description of the method and technical information

Triplicates of water samples were incubated at room temperature with 20 nM (methyl-^{3}H)thymidine (86-90 Curies/mmol, Amersham) for 1 hour. The cold trichloroacetic (5%) insoluble material was collected through filters of 0.2 μm pore size. The effect of pollutants on thymidine incorporation was examined after the water samples were exposed for 60 minutes at room temperature to toxicants at 4 or 6 different concentrations. A positive control was prepared without toxicant.

Time: from the sampling to the result - around 5 hours for a first estimation.
Equipment: Epifluorescence microscope, filter, scintillation counter.
Material: (methyl-^{3}H)Thymidine, trichloroacetic acid, membrane filters (0.22 μm), scintillation liquid, vials.

For each concentration of the tested pollutant, the percentage of activity was calculated with reference to the control. EC_{50} (**concentration of the test chemical causing 50% inhibition of the activity measured**) was calculated from the linear regression of the percentage of thymidine incorporation, with respect to the positive control without toxicant, against the logarithm of pollutant concentration (Fig. 1). A weak effect of the toxicant (bacterial assemblages are still able to grow normally) is demonstrated by a high value of EC_{50} (curve a on Fig. 1). At the opposite, a strong inhibitory effect is shown by a drastic decrease of the bacterial growth rate and, consequently, a small value of EC_{50} (curve b on Fig. 1).

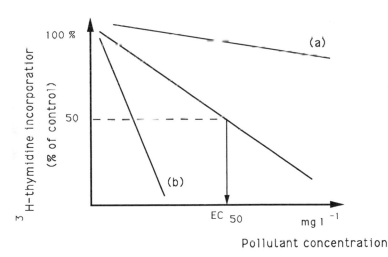

Figure 1. Calculation of EC_{50} from the linear regression of the percentage of thymidine incorporation, with respect to the positive control without toxicant, against the logarithm of pollutant concentration.

Specimen results

First example : fluctuations of EC$_{50}$ values in an estuarine outflow plume
We checked the use of the proposed method as toxic criterion and to assay the toxic effect of Cu^{++}, Ni^{++} and linear alkylbenzene sulfonate (LAS) to the bacterioplankton in the Rhône River plume (Mediterranean Sea). The Rhône River plume constitutes an important nutrient-rich freshwater input (1000 - 2000 m^3 s^{-1}) into the oligotrophic Mediterranean seawater.

Water samples were taken at surface level and along depth profiles (0 - 25 m) during two cruises (spring and fall). Bacterial counts were performed by epifluorescence microscopy (Porter and Faig 1980). Temperature and salinity were measured using a Yellow Spring Instrument meter and probe.

The salinity gradient was well correlated with a decrease in bacterial density. The biomass declined when the salinity increased. However, the correlation between salinity and incorporation of (methyl-^{3}H)thymidine was not significant, indicating that there were not activity changes with the dilution of the freshwater, as previously described (Kirchman *et al.* 1989).

Ecotoxicity was assayed in water samples ranging from 0 to 38 parts per thousand salinity at different stations of the plume. The dose-response regression lines corresponding to each individual assay gave a correlation index superior to 0.90, indicating that the obtained EC$_{50}$ are statistically significant. In general, the toxicity exhibited by Cu^{++} was very high, whereas Ni^{++} and LAS showed a lower toxic effect at any time (Table 1). According to the mean EC$_{50}$ obtained in the experiments, the relative toxicity was in decreasing order : Cu^{++}, Ni^{++}, and LAS.

Along the plume, the toxicity of pollutants varied. Considering natural bacterial populations, we were able to observe : (i) the toxicity of Cu^{++} remained high throughout dilution of the river water in the seawater; (ii) in the plume water (20 ppt combined with high suspended matter load) the toxicity of Ni^{++} decreased by about three times; (iii) the toxicity of LAS increased strongly in the plume water and in the seawater (Table 1).

Table 1. First example : fluctuations of the EC$_{50}$ from a river water to the seawater (Rhône River plume).

EC$_{50}$ (mg l^{-1})	River water	River mouth	From the coast in the seawater 2 km	10 km
Ni^{++}	3.10	3.55	21.95	4.00
LAS	17.50	21.40	4.65	1.15
Salinity (ppt)	0.00	0.15	20.00	37.30
Suspended matter (mg l^{-1})	15.7	19.4	14.7	6.42

Table 2. Second example : fluctuations of the EC_{50} in different geographical areas of the Mediterranean Sea.

$EC_{50}(mg\ l^{-1})$	Cu^{++}	LAS
Rhône Delta	0.032	10.80
Ebro Delta	0.091	1.25
Mediterranean seawater	0.80	1.06

Second example : fluctuations of EC_{50} in different geographical areas of Mediterranean Sea

Considering the toxic effect of Cu^{++} and LAS on the microbial assemblages of a non-polluted Mediterranean seawater, and the water of two Mediterranean river plumes (Rhône and Ebro), Table 2 shows two contradictory results. From the Rhône plume to the Ebro plume and then the Mediterranean seawater, the toxicity effect of Cu^{++} decreased while the toxicity of LAS increased strongly.

Comparing several lakes and marine localities (island of Zealand, Denmark) Riemann and Lindgaard-Jorgensen (1990) also found fluctuations in EC_{50} values established for 3.5-dichlorophenol, 2.4-dinitrophenol and potassium dichromate.

Applications: actual and potential

The (methyl-^{3}H)thymidine incorporation represents a sensitive method in evaluating the impact of pollutants on bacterial populations in aquatic environments. This method is particularly valuable in environmental sites where abiotic parameters can vary greatly.

In conclusion, this method gives an evaluation of the toxic effect of pollutants on natural environments. But, in addition to such a microbial measure, it would be necessary to evaluate exposure of localities to agricultural and/or industrial pollutants in order to determine which factor could have influenced adaptation of the bacterial assemblages. This technique still needs to be used more intensively in environmental monitoring.

References

Bauer, J. and Capone, D.G. (1985) Effect of four aromatic organic pollutants on microbial glucose metabolism and thymidine incorporation in marine sediment. *Applied and Environmental Microbiology,* **49**, 828-835.

Bianchi, M. and Colwell, R.R. (1985) Microbial indicators of environmental water quality. In J. Salánki (ed.) *Biological monitoring of the state of the environment : Bioindicators,* 5-15. IRL Press, Oxford.

Bitton, G. and Dutka, B.J. (1986) *Toxicity testing using microorganisms.* CRC Press Inc. Florida.

Fuhrman, J. and Azam F. (1982) Thymidine incorporation as a measure of heterotrophic bacterioplankton production in marine surface waters : evaluation and field results. *Marine Biology*, **66**, 109-120.

Jonas, R.B., Gilmour, C.G., Stoner, D.L., Weir, M.M. and Tuttle, J.H. (1984) Comparison of methods to measure acute metal and organometal toxicity to natural aquatic microbial communities. *Applied and Environmental Microbiology*, **47**, 1005-1011.

Kirchman, D., Soto, Y., Van Wambeke and Bianchi, M. (1989) Bacterial production in the Rhône River plume : effect of mixing on relationships among microbial assemblages. *Marine Ecology - Progress Series*, **53**, 267-275.

Martinez, J. and Vives-Rego, J. (1988) Thymidine incorporation and bacterial mortality as ecotoxicological assessment in aquatic habitats. *Toxicity Assessment*, **3**, 33-40.

Porter, K.G. and Faig, Y.S. (1980) The use of DAPI for identifying and counting aquatic microflora. *Limnology and Oceanography*, **29**, 943-948.

Rieman, B. and Lindgaard-Jorgensen, P. (1990) Effects of toxic substances on natural bacterial assemblages determined by means of ^{3}H-Thymidine incorporation. *Applied and Environmental Microbiology*, **56**, 75-80.

Vives-Rego, J., Billen, G., Fontigny, A. and Somville, M. (1985) Free and attached proteolytic activity in water environments. *Marine Ecology - Progress Series*, **21**, 245-249.

Bacterial Bioindicators: Biomass Determinations of Methanogenic Bacteria

G.C. Smith and G.D. Floodgate

School of Ocean Sciences, University College of Wales, Bangor, Menai Bridge, Gwynedd, LL59 5EY, U.K.

Key words: methanogenic bacteria, membrane lipids, biomass, detection limit.

Abstract

Methane-forming bacteria belong to the recently devised archaebacterial kingdom and as such possess a number of unique biochemical features. The potential of these attributes with respect to methanogenic bioindication is reviewed. Characteristic membrane lipids found in methanogens are one such feature where phytanyl ether linked phospholipids replace the more usual ester linked analogues. The most common methanogenic membrane ether lipid (Di-Phytanyl Glycerol Ether; DPGE) can be extracted using solvents, chemically broken down, derivatized, purified using thin layer chromatography and finally analysed quantitatively by capillary gas chromatography. In order to evaluate the potential of DPGE as a means of estimating the biomass of methanogens, it was compared with cell numbers, methane production and turbidity at 578 nm in a controlled growth experiment of a marine methanogenic monoculture of *Methanolobus tindarius*. Good correlation between lipid concentration and the other parameters used in the growth experiment was achieved, tracing relatively subtle growth phase changes. It also has the added advantage over direct methane measurements of being more applicable and representative in environmental analysis. Preliminary environmental samples taken from both a marine inter-tidal and a freshwater site were also analysed for the DPGE lipid. The detection limit of this method using a flame ionization detector is at 2.35×10^{-12} moles DPGE lipid with a possible extra sensitivity of 2×10^{-15} moles DPGE yet to be achieved by the use of electron capture detection.

Introduction

The observation and interpretation of biological changes associated with environmental perturbations is the essence of the rationale of bioindicators.

Bioindicators and Environmental Management
ISBN 0-12-382590-3

Such biological changes can give integrated information over a period of time that cannot otherwise be easily obtained. Organisms can also provide an early warning of environmental changes providing that they respond rapidly in a way that is easy to detect. Micro-organisms, particularly bacteria, are excellent organisms for both these purposes. This is because their response time to environmental disturbances is generally short, and usually observed in an increase in certain bacterial numbers. The typical generation time of bacteria in the natural environment is measured in hours or days depending on the conditions. Hence a relatively small environmental change can result in a large change in bacterial biomass within a short time.

The major drawback to the use of bacteria as indicators is that their small size makes them difficult to count, and that examination under the microscope does not usually allow differences in biological function and activity to be expressed. This can be overcome in many instances by growing the organisms on suitable media, some of which are to be quite specific for the growth of a particular physiological group of bacteria. Unfortunately very many species cannot be cultivated easily or at all, so that changes in numbers cannot be fully recognized.

The development over the last decade of new and sensitive methods of biochemical analysis has, however, opened up another avenue of approach. Prokaryotes are distinguished from eukaryotes not only by a different organisation of the cell, but by their possession of a number of macromolecules which are unique to them, and in some cases, to certain groups within the bacteria. Chemical detection and estimation of the variation in the amounts of these substances using the new chemical capabilities then provides a way of detecting changes in the composition of bacterial communities and therefore a means of monitoring environmental disturbance. A list of the substances in common use is shown in Table 1.

The review and data presented in this paper shall be concerned with the bioindication of a diverse group of bacteria that are able to biologically generate methane and are known as methanogens. These together with certain thermoacidophilic and halophilic bacteria constitute a separate kingdom of bacteria known as the archaebacteria; a distinction that is made due to the possession of a number of unique biochemical features that clearly differentiate them from all other prokaryotic and eukaryotic life (Balch *et al.* 1979).

As the name would suggest the archaebacteria are thought to have diverged from the other lines of descent at least by the time when the prokaryotic and eukaryotic groups became phylogenetically distinct kingdoms. This distinction has led to the renaming of the traditional prokaryotic kingdom into the eubacterial and archaebacterial primary kingdoms.

Methanogens are mesophilic or thermophilic bacteria that exhibit extreme habitat diversity; species have been isolated from virtually every habitat in which anaerobic biodegradation of organic compounds occurs. Such environments are of low redox potential and include the guts of ruminant animals and termites (Smith and Hungate 1958), paddy fields (Holzapfel-Pschorn and Seiler 1986), swamps (Archer and Harris 1986), marshes (Jones and

Table 1. General biomarkers of microbial biomass (Parkes 1987).

Biomarker	Measures	Reference
Phospholipid 16:0	Total biomass, eukaryotes and prokaryotes	White *et al.* (1979) Harwood & Russell, 1984
Glycerol or fatty acids released from phospholipid	The biomass of some bacteria	Gheron and White (1983)
Cyclopropyl fatty acids	The biomass of some bacteria	Harwood and Russell (1984)
Lipopolysaccharide (LPS)	Gram negative bacteria	Rogers (1983)
Teichoic acid	Most gram positive bacteria	Rogers (1983)
Muramic acid	All bacteria except archaebacteria	Rogers (1983)
Plasmalogens	Some anaerobic bacteria	Goldfine & Hagen, 1972
Hopanoids	Some bacteria	Rohmer, Bouvier-Nave & Ourisson (1984)
Phytanyl ether lipids	Archaebacteria	Harwood & Russell, 1984
Polyenoic fatty acids	Eukaryotes and Gliding bacteria	Harwood & Russell (1984)
Phytopigments	Photoautotrophs	Gillan & Johns (1983)

Paynter 1980), sewage digesters (Henson *et al.* 1985), marine snow (Conrad and Seiler 1988), landfill sites, marine and freshwater sediments (Bernard 1979; Harris *et at.* 1984). Hence any bioindication technique must be flexible enough to function in a wide range of circumstances. Methanogenic bacteria can utilize only a limited range of substrates for growth and methanogenesis (Smith *et al.* 1980). These substrates are products of previous consecutive oxidation reactions and have relatively little free energy available for ATP formation compared to other fermentation reactions which occur at much higher redox potentials. The energy yields for the substrates typically available to methanogens are given in Table 2.

Extreme halophilic bacteria require habitats of up to 5M sodium chloride and are typically capable of photosynthetic carbon dioxide assimilation. Thermoacidophiles either utilize organic substrates aerobically or assimilate carbon dioxide through the oxidation of ferrous iron or sulphur and tend to be confined to habitats of low pH and high temperature (pH=0.5, T=90°C) such as hot springs and hydrothermal vents.

The process of biological methane formation is of considerable current interest since methane emitted to the atmosphere may affect the earth's climate (Burke and Sackett 1986). The concentration of atmospheric methane has been calculated to be increasing by approximately 1% per year (Blake *et al.* 1982; Rasmussen and Khalil 1984). The methane molecule, which absorbs radiation

Table 2. Methanogenic Substate Energy Yields (Archer and Harris, 1986).

Reaction	$dG^{o\prime}$ KJ/reaction
Acetate + H_2O ----> CH_4 + HCO_3^-	-31.0
4 Formate + H_2O + H^+ ----> CH_4 + $3HCO_3^-$	-130.7
$4H_2$ + HCO_3^- + H^+ ----> CH_4 + $3H_2O$	-135.6
$4CH_3NH^+$ + $3H_2O$ ----> $3CH_4$ + HCO_3^- + $4NH_4^+$ + H^+	-225.4
$4CH_3OH$ ----> $3CH_4$ + HCO_3^- + H^+ + H_2O	-315.1

between 700 and 1400 cm^{-1}, is a "greenhouse gas" and as such is directly related to global warming. The contribution of methane is approximately 20-40 % of that attributed to the increases in atmospheric carbon dioxide (Craig and Chou 1982; Lacis *et al.* 1981).

More than 80% of the earth's annual output of atmospheric methane is produced biologically in anaerobic environments already mentioned. Most of which is attributed to the enteric fermentations of animals, paddy fields, swamps and marshes (Archer and Harris 1986).

Besides the atmospheric concern of biological methane formation it has also been observed at high pressure in landfill sites which can be dangerous if not released in a controlled manner. Again methane formed close to the surface in marine sediments has the effect of blanketing out seismic stratigraphical information as well as presenting drilling operations with potentially dangerous high pressure sources of methane known as "shallow gas" (Hovland and Judd 1988).

However, not all the effects of methane are detrimental to mankind. The biogenesis of methane under controlled conditions has the potential to become a significant source of renewable energy, which is an important step in a future of predicted limited resources. The synthesis of diesel from methane (Stat-Oil Newsletter, 1989) only acts to emphasise the future benefits of biogenic methane.

For these reasons stated there is a need to detect and quantify the biomass of the methanogenic bacteria responsible for methane production. This can be achieved by using the biochemically distinct and unique features possessed by methanogens, which together account for their incorporation in the archaebacterial kingdom. Such features currently employed to determine the methanogenic biomass are reviewed below together with experimental data to support the use of unique membrane phospholipids.

1. Ribosomal RNA

One of the most significant biochemical features possessed by the archaebacteria is the structurally distinct and unique 16S ribosomal RNA's as discovered by partial sequencing (Woese 1981). Such features were instrumental in the formulation of the archaebacterial kingdom since ribosomal RNA molecules appear to remain constant in function over great evolutionary periods. This as a technique of estimating methanogenic biomass has not been tested as yet partly due to the difficulty of sample workup and the instrumental expense.

2. Methane Formation

One of the most obvious indications of the presence of methanogenic bacteria in sediments is their ability to produce methane, a process directly linked to energy generation and found to mainly occur relatively close (<1000 m) to the marine sediment surface. Methane synthesis is not unique to methanogenic bacteria since some anaerobic eubacteria evolve small amounts of methane during their normal metabolism (Postgate 1969). However, this is not coupled to energy production and is also not considered to be a major input of methane into the sedimentary system. Besides biogenic methane, thermogenic sources of methane also exist, formed by the cracking of long chain hydrocarbons deep in the sediment column (>>1000 m). Upward migration of thermogenic methane can often lead to mixing with shallower biogenic sources, therefore confusing the scientific interpretation of such gaseous origins. Although biogenic methane has its own characteristic isotopic ratio with respect to carbon 12/13, chemical and biological redox reactions on both biogenic and thermogenic methane can often confuse the isotopic ratio and render it ambiguous as a means of estimating how much of the methane present is produced by methanogens (Hovland and Judd 1988). Recent work by Jorgensen *et al.* (1990) has shown that a ratio of both hydrogen and carbon isotopic compositions is better at discriminating between biogenic and thermogenic sources.

Although such *in vivo* measurements of methane are not suitable indicators of the abundance of methanogenic organisms, *in-vitro* analysis can prove more productive as a direct indicator of biomass. This can be achieved in the use of most probable number (MPN) estimations of methanogenic bacteria in a particular sediment sample. The method involves serial dilutions of the sediment sample in a suitable liquid growth media to the point of methanogenic extinction. This point is realized by the lack of methane production upon incubation. From sample replication, dilution rates and statistical tables the number of methanogens can be estimated usually to well within an order of magnitude (Rowe *et al.* 1977). Under-estimation of the microbial population is common, an error encountered when the liquid growth media exerts selective pressures. This can causes only part of the methanogenic population to grow (Jones 1979). The confidence limits of the MPN method are relatively wide and the statistical reliability is quite low Not withstanding these inherent shortcomings, the MPN method can provide an indication of the order of magnitude of methanogens in a particular sediment sample (Seyfried and Owen 1978).

3. Coenzymes

Methanogenic bacteria can be microscopically identified by their strong autofluorescence under oxidizing conditions at ultra-violet wavelengths. The contributing components to this phenomenon are several unique coenzymes that act as electron carriers and include coenzyme $F_{420'}$ $F_{350'}$ coenzyme M and some methanopterin derivatives (Moura *et al.* 1983). F_{420} shows the strongest fluorescence and is found in all methanogenic cells having an absorbtion maximum at 420 nm with an associated emission fluorescence at 470 nm. It acts as a low potential electron carrier (Zeikus 1977) and was recently shown to participate directly in the reduction of carbon dioxide to methane (Hartzell *et al.* 1985).

Although tentative identification of methanogenic bacteria can be made using ultra violet microscopy Doddema and Vogels (1978) found that fluorescence decreased with both the aging of cells and also over exposure to UV light. *Escherichia coli* and *Pseudomonas aeruginosa* have also been shown to weakly fluoresce at 420 nm, though the enzyme responsible is different in structure, but no fluorescence is observed at at 350 nm. The F_{420} chromophore has recently been discovered in the light activated DNA repair enzyme of *Streptomyces griseus* which has led to the prediction that the biochemical involvement of coenzyme F_{420} is more widespread in nature than was previously suspected (Jones *et al.* 1987). Although the microscopic enumeration of methanogenic bacteria using F_{420} would appear both difficult and not necessarily specific to the total methanogenic population in environmental samples, the fluorescence of coenzyme F_{420} has proved an important screening feature for new methanogenic isolates in addition to checking monoculture purity (Edwards and McBride 1974).

Attempts to use these unique coenzymes as indicators of methanogenic biomass have been made though not through microscopic means. Complicated extraction and chromatographic techniques performed under anaerobic conditions have led to the quantitation of the levels of coenzyme F_{420} present in methanogenic cultures (Heine-Dobbernack *et al.* 1988). Comparisons with cell protein, cell dry weight, optical density and specific methane production rates have shown that the intracellular F_{420} content approximately followed methanogenic biomass.

4. Lipids

Since the original definition of archaebacteria by Woese (1977) a number of archaebacterial traits have been catalogued. The membrane lipids of the archaebacteria are one such distinguishing feature where the usual ester linked phospholipids are replaced by ether linked analogues. These lipids consist of either two 20 carbon saturated isoprenoid hydrocarbons ether linked to a glycerol backbone (di-*O*-phytanyl glycerol ether; DPGE) (Fig.1.) or two 40 carbon isoprenoid hydrocarbons ether linked to two glycerol molecules (tetra-*O*-di(biphytanyl glycerol ether); biDPGE) in a "monolayer" type membrane configuration. Greater chemical stability is afforded by the ether linkage of the alkyl chains to the glycerol molecules (Jones *et al.* 1987).

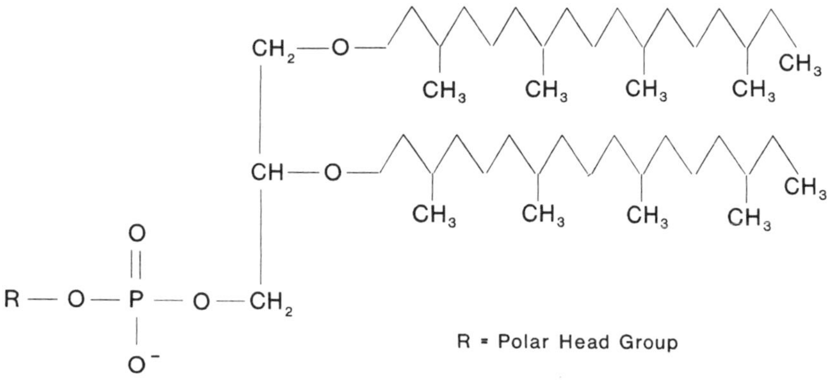

Figure 1. Di-Phytanyl Glycerol Ether membrane phospholipid found in methanogenic bacteria.

Thermoacidophilic archaebacteria, living in what is believed to be a more extreme environment, tend to possess a significant proportion of the stronger tetraether membrane constituent. Woese *et al.*, (1978) postulated that such archaebacteria are similar to those early life forms when primordial earth was a hot, acidic, anaerobic environment. Halophiles and methanogens occupy relatively less extreme environments and as a result contain mainly, if not all, diether phospholipid components (Fig.1.).

All of the methanogenic species analysed to date found in marine and freshwater sediments contain either all diether (DPGE) polar membrane lipids or a significant proportion of this relative to the tetraether (biDPGE) (Konig and Stetter 1988). Therefore since the thermoacidophiles and halophiles are unlikely to form a significant biomass in such mesophilic sediments the diether lipid (DPGE) could be an ideal signature lipid for the determination of methanogenic biomass and growth.

In order to assess the significance of this lipid as an indicator of methanogenic biomass it was considered best to compare it to biomass associated parameters in a controlled growth experiment.

Materials and methods

All solvents used were purchased from Rathburn Chemicals Ltd (Walkerburn, Scotland). Other chemicals came from Sigma Chemicals Ltd. (Dorset, England). Spectrophotometric determinations were made on a diode array spectrophotometer (Hewlett Packard, Vectra ES/12).

Growth of the bacterium
One hundred millilitres of monoculture was grown in 200 ml medical flasks fitted with modified suba seals and a metal screw top to facilitate head space

sampling. *Methanolobus tindarius* is a marine, mesophilic, coccoidal or lobal, monotrichous flagellated methanogen isolated from a marine sediment at Tindari in Sicily (Konig and Stetter 1982), and was a gift from Dr Michael Blaut (Institut für Mikrobiologie, Gottingen, Germany). The culture was grown at 25°C and pH=6.2 on Medium no.3 (Balch *et al*. 1979) without acetate, yeast extract or trypticase and was supplemented with 123 mmol l^{-1} methanol and 59 mmol l^{-1} trimethylamine.

Bacterial direct counts
Cells were first immersed in tetrasodium pyrophosphate (0.001M), a deflocculent used in conjunction with a sonic probe to gently disperse the cells evenly (Velji and Albright 1986). Using the method of Parsons *et al*. (1984) the bacteria were stained with acridine orange and diluted with formalin (40%) before being filtered on to a prestained irglan black nucleopore filter. Enumeration was made with a Leitz (Ortholux, 8667-51) UV microscope using BG38 and BG12 excitation filters and a K510 barrier filter on setting 2.

Lipid extraction
The gas headspace, cell number and turbidity analyses were made before the lipid extraction commenced. Each 100 ml of growth culture was then centrifuged at 30000 rpm (10000 g) for 20 minutes before discarding the supernatant.

Environmental samples (approx. 90 g wet weight) were collected using a hand held corer from approximately 40 cm below the surface for the marine intertidal sample (Menai Straits, N. Wales) and 15 cm for the freshwater sample (Llyn Maelog, Rhosneigr, N. Wales).

The lipid extraction technique of Bligh and Dyer (1959) was completed for all samples and modified with 5% trichloroacetic acid aqueous phase for increased yield from bound proteins (Nishihara and Koga 1988). Sonication also helped to break up particles and aid in lipid-solvent interaction. An addition of 50 μg internal standard of 1,2-di-O-hexadecyl *rac* glycerol ensured that extraction losses were accounted for. Each single phase organic/aqueous extraction was maintained for 2 hours for the growth culture experiments and 6 hours for the environmental samples. The latter samples were filtered and the residue re-extracted and filtered before the chloroform fraction was rotary evaporated to dryness in a reacti-vial.

Cleavage and substitution of the ether lipids
Approximately 2.5 ml of 47% hydriodic acid was added to each rotary evaporated sample and refluxed in the reacti-vial for 12 hours at 100°C with intermittent shaking. Addition of deionized water facilitated the transfer of lipid into a petroleum ether (40-60°) phase which was subsequently washed with saturated sodium thiosulphate solution. The alkyl iodides were substituted with an acetate group by further blowing down of the organic phase in a reacti-vial and adding 2.5ml glacial acetic acid in an excess of silver acetate. This was refluxed at 100°C for 24 hours with intermittent shaking (Guyer *et al*. 1963). Lipid removal and clean up was again repeated also using saturated sodium

hydrogen carbonate solution followed by anhydrous sodium sulphate to dry the sample.

Purification of the Alkyl Acetates
The dried sample was blown down and taken up in 100 µl of chloroform and spotted on a TLC plate. A hexane/ether (95:5) mobile phase was used in a continuous coelution tank for 50 minutes. Coelution of the phytanyl acetate standard together with exposure to iodine vapour showed the location of the alkyl acetates, which were scraped off and eluted with chloroform and methanol (1:1) through a glass wool plug. The sample was blown down under nitrogen and taken up in 200 µl of n-hexane. The phytanyl acetate standard was prepared from phytol using platinum dioxide (Adams' Catalyst) and the method of Guyer et al.(1963).

Gas-liquid chromatography
Phytanyl acetate was analysed on a Carlo Erba (HRGC 5160) gas chromatograph equipped with an SE30 capillary column (0.3 mm x 25 m) (Alltech Associates Inc. Lancashire), using nitrogen carrier gas and a flame ionization detector (FID). A temperature program from 50°C to 300°C at 10°C/minute gave a phytanyl acetate retention time of approximately 20 minutes for a 70KPa carrier gas pressure. The internal standard gave a retention time of 19 minutes (Hexadecanyl acetate; $C_{16}H_{33}$-O-CO-CH$_3$). Sample and standard peaks were resolved both from other peaks of similar retention time and also to the baseline in every analysis made.

Headspace gas was analysed for methane using a DANI (3800) gas chromatograph with a 1.5 m x 3 mm packed (Porapak Q) column, nitrogen carrier gas and an FID.

Results and discussion

Growth experiment
The cell number growth curve (Fig.2) shows negligible lag phase immediately after culture inoculation. Over the 8 to 56 hour period the bacterial cell numbers increased in an exponential manner reaching the final stationary phase after approximately 56 hours. Observations made using epifluorescent microscopy over the 8 to 56 hour time period showed that *M. tindarius* tended to devote more energy towards cell division rather than individual cell growth, with mean cell diameters of approximately 0.3 µm. During the stationary phase after 56 hours individual cell growth became the more dominant process, with bacterial cells reaching 1µm in diameter. Turbidity, which was measured as absorbance at 578 nm, also shows this change in the growth phase after approximately 48/56 hours (Fig. 2).

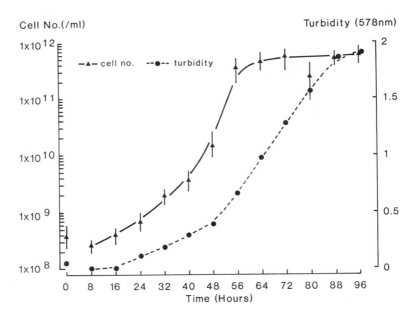

Figure 2. Growth experiment showing *M.tindarius* cell number and turbidity at 578 nm over time.

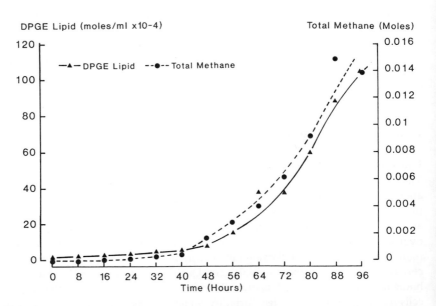

Figure 3. Growth experiment showing DPGE lipid concentration and total methane produced over time for *M. tindarius*.

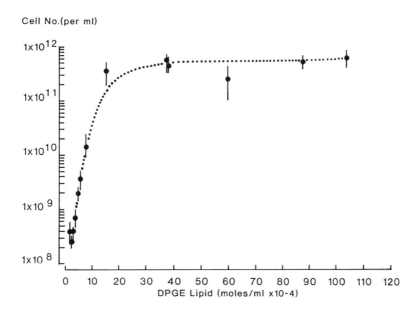

Figure 4. Relationship between cell number and diether (DPGE) phospholipid during the growth experiment of *M. tindarius*.

Good correlation was found between the DPGE lipid concentration and total methane produced over time (Fig.3). This result might be expected in such a closed system growth experiment during relatively early growth phases since methane production is directly related to the cell energy. During this log phase the energy produced would be used both for cell division and also for increasing the cell size, both of which will contribute to the membrane lipid concentration.

Figure 4 again shows this relatively abrupt change in the growth experiment after approximately 48 hours. This result could either be totally due to the noticeable change in the cell size during the growth phase or also due to a change in the diether lipid composition. A further experiment will need to be made using the additional parameter of cell volume in order to conclude whether the phospholipid composition does change during the growth phase of *Methanolobus tindarius*.

Preliminary environmental results

Table 3 shows that the concentration of DPGE lipid in the freshwater sample is an order of magnitude greater than that of the marine sample, a result that is expected due to the higher concentration of sulphate in the marine environment. This allows the sulphate reducing bacteria to predominate by outcompeting the methanogens for electron acceptors (Lovely *et al.* 1982).

Table 3. Diether (DPGE) lipid concentrations for environmental samples.

	Concentration DPGE lipid per dry gram sediment.
Freshwater Sample (Llyn Maelog)	2.20 µg DPGE / dry g 3.23 nmoles DPGE / dry g
Marine Sample (Menai Straits)	303 ng DPGE / dry g 454 pmoles DPGE / dry g

The only other work on environmental DPGE lipid concentrations was reported by Martz, Sebacher and White (1983) using a different method with final analysis by high performance liquid chromatography (HPLC). They found a freshwater value of 8.7 nmoles DPGE/dry g sediment which is comparable to the 3.23 nmoles DPGE/dry g measured from the single site and depth in Maelog Lake. Martz *et al.* (1983) also quote an estuarine concentration of 70 pmoles DPGE/dry g sediment. This is almost an order of magnitude less than the Menai Straits' sample (454 pmoles), a difference possibly attributed to the greater sewage loading of the Menai Straits'. A combination of both a high organic loading and a high sedimentation rate has the effect of causing a more rapid sediment redox change, due to the scavenging of oxygen associated with aerobic organic matter breakdown. This process provides the anaerobic microbiota with more organic substrates which are either directly or indirectly utilizable by the methanogenic bacteria.

Detection limit

By using capillary gas chromatography with a flame ionization detector a detection limit of 1.6×10^{-9} g phytanyl acetate was achieved. This corresponds to 2.35×10^{-12} moles DPGE lipid. From dry weight measurements it was calculated that the detection limit corresponds to 1.21×10^{-7} g bacteria.

This sensitivity in terms of the number of bacteria clearly depends very closely on the mean size of the organism. The methanogen *M. tindarius* used in the growth experiment had a mean size of 0.7 µm diameter (0.17 µm^3). Hence 10^7 cells would be required to detect this small methanogen. However, other methanogens found in marine sediments, such as *Methanosarcina, Methanogenium, Methanomicrobiales* and *Methanobacterium* species tend to be much larger in size with cell volumes ranging from 0.52 µm^3 up to 14.2 µm^3 (3.0µm dia.) and therefore the number required for detection will decrease. By using a typical bacterial size at 3x1 µm, with a cell volume of 2.36 µm^3 and a dry weight of 1.1×10^{-12} g/cell, a total of 110,000 cells would be required for detection and quantification. Greater sensitivity of up to 3 orders of magnitude (2×10^{-15} moles DPGE) can also be achieved in this method by exchanging the flame ionization detector for an electron capture detector (ECD). The revised method would require an extra TLC purification stage (Tornabene and Langworthy 1979) before the acid methanolysis stage with final analysis of the phytanyl iodide derivatives. Martz *et al.* (1983) quote a sensitivity of $7x10^{-14}$ moles DPGE but the errors relative to baseline noise are much greater and peak resolution is not fully achieved.

Conclusions

All parameters used to monitor the growth experiment show an interesting change in the growth curve of *Methanolobus tindarius* where the rate of cell division and the growth of individual cells appears to be dominant in turn. The relationship between the total methane produced and the DPGE lipid concentration also illustrates this change in the growth phase as well as possibly suggesting a change in the DPGE lipid composition early in the growth phase of *M. tindarius* though this appears unlikely.

The DPGE lipid concentration appears to be an adequate bioindicator for methanogenic biomass in a controlled closed system growth experiment. The total methane produced is also appropriate which is expected in such an experiment where methane is produced directly from all energy associated processes, and is subsequently not chemically or biologically oxidized. The DPGE lipid concentration however, together with total substrate availability, has the added advantage of possibly differentiating between methane produced at a particular point in a sediment and methane migration to that point. Therefore the potential of using the DPGE lipid bioindicator in environmental samples is far greater.

The preliminary environmental results show that the DPGE lipid can be both detected and quantified at levels as low as $2.35x10^{-12}$ moles DPGE using the current method. But also there is an extra sensitivity yet to be achieved using electron capture detection to achieve $2x10^{-15}$ moles of CDPGE as a detection limit.

Acknowledgements

The authors wish to thank Dr M. Blaut for the methanogenic monoculture of *Methanolobus tindarius*. This work was supported by grant No. GR/E/9746.3 from the Science and Engineering Research Council and also by a case award from ESSO Petroleum Company Ltd.

References

Archer, D.B. and Harris, J.E. (1986) Methanogenic bacteria and methane production in various habitats, In E.M. Barnes and G.C. Mead (eds.). *Anaerobic Bacteria in Habitats Other Than Man*. The Society of Applied Bacteriology, Symposium Series No. 13, 185-208. Blackwell Scientific

Publication, London.

Balch, W.E., Fox, G.E., Magrum, L.J., Woese, C.R. and Wolfe, R.S. (1979) Methanogens: Re-evaluation of a unique biological group. *Microbiological Reviews*, **43**, 260-296.

Bernard, B.B. (1979) Methane in marine sediments. *Deep Sea Research*, **26**, 429-443.

Blake, D.R., Mayer, E.W., Tyler, S.C., Makide, Y., Montague, D.C. and Rowland, F.S. (1982) Global increase in atmospheric methane concentrations between 1978 and 1980. *Geophysical Research Letters*, **9**, 477-480.

Bligh, E.G. and Dyer, W.J. (1959) A rapid method of total lipid extraction and purification. *Canadian Journal of Biochemistry and Physiology*, **37**, 911-917.

Burke, R.A. and Sackett, W.M. (1986) Stable hydrogen and carbon isotopic compositions of biogenic methanes from several shallower aquatic environments. In M.L. Sohn (ed.) *Organic Marine Geochemistry*. ACS Symposium Series 305, 297-313. American Chemical Society, Washington.

Conrad, R. and Seiler, W. (1988) Methane and hydrogen in sea water (Atlantic Ocean). *Deep Sea Research*, **35**, 1903-1917.

Craig, H. and Chou, C.C. (1982) Methane: the record in polar ice cores. *Geophysical Research Letters*, **9**, 1221-1224.

Doddema, H.J. and Vogels, G.D. (1978) Improved identification of methanogenic bacteria by fluorescence microscopy. *Applied and Environmental Microbiology*, **36**, 752-754.

Edwards, T. and McBride, B.C. (1974). New method for the isolation and identification of methanogenic bacteria. *Journal of Applied Microbiology*, **29**, 540-545.

Gehron, M.J. and White, D.C. (1983) Sensitive assay of phospholipid glycerol in environmental samples. *Journal of Microbiological Methods*, **1**, 23-32.

Gillan, F.T. and Johns, R.B. (1983) Normal-phase HPLC analysis of microbial carotenoids and neutral lipids. *Journal of Chromatographic Science*, **21**, 34-38.

Goldfine, H. and Hagen, P.O. (1972) Bacterial plasmalogens. In F. Snyder (ed.) *Ether Lipids: Chemistry and Biology*, 329-350. Academic Press, New York.

Guyer, K.E., Hoffman, W.A., Horrocks, L.A. and Cornwell, D.G. (1963) Studies on the composition of glycerol ethers and their preparation from diacyl glycerol ethers in liver oils. *Journal of Lipid Research*, **4**, 385-391.

Harris, J.E., Pinn, P.A. and Davis, R.P. (1984) Isolation and characterization of a novel thermophilic freshwater methanogen. *Applied and Environmental Microbiology*, **48**, 1123-1128.

Hartzell, P.L., Zvilius, G., Escalante-Semerena, J.C. and Donnelly, M.I. (1985) Coenzyme F_{420} dependence on the methylene tetrahydro methanopterin dehydrogenase of *Methanobacterium thermoautotrophicum*. *Biochimica et Biophysica Resume (Commun.)*, **133**, 884-890.

Harwood, J.L. and Russell, N.J. (1984) *Lipids in Plants and Microbes*. George Allen and Unwin, London.

Heine-Dobbernack, E., Schoberth, S.M. and Sahm, H. (1988) Relationship of intracellular coenzyme F_{420} content to growth and metabolic activity of *Methanobacterium bryanti* and *Methanosarcina barkeri*. *Applied and Environmental Microbiology*, 54, 454-459.

Henson, J.M., Smith, P.H. and White, D.C. (1985) Examination of thermophilic methane producing digesters by analysis of bacterial lipids. *Applied and Environmental Microbiology*, 50, 1428-1433.

Higgins, I.F., Best, D.J., Hammond, R.D. and Scott, D. (1981). Methane oxidising microorganisms. *Microbiological Reviews*, 45, 556-590.

Holzapfel-Pschorn, A. and Seiler, W. (1986) Methane emissions during a cultivation period from an Italian rice paddy. *Journal of Geophysical Research*, 91, 11803-11814.

Hovland, M. and Judd, A.G. (1988). *Seabed Pockmarks and Seepages*. Graham and Trotman Ltd, London.

Jones, J.G. (1979). *A Guide to Methods for Estimating Microbial Numbers and Biomass in Freshwater*. Freshwater Biological Association, Scientific Publication No. 39. Titus Wilson and Son Ltd, Kendal.

Jones, W.J., Nagle, D.P. and Whitman, W.B. (1987). Methanogens and the diversity of archaebacteria. *Microbiological Reviews*, 51, 135-177.

Jones, W.J. and Paynter, M.J.B. (1980) Populations of methane producing bacteria and *in-vitro* methanogenesis in saltmarsh and estuarine sediments. *Applied and Environmental Microbiology*, 39, 864-871.

Jorgensen, N.O., Laier, T., Buchardt, B. and Cederberg, T. (1990) Shallow hydrocarbon gas in the northern Jutland-Kattegat region, Denmark. *Bulletin of the Geological Society of Denmark*, 69-76.

Konig, H. and Stetter, K.O. (1982) Isolation and characterization of *Methanolobus tindarius* spp.nov., a coccoid methanogen growing only on methanol and methylamines. *Zentralblatt für Bakteriologie, Mikteriologie und Hygeine*, 1. Abt. Orig., C3, 478-490.

Konig, H. and Stetter, K.O. (1988) Archaebacteria. In J.T. Staley and M.P. Bryant (eds.) *Bergley's Manual of Systematic Bacteriology* Vol.3, section 25, 2171-2220.

Lacis, A., Hansen, J., Lee, P., Mitchell, T. and Lebedeff, S. (1981) Greenhouse effect of trace gases 1970-1980. *Geophysical Research Letters*, 8, 1035-1038.

Lovely, D.R., Dwyer, D.F. and Klug, M.J. (1982) Kinetic analysis of competition between sulphur reducing bacteria and methanogens for hydrogen in sediments. *Applied and Environmental Microbiology*, 43, 1373-1379.

Martz, R.F., Sebacher, D.I. and White, D.C. (1983) Biomass measurement of methane forming bacteria in environmental samples. *Journal of Microbiological Methods*, 1, 53-61.

Moura, I., Moura, J.J.G., Santos, H., Xavier, A.V., Burch, G., Peck, H.D. and LeGall, J. (1983) Proteins containing the factor F_{420} from *Methanosarcina barkeri* and *Methanobacterium thermoautotrophicum*: Isolation and properties. *Biochimica et Biophysica Acta*, 742, 84-90.

Nishihara, M. and Koga, Y. (1988) Quantitative conversion of diether or tetraether phospholipids to glycerophosphoesters by dealkylation with BCl_3: a tool for structural analysis of archaebacterial lipids. *Journal of Lipid*

Research, **29**, 384-388.

Parkes, R.J. (1987) Analysis of microbial communities within sediments using biomarkers. In M. Fletcher, T.R.G. Gray and J.G. Jones (eds.) *Ecology of Microbial Communities*, Forty First Symposium of the Society for General Microbiology, 147-178. Cambridge University Press, London.

Parsons, T.R., Maita, Y. and Lalli, C.M. (1984) *A Manual of Chemical and Biological Methods for Seawater Analysis*. Pergamon Press, Oxford.

Postgate, J.R. (1969). Methane as a minor product of pyruvate metabolism by sulphur reducing and other bacteria. *Journal of General Microbiology*, **57**, 293-302.

Rasmussen, R.A. and Khalil, M.A.K. (1984) Atmospheric methane in the recent and ancient atmospheres: Concentration, trends and interhemispheric gradient. *Journal of Geophysical Research*, **89**, 11599-11605.

Rogers, H.J. (1983) Bacterial cell structure. In J.A. Cole, C.J. Knowles and D. Schlessinger (eds.). *Aspects of Microbiology*, Van Nostrand Reinhold (UK) Co. Ltd, Wokingham.

Rohmer, M., Bouvier-Nave, P. and Ourisson, G. (1984) Distribution of hopanoid triterpenes in prokaryotes. *Journal of General Microbiology*, **130**, 1137-1150.

Rowe, R., Todd, R. and Waide, J. (1977) Microtechnique for the most probable number analysis. *Applied and Environmental Microbiology*, **33**, 675-680.

Seyfried, P.L. and Owen, A.R.G. (1978) Evaluation of the most probable number technique for the enumeration of feacal coliforms and *Pseudomonas aeruginosa* in sediments. In C.D. Litchfield and P.L. Seyfried (eds.) *Methodology for Biomass Determinations and Microbial Activities in Sediments, ASTM STP 673*, 52-63. American Society for Testing Materials, Philadelphia.

Smith, P.H. and Hungate, R.E. (1958) Isolation and characterization of *Methanobacterium ruminantum* n. spp. *Journal of Bacteriology*, **75**, 713-718.

Smith, M.R., Zinder, S.H. and Mah, R.A. (1980) Microbial methanogenesis from acetate. *Processes in Biochemistry*, 34-39.

Tornabene, T.G. and Langworthy, T.A. (1979) Diphytanyl and Dibiphytanyl glycerol ethers in methanogenic bacteria. *Science*, **203**, 51-53.

Velji, M.I. and Albright, L.J. (1986) Microscopic enumeration of attached marine bacteria of seawater, marine sediment, faecal matter and kelp blade samples following pyrophosphate and ultrasound treatments. *Canadian Journal of Microbiology*, **32**, 121-125.

White, D.C., Davis, W.M., Nickels, J.S., King, J.D. and Bobbie, R.J. (1979) Determination of the sedimentary microbial biomass by extractable lipid phosphate. *Oecologia*, **40**, 51-62.

Woese, C.R. (1977) A comment on the methanogenic bacteria and the primitive ecology. *Journal of Molecular Evolution*, **9**, 369-371.

Woese, C.R. (1981) Archaebacteria. *Scientific American*, **244**, 94-107.

Woese, C.R., Magrum, L.J. and Fox, G.E. (1978). Archaebacteria. *Journal of Molecular Evolution*, **11**, 245-252.

Zeikus, J.G. (1977) The biology of methanogenic bacteria. *Bacteriological Reviews*, **41**, 514-541.

Use of Phosphatase Assays with Algae to Assess Phosphorus Status of Aquatic Environments

B.A. Whitton

Department of Biological Sciences, University of Durham, Durham DH1 3LE, U.K.

Key words: bioassay, phosphatase, cyanobacteria (blue-green algae), bloom, filamentous alga, rivers, rice-fields.

Abstract

The literature on use of phosphatase activity as an environmental indicator is reviewed and the reasons why previous authors have differed markedly in their assessment of its value are discussed. The technique depends on the ability of 'surface' phosphatases to hydrolyze organic phosphates in the environment and the fact that this ability is 'inducible' in many species under conditions of moderate phosphorus limitation. The presence of phosphatase in populations where such activity is known to be inducible therefore indicates that phosphorus supply is probably a limiting factor influencing the growth rate. A range of surface phosphatase activities have been demonstrated (e.g. phosphomonoesterase, phosphodiesterase and phytase), but for most practical purposes an assay for phosphomonoesterase (PMEase) is sufficient, using the substrate p-nitrophenylphosphate. The influence of pH on PMEase activity measured in laboratory assays (using millimolar concentrations of substances) differs markedly between different populations and the pH at which maximum activity occurs shows only a weak correlation with the pH of the environment from which the organism was taken. The paper includes the results of several practically orientated studies.

Provided suitable precautions are taken, measurement of phosphomonoesterase activity provides a rapid and robust means of assessing the P status of aquatic environments. The method is easiest to apply with filamentous algae and is especially useful for flowing waters or markedly heterogeneous environments such as flooded rice-fields. Sufficient information is available to present it as a standard method and the procedure can be miniaturized to permit assays on a large number of samples. The main precautions required are that tests should be made at several pH values and that care is taken in the choice of buffers; for organisms taken from natural environments at or

Bioindicators and Environmental Management
ISBN 0-12-382590-3

above neutral, the pH values recommended are 7.6 and 10.3.

Introduction

Cells involved in uptake of nutrients from the external environment often show 'surface' (= cell-bound) phosphatase activity, if the organisms are phosphorus-limited. In the great majority of cases the activity is 'inducible', perhaps arising as a result of depression of enzyme activity in the absence of inorganic phosphate. Because of this, various researchers have suggested that measurement of this activity in organisms at the time of harvesting is a useful indicator of the phosphorus status of the environment and the manner in which the organisms are likely to respond to changes in environmental phosphorus concentration. The use of freshwater algae in this way was first suggested by Fitzgerald and Nelson (1966), who used the p-nitrophenylphosphate assay technique developed by Torriani (1960) to show that cultures of three green algae, four cyanobacteria (blue-green algae) and two diatoms all showed inducible 'alkaline phosphatase' activity. Others who have suggested that this activity can be used as an indicator of phosphorus status in natural populations include Reichardt *et al.* (1967), Healey (1978) and Rivkin and Swift (1982).

The older literature was sometimes muddled by attempts to relate the phosphatase activity of natural populations to phosphorus levels in the environment rather than to levels in the organisms. However, it is the latter which influences the concentration of enzyme and hence activity measured in most assays (Chróst and Overbeck 1987), though inorganic phosphate in the assay medium may be a competitive inhibitor of the phosphatase activity.

PMEase and PDEase activities can be measured colorimetrically or fluoro-metrically. It is also possible to demonstrate localization of activity by staining. Although this is of relatively little practical importance for assessing environmental phosphorus status, it helps to establish whether the activity of field samples is due to the algae or to associated organisms.

The measurement of surface phosphatase activity is only one of several biological methods for assessing the phosphorus status of water bodies. Other methods which have been applied to the natural populations are the measurement of 'surplus' cellular phosphorus (Fitzgerald and Nelson 1966), measurement of cellular N : P ratios (Chiaudani and Vighi 1974) and determination of V_{max} for phosphorus uptake (Gotham and Rhee 1981; Pettersson 1980). Overall the approach is most effective when several different methods are applied at the same time (Pettersson 1980; Vincent 1981a,b). Ideally these methods would be applied at the same time as an algal bioassay of the water (see Klapwijk *et al.* 1989 for references). However, the phosphatase assay technique is the simplest and quickest of the various methods.

The aim of this review is to summarize the literature on algal 'surface' and extracellular phosphatases, comment on the advantages and limitations of phosphatase measurements for monitoring purposes and describe a few practical examples of their use. The term 'phosphatase' in the review refers only to activity detectable by hydrolysis of organic phosphates in the environment. Most intracellular phosphatases are unlikely to influence such activity, although several authors have suggested (e.g. Berman 1970) that organic phosphates are sometimes taken up and hydrolyzed inside the cell.

The terms alkaline and acid phosphatase have been applied to phosphomonoesterase (PMEase) activities in the respective pH regions (Feder 1973). Other types of surface phosphatase are widespread, but an easy assay method has been developed only for phosphodiesterase (PDEase).

Occurrence of phosphatases in algae

Phosphatases able to hydrolyze organic phosphate in the environment have been reported from a wide range of algae (Aaronson and Patni 1976). The earlier studies were mostly based on laboratory strains, such as *Chlorella vulgaris* (Brandes and Elston 1956), *C. pyrenoidosa* (Overbeck 1962) and five marine diatoms and two chrysophytes (Kuenzler and Perras 1965) These and most subsequent algal studies have been carried out at pH values above 7.0, and often much higher, and the enzymes under study have frequently been termed alkaline phosphatases, following the numerous studies on such enzymes in bacteria and mammalian tissues. These are the algal phosphatase(s)', which have been used to assess environmental phosphorus status. They have a broad specificity against different substrates: their activity is only restricted to the P-0 bond of phosphomonoesters.

The great majority of cyanobacteria reported possess alkaline phosphatase activity and have probably received the most detailed study (Doonan and Jensen 1979; 1980; Healey 1982). All 50 strains, representing 10 genera, tested by Whitton *et al.* (in press) showed obvious PMEase activity and most of them also phosphodiesterase (PDEase) activity (Table 1). 35 strains grew with phytic acid, but as there is no simple method for measuring the hydrolysis of phytic acid, the development of phytase activity is of little practical use for monitoring. In all 50 strains, PMEase activity was inducible and in the two strains where detailed observations have been made (*Calothrix parietina* D550: Livingstone *et al.* 1983; *Nostoc commune*: Whitton *et al.* 1990), activity is non-detectable in P-rich filaments. In contrast to these studies, no activity was found associated with P-limited *Coccochloris* spp. (Kuenzler and Perras 1965), *Synechococcus* spp. (Kuenzler 1965) or *Pseudanabaena catenata* (Healey and Hendzel 1979a, b). Cyanobacteria with trichomes ending in multicellular hairs apparently nearly always have high PMEase activity associated with these structures (Whitton 1987), so any organisms with hairs are likely to be P-limited. The only other factor which can lead to hair formation in some hair-forming strains is Fe limitation (Sinclair and Whitton 1977).

Alkaline PMEase activity has been reported from most eukaryotic algal phyla. Among the multicellular algae it is again often, though not necessarily, particularly marked in hair-forming species (red, brown and green algae:

Table 1. Influence of pH on ability to detect phosphatase activities of 50 cyanobacterial strains using standard p-nitrophenylphosphate (PMEase) or bis-p-nitrophenylphosphate (PDEase) methodology.

	pH	No. of strains in which activity was detectable	max. activity found in any strain (mM pNPP hydrolyzed mg d. wt^{-1} h^{-1})
PMEase	7.6	46	2.72
	10.3	50	5.23
PDEase	7.6	32	0.49
	10.3	34	1.07

Whitton 1988). Many diatoms show surface PMEase activity, with the most detailed studies in the marine *Phaeodactylum tricornutum* (Chiaudani and Vighi 1982; Flynn *et al.* 1986). PMEase has been found to be inducible in all hair-forming species showing activity (Whitton 1988) and most, but not all, of the diatoms and green algae showing activity (e.g. Fitzgerald and Nelson 1966; Hino 1988). Many species of potential use as indicators remain to be investigated.

Acid phosphatases with the ability to hydrolyze environmental organic phosphate have also been recorded for a range of organisms (Aaronson and Patni 1975) from acidic sites and also ones with pH values above neutral. Like alkaline phosphatases they have a broad specificity. However, little attempt has been made to use this ability for assessing the phosphorus status at field sites. In many cases the enzymes appear to be constitutive (Siuda 1984), their synthesis not being inhibited by high phosphate (e.g. *Peridinium* spp.: Wynne 1977). However, acid phosphatases of *Euglena gracilis* (Price 1962) and *Chlamydomonas acidophila* (Boavida and Heath 1986) have been shown to be inducible, so it should prove possible to develop a phosphatase assay methodology suited for environments with low pH values.

Previous literature on phosphatase as an environmental indicator

Seasonal fluctuations in PMEase activity have been followed in samples of the whole water column from various lakes and oceanic sites and/or in their phytoplankton populations. Based on PMEase activity and analysis of total phosphorus in Lake Kinneret during 1969, Berman (1970) concluded that phosphate was probably adequate during the main season of *Peridinium westii* bloom, but that it probably became limiting from the end of summer until the time of the overturn. PMEase activity of the *Peridinium* was followed in more detail from March to June 1976 (Wynne 1977) and showed maximum activity at the end of the period after the peak of the bloom. Some other authors have also concluded that PMEase activity provides a useful indication of the phosphorus status of particular lakes. Pettersson (1980), who studied Lake Erken, Sweden, from 1975-1978, found that PMEase activity and a range of other phosphorus deficiency indicators all showed severe phosphorus limitation in May to mid-June. Stevens and Parr (1977) also found an increase in PMEase activity in Lough Neagh, N. Ireland, when the algae were P-limited, but concluded that this could not be used as reliable indicator in this lake because of a significant allochtonous contribution to the total activity.

In contrast to the above studies, Pick (1987) concluded that PMEase activity was not a sensitive indicator of phosphorus deficiency in Lake Ontario. Unlike most other researchers, Pick used unbuffered water, although this remained in the pH range 8.0 - 8.7 during the assays. Addition of orthophosphate at concentrations well above those occurring in the lake took some 18 h to bring about a 50% reduction in PMEase activity. The author suggested that this might be due in part to some phytoplankton species possessing constitutive phosphatase activity. Orthophosphate was however, a strong competitive inhibitor of PMEase activity of phytoplankton in the size fraction > 3 μm in the Plusssee, Germany, although only a slight inhibitor for the predominantly bacterial fraction < 3 μm (Chróst and Overbeck 1987).

A study of *Stigeoclonium* populations from streams in northern England (Gibson and Whitton 1987a) showed that there was a highly significant positive relationship between phosphatase activity and extent of formation of long colourless multicellular hairs and a highly significant negative relationship between the concentrations of phosphate in both the ambient water and the alga (Table 2). This relationship with the water probably reflects the fact that environmental conditions had shown little change in the week prior to sampling.

Table 2. Correlation between surface phosphatase activity in *Stigeoclonium* and values for selected variables (n = 30).

variable	r	p
$\log_{10}$ aqueous filtrable reactive phosphate	- 0.622	< 0.001
algal P	- 0.613	< 0.001
algal N : P ratio	+ 0.485	< 0.01

Influence of environmental factors on development of phosphatase activity

It is clear from above comments that the internal phosphorus concentration is the most important factor influencing the development inducible phosphatase activity. However, the influence of other factors sometimes obscures this relationship. Light in the period shortly before assay may influence activity. Rivkin and Swift (1982), for instance, found that activity in *Pyrocystis noctiluca*

was correlated positively with preconditioning light conditions. Wynne and
Rhee (1988) reported that the light regime had a marked influence on PMEase
activity of P-limited cells of four test cultures.

There are several records of high PMEase activity in metal-contaminated
environments. In some cases this may be due simply to the low solubility of
inorganic phosphate in the presence of high concentrations of Pb or some other
heavy metals, but Jansson (1981) suggested a somewhat different explanation for
the elevated PMEase activity during summer in Lake Gårdsjön, an acidified (pH
4.5) lake in Sweden. This is that high concentrations of Al in the water block the
phosphatase substrates and so the plankton must increase phosphatase
production to exceed the ambient Al concentration in competition for the
substrates. Small Chrysophyta (dominated by *Chromulina* in the < 5μm fraction)
were especially important in production of phosphatase (Olsson 1983) in this
lake.

Influence of environmental factors during assays

General comments
Although the general methodology for phosphatase assays has been used
widely, several environmental factors which can influence results markedly have
not been standardized. In addition there is a need to increase the monoester
concentration used for assay purposes, which complicates the ecological
interpretation of the data (Taft *et al.* 1977). Ideally all studies would include
determination of values for the half-saturation constant and V*max*, but this is
impractical for routine purposes. The more eutrophic the environment, the less
the difference between the concentration of substrate needed for assay and the
maximum concentration likely to occur in the water. It therefore seems
probable that the phosphatase assay methodology is most suited for routine
approach in eutrophic waters.

Alkaline phosphatase (PMEase) assays at any particular pH value may be
influenced by a range of factors (Doonan and Jensen 1980), including pH, buffer,
various cations, phosphate and chelators such as EDTA (Whitt and Savage 1988,
dealing with mouse tissue; Grainger *et al.* 1989; Whitton *et al.* 1990).

The choice of pH for assay purposes is especially important. The value
most widely adopted for assays of alkaline phosphatase activity in microbiological studies not directly related to environmental questions is pH 10.3 - 10.4,
the value recommended in Sigma Technical Bulletin No. 104. This is well above
the pH of the ambient environment for most algae, but near the optimum for
PMEase activity of some strains (e.g. *Calothrix parietina* D550: Grainger *et al.* 1989;
Chaetophora spp., *Draparnaldia plumosa*: Gibson and Whitton 1987a). However,
other cyanobacteria have a pH optimum well below this (pH 7.0 in *Nostoc
commune* UTEX 584: Whitton *et al.* 1990) or above this (pH 12.3 in *Calothrix
viguieri* D253: Mahasneh 1990). By definition, acid phosphatases (see above) all
have pH optima below 7.0.

A recent study of human fibroblasts (Fedde and Whyte 1990) indicates a possible explanation for at least part of the weak correlation between pH optimum found in assays and the pH of the microhabitat from which the organism was isolated. The pH curves using three different substrates assayed with millimolar concentrations were identical, with optima at pH 9.3. When two substrates were assayed with micromolar concentrations, the pH curves differed markedly from the others, but were identical with each other, with optima at pH 8.3. The authors commented that the high affinity response was much nearer normal physiological conditions than the low affinity response. Because of the difficulties in assay methods there have apparently been no similar comparisons with other organisms, nor was any information given by Fedde and Whyte as to whether the same level of phosphate limitation led to the expression of both types of activity in the fibroblasts. Although it seems unlikely to influence the practical methodology proposed in this review, it is nevertheless important that the possibilty of a differing high affinity activity should be investigated in representative algae.

Some field orientated studies with algae have used the natural pH of the site from which the sample was taken (e.g. Livingstone and Whitton 1984; Pick 1987). Those requiring standardized assays have mostly either used pH values between 8 and 9, e.g. pH 8.3 (Pettersson 1980), pH 8.5 (Fitzgerald and Nelson 1966; Siuda *et al.* 1982; Hino 1988), pH 8.6 (Berman 1970), or pH 10.3 (Gibson and Whitton 1987a, b; Grainger *et al.* 1989).

For many purposes it probably makes little difference what pH value is chosen (see Table 1), providing that a consistent approach is used in any particular survey; nevertheless there is always a risk that activity may be overlooked altogether in organisms with atypical pH optima. It is therefore recommended that assays are carried out routinely at two widely differing pH values (7.6, 10.3: see below), but if it is essential to restrict the study to a single value, it is probably better to use pH 8.5. The relative influence of buffers differs between organisms, though the glycine NaOH system is usually among the best for assays in the range 9.0 - 10.5.

Providing that a standard algal medium is used, small differences in cation concentration are unlikely to have much influence. However, excess chelating agent may cause inhibition, probably by removing part of the zinc from the enzyme site. Although high concentrations of phosphate are known to inhibit PMEase and PDEase activities (see above), relatively few quantitative studies have been made. In *Calothrix parietina* (Grainger *et al.* 1989), the presence of 0.05 mM phosphate caused a 50% reduction in PMEase activity during a 15-min assay with 0.25 mM substrate.

Acidic phosphatases appear to have no cation requirement, but are specifically inhibited by fluoride (Cembella *et al.* 1984). As with studies at higher pH values, PMEase assays may be carried out colorimetrically using p-nitrophenylphosphate (Patni *et al.* 1974) or fluorometrically (see below).

As mentioned earlier, the light regime influencing growth of the organisms shortly before assays sometimes has a marked influence on activity of cells with the same degree of P-limitation. There are apparently no reports of light influencing activity during short-term assays and no difference in PMEase activity occurred between light and dark conditions during assays on a wide range of cyanobacteria an filamentous green algae (author, unpublished data). Nevertheless the possibility should be borne in mind that light might sometimes influence the results, when developing assay procedures for a particular purpose. It is easy to suggest ways in which light might have an indirect influence. For instance, while the initial activity might be the same in light and dark, transport of phosphate ions away from the site of hydrolysis might be slower in the dark and accumulated phosphate might start to inhibit phosphatase activity.

Stability of phosphatases
Alkaline surface PMEases and PDEases are robust enzymes with high temperature optima and which can persist for relatively long periods. For instance, Kobori and Taga (1978) reported that activity dropped over one week by only 30%. Dried *Nostoc commune* showed no loss of PMEase activity on re-wetting, but a 40% gain in PDEase activity (Whitton *et al.* 1990). 10-yr old dried colonies of stream *Rivularia* showed similar PMEase activity to fresh ones, suggesting that there had been little loss of activity over the period (unpublished data).

Berman (1970) added chloroform to prevent continued microbial growth after collection and found that 1.7 % chloroform by volume had no influence on PMEase activity. Details of storage time in the presence of chloroform were not given, but conditions suited for long-term storage of samples for assay purposes should be easy to achieve. It would be useful to test preservatives such as buffered formaldehyde.

Coulorimetric method
This is derived from Torriani (1960), who measured PMEase activity by the product of the reaction:

$$\text{p-nitrophenyl phosphate} \longrightarrow \text{p-nitrophenol} + \text{phosphate}$$

PDEase activity may be measured by substituting bis-p-nitrophenyl phosphate in the reaction. Both substrates are relatively cheap and solutions are stable over most of the pH range for at least a few days, provided they are sterilized (by membrane filtration). However, p-nitrophenylphosphate hydrolyzes spontaneously at high pH values, though this is only likely to interfere with short-term assays above about pH 12. Controls should, however, be included with all assays and particular care taken to check for spontaneous hydrolysis if unusual environmental conditions are used.

It is clear that further research is needed before a standard methodology can be recommended, but the following procedure is satisfactory for most purposes. The assay should be carried out in a standard algal medium (see Stein 1973), but omitting phosphate; zinc must be included in the trace element stock. A HEPES - NaOH buffering system is suitable for *Calothrix parietina* at pH 7.6 and glycine - NaOH at pH 10.3 (Grainger *et al.* 1989), but further studies are required (see above). If the assay is to be carried out at only a single concentration of the substrate (rather than a more detailed kinetic study), it is recommended that the final concentration in the medium should be 0.25 mM. Suitable incubation conditions for samples from temperate environments are 30 min at 25°C and from tropical environments, 20 min at 30°C.

The reaction must be terminated. Even if the alga is removed, the presence of extracellular enzyme or bacteria may lead to continued phosphatase activity. In addition the solution must be raised to pH c 12.5 to obtain the maximum colour. Both aims are achieved by adding 5 M NaOH to make the final concentration 0.5 M. Activity is measured at 405 nm.

Fluorometric method
Different substrates are required for the more sensitive fluorometric method, which is needed for the study of natural (unconcentrated) populations from oligotrophic and mesotrophic lakes. Perry (1972) introduced the use of 3-o-methylfluorescein phosphate for oceanic water and the practical problems in its use have been discussed by Pettersson and Jansson (1978) and Pettersson (1980). Jansson *et al.* (1981) employed 3-methylumbelliferryl phosphate as a fluorogenic substrate over the pH range 3.3 to 8.0 and provided a calibration curve to compensate for the decrease in fluorescence of this substrate with rise in pH. Related fluorochromes have been used to evaluate phosphatase and other enzyme activities in the humic-rich Lake Mekkojärvi, Finland (Münster *et al.* 1989).

Staining procedures
Several azo-dye substrates have been adopted to demonstrate localization of PMEase activity by staining visible with the light microscope In many cases they all give similar results, but sometimes only one is successful or, rarely, tissues which almost certainly possess PMEase activity fail to show this by staining. The use of staining should therefore always be combined with enzymatic assays.

The use of 5-bromo-4-chloro-3-indolyl phosphate (BCIP) was introduced for bacteria by Holt and Withers (1958) and Coston and Holt (1958) and since been applied widely in bacteriological studies, but there are only passing references to its use with algae. The hydrolysed product is bright blue. Localization of PMEase activity has been studied with several *Calothrix* spp. (Livingstone *et al.* 1983; Mahasneh 1990) with other substrates. The use of 3-hydroxy-2-naphthoic acid 2,4-diamethylanilide phosphate (napthol AS-MX phosphate) as the organic-P source and diazotized-4-benzolyamino-2,5-dimethoxyaniline zinc chloride (Fast Blue RR diazonium salt) as the coupling agent (see Sigma technical bulletin No. 85) is probably the most successful. The hydrolysed product is violet.

Localization of phosphatase has also been demonstrated using the lead capture technique, as on the hairs of *Chaetophora* (Gibson and Whitton 1987a). In this case it is the phosphate released which is precipitated as lead phosphate in the presence of a soluble lead salt.

Practical examples

In addition to published studies mentioned above, the author has carried out several different types of field survey, examples of which are given here. Broadly these fall into two types, those based on simple visual comparisons of activity and those involving quantitative measurements. The former are of use where rapid surveys of a large number of sites are involved, such as a preliminary survey of water-bodies with cyanobacterial blooms or for situations where there is no access to a spectrophotometer. In neither case can the possibility be ruled out that part of the activity is due to bacteria. For assessing environmental conditions, this is often unimportant, but it needs to be considered when making comparisons with laboratory data for axenic strains of algae. In practice bacteria probably make a substantial contribution to the phosphatase activity measured in some cyanobacterial blooms, but are of much less importance with filamentous algae or gelatinous colonies, provided the material appears healthy.

An example of the results of a visual survey is shown in Figure 1. Samples of algae and *Azolla pinnata* from rice-fields and shallow pools in the Philippines were taken for p-nitrophenylphosphate assay and the results compared against a colour chart. Standardization of the amount of alga was based on visual comparison and is not quantitative, so the results are summarized on a 1 - 5 scale. They show clearly, however, that some natural populations of some species or genera of cyanobacteria generally show high PMEase activity, whereas the filamentous green algae *Cladophora* and *Rhizoclonium* generally show low activity.

Similar results to those in Figure 1 have been obtained in the U.K. for all the organisms except *Azolla pinnata*, which does not occur there. *Gloeotrichia* and *Rivularia* colonies, for instance, almost always show very high activity, although isolates can be grown in the laboratory where the activity is suppressed due to growth in P-rich conditions. This suggests that the natural environment for these genera is one where the organisms occur for most of the time under conditions of at least moderate P-limitation. In contrast, *Cladophora glomerata* taken from the field usually shows no or very little PMEase activity, although activity can be induced in the laboratory by incubation in high-N, low-P medium. This suggests either that *Cladophora* only develops PMEase activity when it is severely P-limited, or that it tends to occur abundantly in the field only when the water has a relatively low N : P ratio.

Long-term monitoring at a particular site requires quantitative results. Table 3 provides an example of data collected on one day at stream and river sites in northern England, all sites where the pH of the water is above 7.0. Half the samples show detectable activity when assayed at pH 7.6 and one more becomes detectable when assayed at pH 10.3. There are considerable differences between the ratios of activity at pH 7.6 and 10.3, ranging from slightly greater activity at 7.6 to about three times as much activity at pH 10.3.

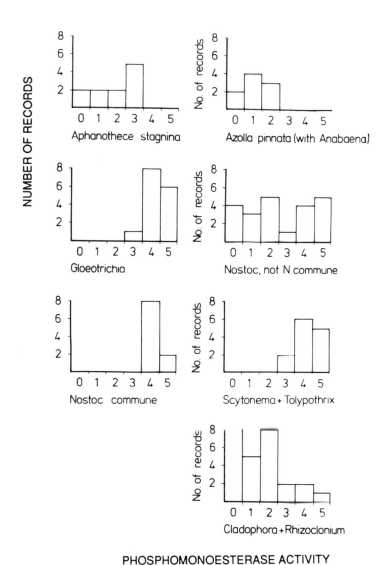

PHOSPHOMONOESTERASE ACTIVITY

Figure 1. Survey of PMEase activity in populations of algae and *Azolla* in rice-fields and small pools near Los Baños, Philippines, in January 1982. Each sample of a particular genus represents a different site. Activity measured by standard p-nitrophenylphosphate assay for 20 min at *c.* 30°C; results expressed on 1-5 scale, with 5 indicating the highest activity. Most (probably all) of 'Nostoc, not *N. commune*', was *N. linckia*; the one sample showing akinetes had very high PMEase activity.

Table 3. Example of a survey of flowing waters for PMEase activity. Samples taken from sites in the catchments of the Rivers Tyne, Wear and Tees, N-E. England, on 18 September 1990. Distances down R. Wear measured from source; PMEase activity assayed at 25°C and at two pH values.

Organism	site	PMEase activity μmol pNPP mg^{-1} h^{-1}	
		pH 7.6	pH 10.3
Batrachospermum moniliforme	Station Quarry	0.123	0.162
Chara vulgaris	Station Quarry	0.012	0.023
Cladophora glomerata	R. Wear km 24.3	<0.005	<0.005
Cladophora glomerata	R. Wear km 44.7	<0.005	<0.005
Cladophora glomerata	R. Gaunless	<0.005	<0.005
Cladophora glomerata	Percy Beck	<0.005	<0.005
Enteromorpha flexuosa	R. Wear km 44.7	<0.005	<0.005
Enteromorpha flexuosa	R. Gaunless	<0.005	<0.005
Mougeotia spp. 8 μm	Caplecleugh	0.009	0.029
Oedogonium spp.	R. Gaunless	<0.005	0.009
Spirogyra spp. 32 μm	Station Quarry	0.035	0.114
Stichococcus bacillaris	R. Nent	0.023	0.026
Zygnema spp. 30 μm	R. Wear km 0.1		
	river	0.018	0.011
	pool	0.026	0.021

Conclusions

In spite of the reservations of some previous authors of lake studies, assays of PMEase activity are of considerable value for environmental monitoring, since they provide a rapid and robust means of assessing the P status of aquatic environments. The method is easiest to apply when relatively uniform samples of a single species can be harvested for assay purposes. Examples are filamentous species or ones with distinct colonies and dense cyanobacterial blooms. Flowing waters and rice-fields are environments where the method is especially useful. Preliminary studies also suggest that the method has considerable potential in assessing the P status of cyanobacterial blooms. Interpretation of results for other types of lake plankton requires further research before their value can be assessed, and this is even more true for dissolved phosphatase in lake waters.

If the method is to achieve its full potential, however, considerably more background information is needed. Important requirements are:
1. To establish the relationship between PMEase activity and P status of individual species, especially ones which are widespread and easy to harvest. Obvious species in freshwaters (in addition to *Stigeoclonium tenue*, for which data

exist already) include *Cladophora glomerata, Enteromorpha flexuosa, Hydrodictyon reticulatum* and the common bloom-forming cyanobacteria. It is also important to establish whether or not most species within a particular genus show a similar quantitative response to P limitation.

2. To quantify the influence of environmental factors on PMEase activity, especially pH.

Acknowledgement

The author is most grateful to Prof. M. Martinez of the University of Philippines at Los Baños for help in collecting algal materials.

References

Aaronson, S. and Patni, N.J. (1976) The role of surface and extracellular phosphatases in the phosphorus requirement of *Ochromonas*. *Limnology and Oceanography*, **21**, 838-845.

Berman, T. (1970) Alkaline phosphatases and phosphorus availability in Lake Kinneret. *Limnology and Oceanography*, **15**, 663-674.

Boavida, M.J. and Heath, R.T. (1986) Phosphatase activity of *Chlamydomonas acidophila* Negoro (Volvocales, Chlorophyceae). *Phycologia*, **25**, 400-404.

Brandes, D. and Elston, R.N. (1956) An electron microscopical study of the histochemical localization of alkaline phosphatase in the cell wall of *Chlorella vulgaris*. *Nature*, **177**, 274-275.

Cembella, A.D., Antia, N.J. and Harrison, P.J. (1984) The utilization of inorganic and organic phosphorus-compounds as nutrients by eukaryotic microalgae. A multi-disciplinary perspective. Part I. *Critical Reviews in Microbiology*, **10**, 317-391.

Chiaudani, G. and Vighi, M. (1974) The N:P ratio and tests with *Selenastrum* to predict eutrophication in lakes. *Water Research*, **8**, 1063-1069.

Chiaudani, G. and Vighi, M. (1982) Multistep approach to identification in Northern Adriatic eutrophied coastal waters. *Water Research*, **16**, 1161-1166.

Chróst, R.J. and Overbeck, J. (1987) Kinetics of alkaline phosphatase activity and phosphorus availability for phytoplankton and bacterioplankton in Lake Plusssee (North German eutrophic lake). *Microbial Ecology*, **13**, 229-248.

Coston, S. and Holt, S.J. (1958) Kinetics of aerial oxidation of indolyl and some of its halogen derivatives. *Proceedings of the Royal Society B*, **148**, 506-510.

Doonan, B.B. and Jensen, T.E. (1979) Effect of ions on the activity of the enzyme alkaline phosphatase from *Plectonema boryanum*. *Microbios*, **25**, 177-186.

Doonan, B.B. and Jensen, T.E. (1980) Physiological aspects of alkaline phosphatase in selected cyanobacteria. *Microbios*, **29**, 185-207.

Fedde, K.N. and Whyte, M.P. (1990) Alkaline phosphatase (tissue-nonspecific isoenzyme) is a phosphoethanolamine and pyridoxal-5'-phosphate ectophosphatase : normal and hypophosphatasia fibroblast study. *American Journal of Human Genetics*, **47**, 767-775.

Feder, J. (1973) The phosphatases. In E.J. Griffiths *et al.* (eds.) *Environmental Phosphorus Handbook*, 474-578. Wiley and Sons, New York.

Fitzgerald, G.P. and Nelson, T.C. (1966) Extractive and enzymatic analyses for limiting or surplus phosphorus in algae. *Journal of Phycology*, **2**, 32-37.

Flynn, K.J., Öpik, H. and Syrett, P.J. (1986) Localization of the alkaline phosphatase and 5′-nucleotidase activities of the diatom *Phaeodactylum tricornutum*. *Journal of General Microbiology*, **132**, 289-298.

Gibson, M.T. and Whitton, B.A. (1987a) Hairs, phosphatase activity and environmental chemistry in *Stigeoclonium*, *Chaetophora* and *Draparnaldia* (Chaetophorales). *British Phycological Journal*, **22**, 11-22.

Gibson, M.T. and Whitton, B.A. (1987b) Influence of phosphorus on morphology and physiology of freshwater *Chaetophora*, *Draparnaldia* and *Stigeoclonium*, (Chaetophorales, Chlorophyta). *Phycologia*, **26**, 59-69.

Gotham, I.J. and Rhee, G.-Y. (1981) Comparative kinetic studies of phosphate limited growth and phosphate uptake in phytoplankton in continuous culture. *Journal of Phycology*, **17**, 257-265.

Grainger, S.L.J., Peat, A., Tiwari, D.N. and Whitton, B.A. (1989) Phosphomonoesterase activity of the cyanobacterium *Calothrix parietina*. *Microbios*, **59**, 7-17.

Healey, F.P. (1978) Physiological indicators of nutrient deficiency in algae. *Mitteilungen Internationalen Vereinigung Limnologie*, **21**, 34-41.

Healey, F.P. (1982) Phosphate. In N.G. Carr and B.A. Whitton (eds.) *The Biology of Cyanobacteria*, 105-124. Blackwell, Oxford.

Healey, F.P. and Hendzel, L.L. (1979a) Fluorometric measurement of alkaline phosphatase activity in algae. *Freshwater Biology*, **9**, 429-439.

Healey, F.P. and Hendzel, L.L. (1979b) Indicators of phosphorus and nitrogen deficiency in five algae in culture. *Canadian Journal of Fisheries and Aquatic Science*, **37**, 442-453.

Hino, S. (1988) Fluctuations of algal alkaline phosphatase activity and the possible mechanisms of hydrolysis of dissolved organic phosphorus in Lake Barato. *Hydrobiologia*, **157**, 77-84.

Holt, S.J. and Withers, R.F.J. (1958) An appraisal of indigogenic reactions for esterase localization. *Proceedings of the Royal Society* B, **148**, 510-526.

Jansson, M. (1981) Induction of high phosphatase activity by aluminium in acid lakes. *Archiv für Hydrobiologie*, **93**, 32-44.

Jansson, M., Olsson, H. and Broberg, O. (1981) Characterization of acid phosphatases in the acidifed Lake Gårdsjön. *Archiv für Hydrobiologie*, **92**, 377-395.

Klapwijk, S.P., Bolier, G. and van der Does, J. (1989) The application of algal growth potential tests (AGP) to the canals and lakes of western Netherlands. *Hydrobiologia*, **188/189**, 189-199.

Kobori, H. and Taga, N. (1978) Phosphatase activity and its role in the mineralization of organic phosphorus in coastal sea water. *Journal of Experimental Marine Biology and Ecology*, **36**, 23-39.

Kuenzler, E.J. (1965) Glucose-6-phosphate utilization by marine algae. *Journal of Phycology*, **1**, 156-164.

Kuenzler, E.J. and Perras. J.P. (1965) Phosphatases of marine algae. *Biological Bulletin of Marine Biological Laboratory Woods Hole*, **128**, 271-284.

Livingstone, D., Khoja, T.M. and Whitton, B.A. (1983) Influence of phosphorus on physiology of a hair-forming blue-green alga (*Calothrix parietina*) from an upland stream. *Phycologia*, **22**, 345-350.

Livingstone, D. and Whitton, B.A. (1984) Water chemistry and phosphatase activity of the blue-green alga *Rivularia* in Upper Teesdale streams. *Journal of Ecology*, **72**, 405-421.

Mahasneh, I.A. (1990) Influence of salinity on hair formation and phosphatase activities of the blue-green alga (cyanobacterium) *Calothrix viguieri*. *British Phycological Journal*, **25**, 25-32.

Münster, U., Einiö, P. and Nurminen, J. (1989) Evaluation of the measurements of extracellullar enzyme activities in a polyhumic lake by means of studies with 4-methylumbelliferyl-substrates. *Archiv für Hydrobiologie*, **115**, 321-337.

Olsson, H. (1983) Origin and production of phosphatases in the acid Lake Gårdsjön. *Hydrobiologia*, **101**, 49-58.

Overbeck, J. (1962) Untersuchungen zum Phosphathaushalt von Grünalgen. II. Die Verwertung von Pyrophosphat und organisch gebundenen Phosphaten und ihre Beziehung zu den Phosphatasen von *Scenedesmus quadricauda* (Turp.) Bréb. *Archiv für Hydrobiologie*, **58**, 281-308.

Patni, N.J., Aaronson, S., Holik, K.J. and Davis, R.H. (1974) Existence of acid and alkaline phosphohydrylase activity in the phytoflagellate *Ochromonas danica*. *Archives of Microbiology*, **96**, 63-67.

Perry, M.J. (1972) Alkaline phosphatase assay in subtropical Central North Pacific waters using a sensitive fluorimetric method. *Marine Biology*, **15**, 113-119.

Pettersson, K. (1980) Alkaline phosphatase activity and algal surplus phosphorus as phosphorus-deficiency indicators in Lake Erken. *Archiv für Hydrobiologie*, **89**, 54-87.

Pettersson, K. and Jansson, M. (1978) Determination of phosphatase activity in lake water - a study of methods. *Verhandlungen internationalen Vereinigung theoretische und angewandte Limnologie*, **20**, 1226-1230.

Price, C. A. (1962) Repression of acid phosphatase synthesis in *Euglena gracilis*. *Science*, **135**, 46.

Reichardt, W., Overbeck, J. and Steubing, L. (1967) Free dissolved enzymes in lake waters. *Nature*, **216**, 1345-1347.

Pick, F.R. (1987) Interpretations of alkaline phosphatase activity in Lake Ontario. *Canadian Journal of Fisheries and Aquatic Sciences*, **44**, 2087-2094.

Rivkin, R.B. and Swift, E. (1982) Phosphate uptake by the oceanic dinoflagellate *Pyrocystis noctiloca*. *Journal of Phycology*, **16**, 486-489.

Sinclair, C. and Whitton, B.A. (1977) Influence of nutrient deficiency on hair formation in the Rivulariaceae (Cyanophyta). *Journal of Phycology*, **13**, 297-313.

Siuda, W. (1984) Phosphatases and their role in organic phosphorus transformation in natural waters. A review. *Polskie Archiwum Hydrobiologii*, 31, 207-233.

Siuda, W., Chróst, R. J., Scislo, R. and Krupka, M. (1982) Factors affecting alkaline phosphatase activity in a lake (short-term experiments). *Acta Hydrobiologica*, 24, 2-20.

Stein, J.R. (ed.) (1973) *Handbook of Phycological Methods. Culture Methods and Growth Measurements.* Cambridge University Press, Cambridge.

Stevens, R.J., Parr, M.P. (1977) The significance of alkaline phosphatase activity in Lough Neagh. *Freshwater Biology*, 7, 351-355.

Taft, J.L., Loftus M.E. and Taylor, W.R. (1977) Phosphate uptake from phosphomonoesters by phytoplankton in the Chesapeake Bay. *Limnology and Oceanography*, 22, 1012-1021.

Torriani, A. (1960) Influence of inorganic phosphate in the formation of phosphatases by *Escherichia coli*. *Biochimica Biophysica Acta*, 38, 460-469.

Vincent, W.F. (1981a) Rapid physiological assays for nutrient demand by the plankton. I. Nitrogen. *Journal of Plankton Research*, 3, 685-697.

Vincent, W.F. (1981b) Rapid physiological assays for nutrient demand by the plankton. II. Phosphorus. *Journal of Plankton Research*, 3, 699-710.

Whitt, D.D. and Savage, D.C. (1988) Stability of enterocytes and certain enzymatic activities in suspensions of cells from the villous tip to the crypt of Lieberkühn of the mouse small intestine. *Applied and Environmental Microbiology*, 54, 2398-2404.

Whitton, B.A. (1987) The biology of Rivulariaceae. In P. Fay and C. Van Baalen (eds.) *Cyanobacteria: Current Research*, 513-534. Elsevier, Amsterdam, Netherlands.

Whitton, B.A. (1988) Hairs in eukaryotic algae. In F.E. Round (ed.) *Algae and the Aquatic Environment*, 446-460. Biopress, Bristol, England.

Whitton, B.A., Grainger, S.L.J., Hawley, G.R.W. and Simon, J.W. (in press) Cell-bound and extracellular phosphatase activities of cyanobacterial isolates. *Microbial Ecology*.

Whitton, B.A., Potts, M., Simon, J.W. and Grainger, S.L.J. (1990) Phosphatase activity of the blue-green alga (cyanobacterium) *Nostoc commune* UTEX 584. *Phycologia*, 29, 139-145.

Wynne, D. (1977) Alternations in activity of phosphatases during the *Peridinium* bloom in Lake Kinneret. *Physiologia Plantarum*, 40, 219-224.

Wynne, D. and Rhee, G.Y. (1988) Changes in alkaline phosphatase activity and phosphate uptake in P-limited phytoplankton, induced by light intensity and spectral quality. *Hydrobiologia*, 160, 173-178.

The Toxicity of Freshwater: Estuarine Bioindicators

James G. Wilson[1] and Bernard Elkaim[2]

[1]Environmental Sciences Unit, Trinity College, Dublin 2, Ireland.
[2]Laboratoire d'Hydrobiologie, Universit P. et M. Curie Paris VI, 12 rue Cuvier, 75005 Paris, France.

Key words: bioindicators, estuaries, salinity, pollution.

Abstract

In estuaries, where the seawater is appreciably diluted by freshwater, species numbers are typically cut to around 10% of those in fully marine conditions, yet the number of individuals may actually increase. The result is that 'polluted' values for commonly used indices are obtained even in uncontaminated locations, so alternative methods for assessing quality must be sought.

Chemical indices alone are inadequate, both on account of the complex nature of the estuarine environment and its effect on the form and speciation of contaminants and on account of the effect of the freshwater itself on organisms.

Accordingly, biological indices must take account of the particular characteristics of the system and the two approaches, community-based schemes and physiological and biochemical indices are here compared with examples.

Introduction

Estuaries are naturally places of paradox: to some they are deserts or literally "waste" lands of no value except to be reclaimed, while to others they are places of abundant life. Gradually, the latter view is now becoming more widely accepted, and it is now recognized that estuaries play major roles in many of the coastal processes and that, for example, estuarine productivity is on a level with that of good agricultural land.

By definition (e.g. Cameron and Pritchard 1963) an estuary is a place where "seawater is measurably diluted with freshwater", and this dilution poses problems for the biota at both ends, freshwater and salt water, of the estuary.

The classic work of Remane (1971) outlined clearly the diminution in the number of species in brackish water as opposed to both freshwater and salt water, and the same sort of situation can be found in the Baltic, where the numbers of species decline with distance from the North Sea. However, two points should be made here, namely that the decline in the number of species is not linear, but operates with greater severity at the extremes (i.e. toward the freshwater and salt water ends of the spectrum) and that in an estuary, unlike the Baltic, the degree of temporal variability in physical conditions, including salinity, can be more of a problem than the actual salinity itself. Some authors, for example Bagge (1969) and Leppakowski (1975), have already considered this problem without arriving at a completely satisfactory solution.

Nevertheless, there appears to be a reasonably well-defined group of species that one can accept as largely "estuarine". In this paper, we shall briefly look at the validity of this group as indicators of estuarine conditions and then examine at some more length their use as pollution indicators.

Estuarine species

There have been several classifications of brackish waters according to their salinity, classifications based loosely on the critical salinities for a particular group of species. For example, truly freshwater species are not found at salinities above 0.5ppt, while the truly marine species are not found below 30ppt (Remane 1971) with more or fewer sub-divisions in-between according to inclination.

This system works particularly well with planktonic or nectonic species, which remain within a specific water body and move around the estuary with it. One of the best studied examples of this are the Gammarids, in which there is a well-defined species gradient from freshwater to salt water, starting with *Gammarus pulex* (freshwater), through *G. duebeni* (4 - 18ppt), and *G. zaddachi* (*G.z. zaddachi* 1 - 15ppt, *G.z. salinus* 4 - 31ppt) to *G. locusta* (>25 - 30ppt) and *Marinogammarus* (marine) (Spooner 1947).

Other estuarine species such as the copepod *Eurytemora affinis* exhibit marked preferences for certain salinities, and in this case it is possible to distinguish between those of the adults and of the juveniles (Soltanpour-Gargari and Wellershaus 1985). *Eurytemora affinis* also exhibits a marked preference for the turbidity maximum, and its distribution is thus structured both longitudinally and vertically within the water column.

For species such as these which move with their chosen water body, the spatial and temporal variability of the salinity has little effect, but species which are fixed at one location, such as the benthos which inhabit the soft sediments, are subject to greatly changing conditions. As a consequence, the limits for the salinity tolerances for the benthos are much less well-defined, and the information available shows that both laboratory tolerances and field results from the distribution in the Baltic, for example, are in general much lower than those at which a particular organism might be found within the estuary.

In terms of the conventional bioindicators and pollution monitoring, then, the estuary poses several problems because of the stresses imposed by the salinity variations, and these will now be systematically considered.

Community indices (Holistic approach)

The combination of low species numbers yet high densities of some organisms in estuaries has meant that many of the conventional indices used to measure pollution, indices such as the Shannon-Weiner which is commonly used in fresh and salt water as well as several others put forward by a variety of workers (e.g. Gray 1979; Bascom 1982; Platt and Lambshead 1985) become difficult to interpret. Wilson (1988 p 120 *et seq.*) has taken a number of these indices and shown that they demonstrate a "stressed" community structure even in unpolluted estuarine conditions.

Leppakowski's (1975) Benthic Pollution Index (BPI) and the Biological Quality Index (BQI) developed by Jeffrey *et al.* (1985) avoid some of these restrictions and have the added advantage that they give an overall score for the whole area tested. The BQI, which allocates quality categories with reference to a number of community criteria, including community type, range and kind of species found and their age distribution, has been tested in a number of European estuaries with encouraging results (Wilson and Jeffrey 1987; Wilson and Elkaim in press). Jeffrey *et al.* (1985) also incorporated a Pollution Load Index (PLI) in which pollution status was ascertained from the contaminant loadings in the sediment and recommended that the two be used in conjunction, with further tests of specific targets or for specific contaminants if indicated.

Ducrotoy *et al.* (1989) have taken this procedure one step further, and have outlined an entire suite of protocols, including the BQI and PLI, to be followed in order to give an accurate assessment of the changes in the estuary over space and time.

Central to the BQI is the concept of "opportunistic" zones, where the normal community type is disturbed and certain kinds of organisms called variously "indicator", "opportunist" or "transgressive" predominate, although Wilson and Jeffrey (1987) have made it plain that such zones need not necessarily be a result of pollution, but can occur naturally particularly in highly dynamic situations. Ducrotoy *et al.* (1989), however, have put forward a rather different idea in that their division is into "target" and "key" species. Key species respond quickly to any disturbance and can be used to detect changes in the quality of the system, while target species, which do not show such variations, can be used for large scale comparisons (Desprez *et al.* 1986; Ducrotoy *et al.* 1989).

Indicator species

All those variously defined in the preceding paragraph can be considered under this heading and Pearson and Rosenberg (1978) have listed some 90 species as opportunist/transgressive and cited the references linking them to pollution. Of these 90 or so species, only a few (e.g. *Capitella capitata, Scolelepis fuliginiosa, Streblospio shrubsolii*) actually belong to a specific enrichment category and the rest respond to any disturbance.

In their (Pearson and Rosenberg 1978) list of such species can be found many of the species commonly found in estuaries. Table 1 lists the bivalves with two other typical estuarine species (*Nereis* and *Corophium*) as reference, and distinguishes those occurrences where their presence has been associated with pollution in fully marine or in estuarine (or other areas of lowered salinity) environments. From the list, it therefore appears that estuarine species (e.g. *Macoma, Nereis*) can act as opportunists elsewhere, but that the converse (opportunists as estuarine species) does not necessarily hold - see for example *Corbula* and *Thyasira*.

Within the list of estuarine species are the key species which are acutely sensitive to pollution, and the target species which are relatively insensitive and whose populations respond to long-term environmental changes proposed by Ducrotoy *et al.* (1989). An example of the first type would be the cockle *Cerastoderma*, whose populations in estuaries can show dramatic oscillations, and of the second, *Macoma* whose population fluctuations are much less extreme (Desprez *et al.* 1986; Ducrotoy *et al.* 1989). While these definitions seem to align the definition of key species with the opportunists, it is perhaps more likely that there exists a continuum along which the species are aligned, such that under a certain set of circumstances *Macoma* may act as an opportunist (see Table 1), while under others it can be considered as a target species (see Deprez *et al.* 1986).

Jeffrey *et al.* (1985) recognized these difficulties in their allocation of opportunistic zones, and broadened the definition beyond the mere presence or absence of certain species (although that was included as one of the criteria) to include the concept of domination of a zone by certain types of organisms and their size (c.f. Gray 1979) and life-style (e.g. number of age classes present). Zones thus defined can be clearly distinguished in terms of pollutant levels (Wilson and Elkaim in press; Wilson and Jeffrey 1987) and the concentrations of a range of contaminants analysed from opportunistic zones in a total of 13 Irish and French estuaries is shown in Table 2.

Table 1. Estuarine species as pollution opportunists: see also text, (from Pearson and Rosenberg 1978).

Species	Estuarine Pollution	Non-estuarine Pollution
Corbula gibba	1	8
Thyasira flexuosa	0	4
Mya arenaria	5	7
Macoma balthica	7	4
Mytilus edulis	3	4
Nereis diversicolor	7	5
Corophium volutator	3	2

Table 2. Contaminant concentrations ($\mu g\ g^{-1}$ dry weight sediment unless stated) in the opportunistic zones of Irish and French estuaries. Organic matter is as %LOI.

Contaminant	Mean	S.D	Range
Organic matter	6.86	3.61	0.65 - 0.51
Total N	3172.0	3118.0	257.0 - 22024
Total P	1007.0	703.0	188.0 - 3512
Cd	1.28	2.32	0.00 - 9.90
Cr	43.8	32.4	6.00 - 130.0
Zn	178.0	122.5	6.82 - 567.0
Cu	42.1	36.33	0.00 - 166.0
Pb	101.0	136.3	3.69 - 741.0
Fe (%)	1.80	0.50	0.37 - 3.10
Mn	563.9	252.0	96.30 - 1127
Co	7.80	2.62	0.50 - 12.4
Ni	38.0	16.6	0.00 - 59.3
Hg	0.41	0.18	0.00 - 0.69
Hydrocarbons	861.5	1091.0	0.00 - 4162

As can be seen from Table 2 there is a wide range in the concentrations of contaminants. While this is no doubt partly due to the different sediment types and in particular the different particle size compositions, some of the zones classified as opportunistic displayed very low contaminant levels across the whole spectrum. The conclusion drawn from this is that there are therefore some areas in estuaries which are naturally opportunistic by virtue of salinity or some other stressor such as strong hydrodynamic action, although these conditions may operate unpredictably e.g. floods or storms (Wilson and Jeffrey 1987; Wilson and Elkaim, in press).

Consequently, it is difficult in practice to separate pollution-induced changes in estuarine communities from natural variation, the more so when the period studied follows an extreme environmental perturbation, such as storms or floods as mentioned above, or very hot summers/cold winters. However, abrupt changes which can be related to specific instances either of pollution or of natural stress are easily detected, and the common problem is one of determining the extent to which low level contamination has affected the estuarine community. Under these circumstances, the division into pollution-sensitive species and long-term environmentally sensitive species as suggested by Ducrotoy *et al.* (1989) above is one which deserves further attention.

Bioconcentration

The choice of which component of the estuarine environment to choose for monitoring purposes depends on the particular goal of the exercise and the advantages and disadvantages of each has been summarized by Wilson (1988, p. 108). There are two major advantages of using the biota to monitor contaminant levels. The first is that, like the sediment, the organisms concentrate many substances up to several thousandfold in some instances above those that are present in the water column. The second is that the levels obtained reflect the availability over time of that substance to the biota.

Table 3 shows an example for Cu concentrations in three estuarine species from various estuaries around Ireland in relation to the sediment Cu concentration.

In contrast to many estuaries, notably those used by Bryan in his studies in the south-west of England (Bryan and Gibbs 1983; Bryan et al. 1980; 1985) the Irish estuaries are in general lightly contaminated with metals, yet there is still considerable interspecific and intraspecific variation in the Cu concentrations in the three animals above (Table 3). Bryan (Bryan et al. 1985) recognized this difficulty and recommended that while some species such as S. plana could be good general indicators, and others useful for specific metals (for example N. diversicolor for Cu - see Table 3), no species was ideal for every element. He therefore recommended including a number of different species with different life-styles and in particular different modes of feeding in the monitoring scheme to assess contamination and differential availability of contaminants as well as the possibility of food-chain biomagnification. It should be stated however, that evidence for biomagnification of metals (excluding As) is lacking, except in the case of organic ligands, and the phenomenon may therefore be an important consideration only where organic or organically-linked substances are concerned (Mance 1987).

Table 3. Copper levels ($\mu g \ g^{-1}$ dry weight) in sediment, *Macoma balthica*, *Scrobularia plana* and *Nereis diversicolor* and sediment organic content (%LOI).

Estuary	Sediment	LOI	M. balthica	S. plana	N. diversicolor
Boyne[1]	19.0	5.2	57.0	49.0	50.0
Malahide[1]	9.0	2.4	n.a.	57.0	48.0
Portmarnock (a)[1]	9.0	2.0	81.0	67.0	51.0
Portmarnock (b)[1]	11.0	2.5	57.0	57.0	38.0
Bull Island[1]	10.0	2.1	282.0	252.0	43.0
Bull Island[2]	4.7	1.5	228.6	165.5	22.9
Tolka[1]	47.0	5.1	191.0	n.a.	40.0
Wicklow[1]	39.0	8.2	71.0	n.a.	25.0
Wexford[1]	19.0	3.6	35.0	111.0	28.0
Bannow[2]	6.8	2.9	n.a	17.3	33.1
Waterford[1]	19.0	7.5	n.a	40.0	27.0
Cork[1]	21.0	6.9	110.0	63.0	24.0
Galway[1]	98.0	13.3	261.0	291.0	101.0

[1]Brennan, 1988. [2]Magennis, 1987a

Table 4. Concentration factors (mean, standard deviation and coefficient of variation) for Cu in *M. balthica, S. plana* and *N. diversicolor* against Cu concentration - a) in whole sediment; b) normalized with respect to %LOI; c) pelite (<63μm) fraction; d) HCl extract; and e) porewater, (from Brennan, 1988).

	Mean	s.d.	c.v.
Macoma balthica			
a) Total sediment	6.78	8.34	1.23
b) normalized	0.24	0.16	0.67
c) pelite	3.84	2.36	0.62
d) HCl	10.48	12.74	1.22
e) Porewater	19.17	11.64	0.61
Scrobicularia plana			
a) Total sediment	6.74	7.17	1.06
b) normalized	0.23	0.14	0.61
c) pelite	3.99	2.28	0.57
d) HCl	16.11	14.67	0.91
e) Porewater	16.38	12.52	0.76
Nereis diversicolor			
a) Total sediment	2.54	1.86	0.73
b) normalized	0.093	0.034	0.37
c) pelite	1.58	0.77	0.49
d) HCL	8.94	12.08	1.35
e) Porewater	7.04	4.58	0.65

The accumulation of contaminants by organisms is dependent on the relative availability of the contaminant to the organism and the length of time of exposure. Consequently, the concentration in the organism will vary not only with the mode of uptake, but also with the partitioning of the contaminant in the environment. Table 4 shows the concentration factor for Cu in *Macoma balthica, Scrobicularia plana* and *Nereis diversicolor* with respect to the sediment for five different contaminant fractions.

All the concentration factors show a considerable amount of variability. It is not perhaps unexpected that those with regard to total sediment concentrations are very variable, as the concentration in the sediment will itself vary according to particle size, organic content etc. Normalizing the sediment concentration with regard to organic content approximately halves the variability, as does the analysis of a standard fraction, in this case the pelite (<63μm) fraction. What is rather unexpected is that an HCl digest of the sediment, which has been considered to reflect more closely bioavailable metals than a HNO_3 digest (Bryan *et al.* 1985; Brennan 1988), still results in very high variation in the concentration factors (Table 4). Extraction of the porewater and its analysis, reduces the variability somewhat, but does not eliminate it entirely, and it seems therefore that the variation stems as much perhaps from the different modes of uptake, which may vary as much from location to location as between species as from variability in the environmental parameter chosen for the comparison. From the above (Table 4) the least variability is consistently displayed with the pelite fraction. Ackerman *et al.* (1983) have demonstrated that the least variable fraction for the comparison of metals in estuarine sediments is that between 20µm and 2µm, so perhaps that should be what is chosen for calculation of organism concentration factors.

As to the problem of the salinity interfering with metal uptake, this can take two forms. Firstly the organism may take steps to isolate itself from unfavourable salinities either by shell closure, burrowing or some other mechanism, such that the exposure of the tissues does not reflect the levels in the water. Secondly, changing salinities can change the rate of many of the physiological processes, most obviously the rate of water and ion-exchange. Brennan (1988) in a comprehensive study of Irish estuaries found that salinity was a significant factor in relatively few combinations of organisms and metals, and it is difficult at this stage to say with any certainty what the action of changing salinity may be in any one case.

Individuals as pollution indicators

When confronted with a change in environmental conditions, the organism will attempt to adapt to them; if the degree of adaptation required is beyond its capacity it dies (or moves away). The result is a change in community structure, and the section above on community indices has discussed their use and their limitations in estuaries. Given these problems at community level, is it possible to use those species which can adapt and apparently flourish?

This approach has been advocated by (among others) Bayne and his co-workers (e.g. Bayne *et al.* 1979) for the mussel *M. edulis* where the degree of adaptation can be measured and correlated with the degree of contamination of the environment. Most estuarine species are fairly tolerant and comply well with the conditions laid down for the selection of indicator species (e.g. Phillips 1980), and there is now a considerable body of results with a variety of organisms and techniques demonstrating the effects of pollution in estuarine situations.

Table 5 shows a range of pollution indices for the same location (Bull Island, Dublin Bay) at different levels of organization to illustrate how the degree of adaptation may be assessed. It also shows not only the degree of variation between species but also the variability in any one species depending on which assay is chosen (see for example *A. marina*) as well as greatly different results in any one case.

There are several reasons for this seeming lack of consistency in the assays, including of course variation inherent at each level of organisation. However, there are two other important sources of variability. The first is temporal variation both in the contaminant delivery or pollution status, as found for example by Magennis (1987b), as well as in the condition and responses of both target and control animals. The relation between pollution stress and response

Table 5. Pollution indices for Bull Island expressed as a percentage of control areas.

Index	Organism	Control	% Response	Ref
Shannon-Weiner	-	Dublin Bay	40%	1
	-	Bannow Bay	112%	1
	-	Dublin Bay	40%	1
P/B ratio	*M. balthica*	Ythan	16%	2
	N. diversicolor	Ythan	63%	2
	S. plana	Lynher	100%	2
	C. edule	Grevelingen	271-73%	2
	A. marina	Grevelingen	383-242%	3
Osmoregulation	*A. marina*	Dublin Bay	90-84%	3
Lysosyme latency	*M. edulis*	Bannow Bay	450-40%	2
	A. marina	Dublin Bay	203-50%	3
Hydroid Assay				
- Growth	*C. flexuosa*	Bannow Bay	106-94%	4
- Gonozoids	*C. flexuosa*	Bannow Bay	2000-75%	4
AEC	*A. marina*	Dublin Bay	94%	3

1 - Wilson (1983); 2 - Magennis (1987a); 3 - Rafferty (1989) 4 - Magennis (1987b)

may be further obscured by the behaviour of the organisms, which can isolate themselves from the surrounding environment - as mentioned previously when discussing bioconcentration. In fact this phenomenon is so well studied in some species, including the mussel *M. edulis* that it is possible to buy a commercial detector system for pollution which makes use of this very response (Kramer *et al.* 1989).

The second reason is that environmental stress can produce the same response as pollution stress, such that for example high temperatures at the control site can obscure any possible pollution effect at the suspected contaminated site (see e.g. Table 5 lysome latency in *M. edulis*). Salinity itself is an important stressor, and an index such as the amino acid ratio (Jeffries 1972) measures the same amino acids used by bivalves for osmoregulation.

The latter poses a severe problem for management, in that it is relatively easy to measure stress with considerable accuracy, but much more difficult to attribute it with certainty. For this reason, Wilson and Jeffrey (1987) (and others) have advocated chemical analyses in combination with biological tests as a standard procedure in pollution investigation. There are some tests which are more specific, for example those which measure the degree of induction of specific enzymes such as aryl hydrocarbon hydroxylase (Lee *et al.* 1981) or of metallotheionines, although with the latter, it has recently been reported that these seem to be lacking in two of the most common estuarine bivalves, *M. balthica* and *S. plana* (Langston and Zhou 1987)

Conclusions

The impact of freshwater on estuarine organisms obstructs much of the conventional selection of indicator organisms, and the current status is summarized in Table 6.

As indicators of estuarine conditions, the best are those planktonic or nectonic species which remain within one fairly well-defined water body. While benthic species do in general show certain obvious characteristics, such as capacity for osmoregulation or tolerance, they are not necessarily confined to one set of conditions. As such, the best alternative is to look for the absence of species or groups, for example echinoderms, as indicators of lowered salinity.

As indicators of pollution, the characteristics which enable them to survive in the estuarine environment may also mask the levels or the effects of contamination. Conversely, the same tolerance or capacity for endurance of adverse conditions may be used and may also prove to be the best method for pollution detection. For contaminant accumulation, a range of species, as recommended by Bryan *et al.* (1985), is indicated the alternative is to use the sediment itself (see Wilson 1988 p108 *et seq.* for discussion).

The choice of *M. edulis* as a stress bioindicator rests on the amount of data on that species and its responses currently available. While the same techniques are available and do work with other species (see Table 5), they lack the volume of information as background. For the same reason, the suggested alternative is the "target" species *M. balthica*, particularly in relation to large scale (temporal and spatial) trends.

Table 6. Suggested indicators and alternatives for estuarine biomonitoring

For	Suggested Indicator	Alternative
salinity	nectonic/planktonic spp.	absences
pollutant loads	range of species	sediment
stress	eurytolerant species (*M. edulis*)	'target' species (*M. balthica*)

Acknowledgements

The authors would like to thank their colleagues, particularly those in the IERP and GEMEL for much interesting discussion and comment on these ideas.

References

Ackerman, F., Bergmann, H. and Scleichart, M. (1983) Monitoring of heavy metals in coastal and estuarine sediment - a question of grain size: <20 um versus <60 μm. *Environmental Technology Letters*, 4, 317-328.

Bagge, P. (1969) Effects of pollution on estuarine ecosystems. *Merentutkimuslait*, 228, 3-118.

Bascom, W. (1982) The effects of waste disposal on the coastal waters of southern California. *Environmental Science Technology*, 16, 226-236.

Bayne, B.L., Moore, M.N., Widdows, J., Livingstone, D.R. and Salkeld, P. (1979) Measurement of the responses of individuals to environmental stress and pollution : studies with bivalve molluscs. *Philosophical Transactions of the Royal Society of London*, B, 286, 563-581.

Brennan, B.M. (1988) *The geochemistry and bioavailability of heavy metals in inshore coastal sediments*. Ph.D. Thesis, University of Dublin, Trinity College.

Bryan, G.W. and Gibbs, P.E. (1983) Heavy metals in the Fal estuary, Cornwall: a study of long-term contamination by mining waste and its effects on estuarine organisms. *Marine Biological Association of the United Kingdom, Occasional Publications*, 2, 1-112.

Bryan, G.W., Langston, W.J. and Hummerstone, L.G. (1980) The use of biological indicators of heavy metal contamination in estuaries. *Marine Biological Association of the United Kingdom, Occasional Publications*, 1, 1-73.

Bryan, G.W., Langston, W.J. and Hummerstone, L.G. (1985) A guide to the assessment of heavy metal contamination in estuaries using biological indicators. *Marine Biological Association of the United Kingdom, Occasional Publications*, 4, 1-92.

Cameron, W.M. and Pritchard, D.W. (1963) Estuaries. In M.N. Hill (ed.) *The Sea*, Vol. 2, 306-324. John Wiley and Sons, New York.

Desprez, M., Ducrotoy, J.-P. and Sylvand, B. (1986) Fluctuations naturelles et evolution artificielle des biocoenoses macrozoobenthiques intertidales des trois estuaries des cotes francaises de la Manche. *Hydrobiologia*, 142, 249-270.

Ducrotoy, J.-P. Desprez, M., Sylvand, B. and Elkaim B. (1989) General methods of study of macrotidal estuaries: The Biosedimentary Approach. In J. McManus and M. Elliott (eds.) *Developments in Estuarine and Coastal Study Techniques*, 41-52. Olsen and Olsen, Fredensorg, Denmark.

Gray, J.S. (1979) Pollution - induced changes in populations. *Philosophical Transactions of the Royal Society of London*, B, 286, 545-561.

Jeffrey, D.W., Wilson, J.G., Harris, C.R. and Tomlinson, D.L. (1985) The application of two simple indices to Irish estuary pollution status. In J.G., Wilson and W. Halcrow (eds.) *Estuarine Management and Quality Assessment*, 147-162. Plenum Press, London.

Jeffries, H.P. (1972) A stress syndrome in the hard clam *Mercenaria mercenaria*. *Journal of Invertebrate Pathology*, 20, 242-251.

Kramer, K.J.M., Jenner, H.A. and De Zwart, D. (1989) The valve movement response in mussels : a tool in biological monitoring. *Hydrobiologia*, 188/189, 433-443.

Langston, W.J. and Zhou, M. (1987) Cadmium accumulation, distribution and elimination in the bivalve *Macoma balthica* : neither metallothionein nor metallothionein - like proteins are involved. *Marine Environmental Research,* **21,** 225 - 237.

Lee, R.F., Stolzenbach, J., Singer, S., and Tenore, K.R. (1981) Effects of crude oil on growth and mixed function oxygenase activity in polychaetes *Nereis* sp. In J. Vernberg, J. Calabrese, F.P. Thurberg and W.B. Vernberg (Eds.). *Biological Monitoring of Marine Pollutants,* 323-334. Academic Press, New York.

Leppakowski, E. (1975) Assessment of the degree of pollution on the basis of macrozoobenthos in marine and brackish-water environments. *Acta Academsis Aboenisis,* B. **35.2,** 1-96.

Magennis, B. (1987a) *The assessment of pollution at the Bull Island, Dublin.* Ph.D. Thesis, University of Dublin, Trinity College.

Magennis, B. (1987b) Hydroids and estuarine water quality. In D.H.S. Richardson (ed.) *Biological Indicators of Pollution'*, 211- 224. Royal Irish Academy, Dublin.

Mance, G. (1987) Pollution Threat of Heavy Metals in the Aquatic Environment. Elsevier Applied Science, London.

Pearson, T.H. and Rosenberg, R. (1978) Macrobenthic succession in relation to organic enrichment and pollution of the marine environment. *Oceanography and Marine Biology Annual Review,* 16, 229-311.

Phillips, D.J.H. (1980) Quantitative Aquatic Biological Indicators. Applied Science Publishers, Barking, Essex.

Platt, H.M. and Lambshead, P.J.D. (1985) Neutral model analysis of patterns of marine benthic species diversity. *Marine Ecology Progress Series,* 24, 75-81.

Rafferty, B. (1989) *Physiological parameters in the lugworm* Arenicola marine *(L.), as indicators of environmental pollution.* Ph.D Thesis, University of Dublin, Trinity College.

Remane, A. (1971) Ecology of Brackish Water. In A. Remane and C. Schlieper *Die Binnengewasser* Band 25, Wiley Interscience, New York.

Soltanpour-Gargari, A. and Wellenshaus, S. (1985) *Eurytemora affinis* - one year study of abundance and environmental factors. *Verofflichten Meeresforschung Bremerhaven,* 20, 183-198.

Spooner, G.M. (1947) The distribution of *Gammarus* species in estuaries. Part 1. *Journal of the Marine Biological Association of the United Kingdom,* 27, 1-52.

Wilson, J.G. (1983) The littoral fauna of Dublin Bay. *Irish Fisheries Investigations,* Series B, **26,** 1-20.

Wilson, J.G. (1988) *The Biology of Estuarine Management.* Croom Helm, London.

Wilson, J.G. and Elkaim, B. (*in press*). A comparison of the pollution status of twelve Irish and French Estuaries. In J.P. Ducrotoy, and M. Elliott (eds.) *Estuaries and Coasts : Spatial and Temporal Comparisons*, Olsen and Olsen, Fredensborg, Denmark.

Wilson, J.G. and Jeffrey, D.W. (1987) "Europe wide indices for monitoring estuarine quality. In D.H.S. Richardson (ed.) *Biological Indicators of Pollution*,225-242. Royal Irish Academy, Dublin.

The Use of Leaf Surface Inhabiting Yeast as Monitors of Air Pollution by Sulphur Dioxide

P. Dowding and J. Peacock

Department of Botany, Trinity College, Dublin 2, Ireland.

Key words: yeasts, leaf surface, sulphur dioxide, air pollution, biomonitor, acid, synoptic survey.

Abstract

Leaf-surface inhabiting yeasts in the Sporobolomycetaceae have been shown to be sensitive to urban and industrial air pollution. They can be specifically isolated onto simple undefined media and the resulting pink colonies can be counted. A large number of successful synoptic surveys of air quality using leaf yeasts as bioindicators have been carried out in northern Europe in the last decade by second-level students. Laboratory and field studies have indicated that leaf yeasts are specifically sensitive to sulphur dioxide, with a relatively rapid response time of >6h. Leaf yeast counts, transformed to natural logarithms, have been shown to be negatively correlated to the average sulphur dioxide concentration during the preceding two to five days over the range 0 - 130 $\mu g\ m^{-3}$. In laboratory simulations of the leaf surface, cell division in pure cultures of *S. roseus* are not inhibited by strong acids down to pH 3.

The practical findings are discussed in relation to current knowledge about sulphur dioxide uptake by plant canopies and about the sensitivity of food spoilage yeasts and other fungi to sulphur dioxide in solution.

Introduction

Leaves of higher plants have evolved as gas-exchange organs and in their relatively exposed position act as effective pollutant traps as well. Organisms living on leaf surfaces are therefore in a very exposed position with regard to the physico-chemical effects of both gaseous and particulate pollutants.

Dowding and Carvill (1980) were the first to report a reduction in counts of *Sporobolomyces* spp. isolated by the sporefall method (Last 1955) which was associated with urban conditions in Dublin, Ireland. They selected *Fraxinus excelsior* as an ubiquitous host with large leaves and a consistent leaf age on

Bioindicators and Environmental Management
ISBN 0-12-382590-3

short shoots and on the first two pairs of leaves on long shoots of different plants. The sample collection for this work, from between 40 and 70 sites in and around Dublin, was done on two consecutive mornings to minimize variations in counts that were expected to arise from changes in weather conditions. Morning sampling allowed rapid processing of samples for an overnight spore-fall phase of 16 h. At this point the method was limited to synoptic surveys that could be accomplished by 8 to 10 h sampling and could only take place in the summer (June to October inclusive). The inability to use the method in the winter when higher pollution emissions from urban rather than industrial sources could be expected was, and remains, a great limitation.

Methodology

The sporefall technique

The spore-fall method for isolating ballistosporous phylloplane yeasts established by Last (1955) and improved by Dickinson (1971) is a robust method. If leaves, or discs cut from them, are stuck to the inside of a petri-plate lid, the only organisms that fall from the leaf to the agar when the plate is incubated lid uppermost are the ballistosporous yeasts. In western Europe there are two common genera, *Sporobolomyces* and *Tilletiopsis*. The former form **pink** glistening colonies on malt extract agar (0.5-2.0% malt) which grow to visible and countable size in 48-72 h at 15-25°C (i.e., uncontrolled summer room temperatures), while the latter genus forms fuzzy semi-mycelial **white** colonies which only become countable after 4-7 days. The occurrence, characteristics and taxonomy of the ballistosporous yeasts are well reviewed in Webster (1970). The ballistospores are violently ejected some 100 μm from the leaf surface and fall to the agar surface, provided that the leaves are suspended <u>above</u> the agar. As there is a circadian rhythm of discharge, a full 24 h incubation with lid uppermost is ideal, and consistent timing of this first phase of the isolation procedure is rather more critical if periods of less than 24 h are used (Pennycook and Newhook 1974; Johnston 1983). For quantitative work all that is necessary to stop effective spore discharge is to invert the plates after 24 h.

It has proved possible to set up the petri dishes for this method in a wide variety of places (including the back seat of a car) as with a few sensible precautions it is remarkably contamination-proof. The plates should be poured at least four days in advance (on Friday for use on Monday, for instance). Any contamination at the pouring stage will be visible after four days at room temperature. A small (30cm x 30cm) alcohol-swabbed plastic surfaced board is used to park the inverted bottom halves of the petri dishes while the leaf discs are being stuck to the lids. This manoeuvre reduces the chances of airborne contamination of the agar surface. The discs are cut on a second wooden board, which need not be sterilized, so long as the discharge surface of the leaves is kept uppermost. Small blobs of vaseline are applied with a finger in a pattern designed to maximize the distance between discs; a predrawn template on paper is useful here. The leaf discs are transferred from the cutting board to petri plate

lid with clean forceps, though fingers will do. If the plates have to be transported
prior to the sporefall phase, then they are kept lid lowermost until they can be
left undisturbed for 24 h with the lid uppermost. The sporefall phase is best done
in the dark to reduce the transfer of water from the medium to the lids and to the
leaves. If these simple precautions are followed, the only significant
contamination will come from discs that have fallen off the lid during the
sporefall phase. Scoring must explicitly exclude such discs. Some drying of the
agar medium (to 33% loss of original weight) does not affect the counts obtained
(Fig. 1).

Selection of host species
Considerable variation in counts from different host species was expected.
Generally, counts are higher on unhairy deciduous leaf surfaces than on hairy or
evergreen leaves. In Figure 1 some idea can be obtained of the variation in mean
counts from individual leaves of *Fraxinus excelsior* collected from several trees in
the same rural location on two occasions a week apart, and from leaves of *Tilia
europaea* collected at the same time and at the same location as the first *F. excelsior*
collection. There was no significant difference between the median of the mean
counts for the two *F. excelsior* and the *T. europaea* collections.

When the host range was widened, and suburban and urban sites
considered as well as rural ones, *F. excelsior* emerged as the host species
returning the highest counts in rural and suburban locations (Figs. 2 and 3), but
it also showed the highest between-tree variability at each site. At the suburban
site, counts per disc from *T. europaea* showed greater within-tree variation than

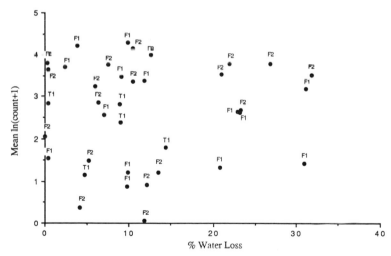

Figure 1. The relationship between drying out of the agar medium and the mean
transformed count (ln(count+1)) of *Sporobolomyces roseus* colonies arising from 8
mm diam. discs cut from leaves of *F. excelsior* (F) and *T.europaea* (T) on 10/7/87
(1) and 21/7/87 (2) taken from trees growing in rural Co. Meath, Ireland.

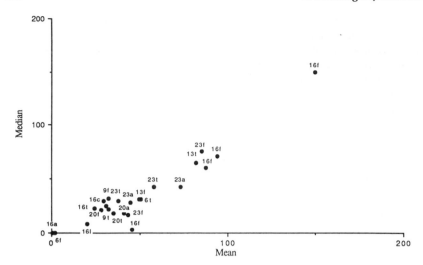

Figure 2. The relationship between the arithmetic mean and median of counts of colonies of *S. roseus* arising from twelve 8 mm discs cut from leaves of *F. excelsior* (f), *T. europaea* (t), *Acer psuedoplatanus* (a), *Crataegus monogyna* (c) and *Ligustrum vulgare* (l) during July 1987. The day of collection and the species code are shown in each point label. The site was rural Co. Meath, Ireland.

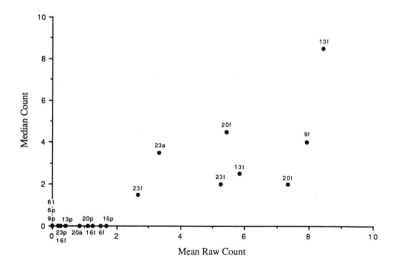

Figure 3. The relationship between the arithmetic mean and median of counts of colonies of *S. roseus* arising from twelve 8 mm discs cut from leaves of *F. excelsior* (f), *T. europaea* (t), *Acer psuedoplatanus* (a), *Platanus orientalis* (p) during July 1987. The day of collection and the species code are shown in each point label. The site was suburban north Dublin.

those from *F. excelsior* (Fig. 3). At both rural and suburban sites, *F. excelsior, T. europaea* and *Acer psuedoplatanus* were suitable hosts (Figs. 2 & 3) supporting disc counts in excess of 50 cm^{-2} d^{-1}. At the urban site, three subsites each with closely neighbouring specimens of *F. excelsior, T. europaea* and *Platanus orientalis* were investigated on seven separate occasions in July 1987 (Fig. 4). The correlation between disc counts from each of the hosts was very poor, even though many of the discs yielded no colonies at all. Discs of *T. europaea* yielded the lowest proportion of zero counts (Fig. 4a). The discs cut from leaves of *P. orientalis* yielded most zeros in urban and suburban locations, and as this species is not planted at all outside cities in Ireland, it was obviously an unsuitable host. However, in much of Europe it would be the most common tree present in both town and country and so might be the only practical choice. This proved to be the case in the survey in Lyon, France in 1987 (Dowding and Richardson 1989; 1990).

There is also a host-mediated difference in the best side to collect discharged spores from: for *F. excelsior* and *A. psuedoplatanus* it is the underside, while for *T. europaea* and *P. orientalis* it is the upperside.

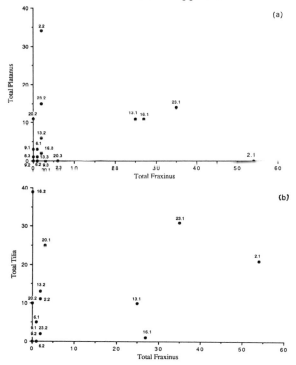

Figure 4. The relationship between the total plate counts of colonies of *S. roseus* arising from twelve 8 mm discs cut from leaves of *F. excelsior, T. europaea* (a), and *Platanus orientalis* (b) during July 1987. The day of collection and the subsite code are shown in each point label. All the subsites were in Trinity College park in the centre of Dublin.

The surface of the thick smooth cuticle of many evergreen leaves is a much harsher environment in terms of variations in water status and in temperature than is the surface of the thinner undulating cuticle of many deciduous leaves (Gates and Papian 1971). Early work in Dublin with common evergreens such as *Hedera helix* confirmed that the highest counts of *Sporobolomyces* spp. on evergreen leaves were so low as to reduce urban/rural differences to non significant levels (Barnes 1979). The relative position of counts from discs cut from *Ligustrum vulgare* in the rural site (Fig. 2) was a more recent confirmation of this view.

The standard protocol states that the same host species be used throughout each sample area.

Nonwoody hosts
The leaves of many herbaceous species and of grasses carry adequate populations of *Sporobolomyces roseus* to be useful as hosts in biomonitoring of air pollution. In addition, many of these nonwoody species are wintergreen, and so offer possibilities for the extension of the biomonitoring season. There are, however, much greater problems associated with obtaining consistent age samples in all herbaceous species, and with identification and standardization of leaf area in grasses in particular. The cultivation and inoculation of individual host plants in pots proved to be too difficult to manage in summer (Bailey and Dowding 1985), but may still be feasible for use in the wetter and cooler winter months. Some preliminary work in Spain by PD has shown that leaves of the ubiquitous weedy species of *Malva* are suitable, and that they showed the expected gradient in counts in a four point transect into Seville.

Other sources of variation in counts
The variations in counts caused by differences in host, in leaf age and physiological status, and by weather events prior to sampling are generally thought to be greater than those caused by microsite variations on the leaf or by the isolation method itself. Bashi and Fokkema (1976) have demonstrated that under favourable controlled conditions not all colonies of *Sporobolomyces* spp. developed sterigmata and spores, so it is likely that the count obtained by the spore-fall method is only a relative measure of the abundance of *Sporobolomyces* spp. cells on a leaf surface. The effect of leaf microsite was investigated using *F. excelsior* leaves from a rural location, and the results are summarized in Figure 5. There were no significant ($p<0.05$) differences in the means of untransformed counts between the different leaflets on the leaf, or between different positions on the midribs of the leaflets.

As the midrib discs were much easier to pick up and because there could be no ambiguity about their position on the leaf, the standard protocol specified midrib discs.

The effects of different disc size were also investigated using leaves of *F. excelsior* collected from a rural location, and the results are shown in Figure 6. Even in the rural location the smaller disc sizes returned more zero counts than the largest used in the trial (1 cm diam). That would imply an even higher

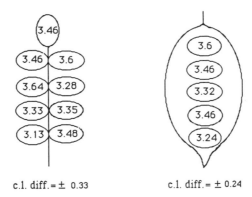

c.l. diff. = ± 0.33   c.l. diff. = ± 0.24

Figure 5. The relationship between the mean (ln(count+1)) of *S. roseus* colonies arising from 5 mm discs cut from different positions on leaves of *F. excelsior* at a rural site in Ireland during August. For comparison of leaflets n = 15, and for midrib positions n = 27.

proportion of zeros from areas the atmosphere of which were inhibitory to leaf yeast sporulation. The most important practical limitation to the upper size of disc used is the fatigue, and therefore error, induced by having to count very large numbers of colonies.

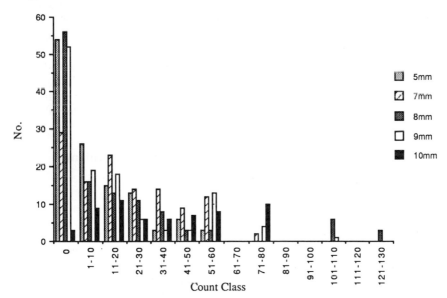

Figure 6. Frequency distribution curves of raw counts of colonies of *S. roseus* arising from different sized discs cut from midribs of *F. excelsior* leaflets during August at a rural site in Ireland.

In large multicentre synoptic surveys, the cost of the cutters becomes significant, and in the EC-funded survey, effective cutters of 1.3 cm internal diameter were made from 5 cm lengths of half inch copper tube, with the bevel cut on the outside of one end on a lathe. For single centre synoptic surveys, a cork borer between 1.0 and 1.2 cm diameter is most comfortable to use.

A schedule suitable for use by Irish schools was described in Richardson *et al.* (1985). The method was further refined and translated into French and German during preparatory work for the EC-funded surveys carried out in five countries in 1987 and 1988 and reported in Dowding and Richardson (1988; 1989; 1990). A description of the method has been published in German by Vietske and Dowding (1988). The schedules comprise a detailed instruction protocol, an explanatory document (for teacher information) and a sample score sheet, and are available on request from the Botany Department, Trinity College, Dublin 2, Ireland.

As the output from synoptic surveys has been described in great detail elsewhere (Dowding and Carvill 1980; Dowding 1986; 1987; Dowding and Richardson 1988; 1989; 1990; Ni Lamhna *et al.* 1983; 1984; 1985; 1986; 1987; 1988) it will not be repeated here. The ideal method of statistical data has not yet been finalized, and remains non-parametric for those without ready recourse to computer assistance (Dowding and Richardson 1990). The exact type of frequency distribution of the data appears to vary with sample size, and examples conforming to a poisson distribution, for which a square-root transformation is appropriate, are given in Figure 7, where n = 24. Dowding and Richardson (1990) suggest that a negative binomial distribution, for which a natural logarithm transformation is appropriate, fits data from sample sizes of 45 and 48, which were used in the EC-funded European surveys. Although the power and complexity of parametric tests are greater than those of the equivalent non-parametric tests, transformations cannot adequately include data from discs giving rise to coalescing colonies, or those from which too many colonies to count accurately have arisen. The median is a much better estimate of the central point of samples where some discs are in the "uncountable" category. These problems were included in the protocols referred to above.

Mechanisms of reduction of yeast populations by air pollutants

Experimental findings

Air humidity and leaf wetness

Phylloplane organisms spend much of the time in a dry and presumably inactive state. Many possess adaptive mechanisms to survive dry periods without damage (Park 1982) as do lichens. It has been found that lichens are much more resistant to the effects of acid gases, in particular sulphur dioxide, when they are dry than when they are wet. There is no direct evidence about the relative sensitivities of phylloplane organisms in the dry and wet states, but it is reasonable to assume that in the cases of water-soluble gases such as sulphur dioxide and ammonia their effects would be greater in the hydrated state.

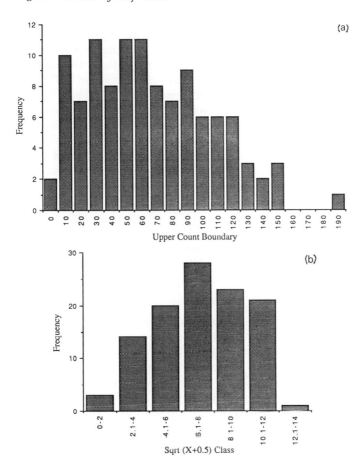

Figure 7. Frequency distributions of untransformed counts (a) and of square-root transformed counts (7b) arising from 10 cm discs cut from the midrib area of *F. excelsior* leaflets on 29/7/87 in rural Co. Meath, Ireland.

Bashi and Fokkema (1977) reported that under controlled conditions populations of *S. roseus* inoculated onto wheat leaves increased for three to eleven days at 95% relative humidity. Vegetative growth rates were observed to increase with the interpolation of an 8 h period of dew in each 24 hours. At lower humidities, without dew, the populations either did not change significantly (75% r.h.) or decreased by 10% d^{-1} (65% r.h.). Mishra and Dickinson (1984) concluded that on holly leaves there was little evidence for active vegetative growth at 95% r.h. and below. Importantly, it does not seem that *S. roseus* requires free water to grow on the leaf surface, though, as Bashi and Fokkema (1977) and Burrage (1971) point out, the conditions at the leaf surface are likely to be more humid than in the air surrounding the sensors responsible for giving the relative humidity measurements. That supposition will only be true so long as the leaf temperature is the same as or lower than that of the air surrounding it (Gates and Papian 1971). In both experiments reported the leaf surface temperatures would have tended to be higher than the air surrounding them as they were under artificial illumination and were shielded from cooler surfaces to which they could radiate. In any case Garland and Branson (1977) estimate that in temperate latitudes in summer, leaf surfaces are wet for about 30% of the time. It is possible that leaf surfaces in towns are drier than those in rural areas but any difference is within the range of humidities tolerated by *Sporobolomyces* spp. and, moreover, the results from the three rural synoptic surveys show that something other than wetness or humidity is exerting a very strong effect on populations of *Sporobolomyces* spp.. Dew certainly occurs in central Dublin but conventional meteorological measurements do not include dew duration or allow its calculation. To make the relationship of leaf yeast count to weather even more confusing, Flannigan and Campbell (1977) observed the highest count of their survey of *S. roseus* (150 000 cells cm^{-2} on wheat flag leaves) immediately after a dry spell in August. Fluctuations in counts caused by natural phenomena have to be taken into account before pollutants can be identified as causative agents in observed diminutions of counts.

Sulphur dioxide in the field

Dowding (1986) reported the results of two simple field experiments which had been conducted by students under his supervision. In these the burning of elemental sulphur in quite small quantities (1.35 kg S in 12 h, giving a mass emission flux of 62.5 mg SO_2 s^{-1}, and 0.25 kg S in 1 h, giving a mass emission rate of 139 mg SO_2 s^{-1}) caused a rapid (as short as 6 h after the end of the exposure) and statistically significant reduction in the counts of leaf yeasts obtained by the sporefall method from trees that had intercepted the ground level based plume. In neither experiment were the ambient concentrations of SO_2 measured.

Dowding and Richardson (1990) were able to calculate a statistically significant regression relationship between the natural logarithm of the medians of samples of leaf yeast counts made by the sporefall method and the ambient SO_2 concentrations measured specifically by UV-absorption at nearby fixed monitors. The field work was carried out by H. Vietske and his students in Hamburg, Germany, where there is an urban network of automated air quality monitors measuring the ambient concentrations of smoke, sulphur dioxide, nitrogen oxides, ozone carbon monoxide and non-methane hydrocarbons. The continuous measurements are reported at 30 min intervals to a central computer and are freely available to the public. More work has to be done with the very large data set generated by the monitors, but the initial analyses are encouraging. Changes in SO_2 concentration are better related to leaf yeast counts than are changes in concentration of any other pollutant, and the range over which the yeasts respond (0 - 100 μg m^{-3}) are in a range suitable to monitor the attainment of EC guide values for SO_2. The best averaging time would appear to be four to five days, but further field work needs to done in conjunction with continuous monitors to separate "knock-down" effects from recovery effects. The relationship is shown in Figure 8.

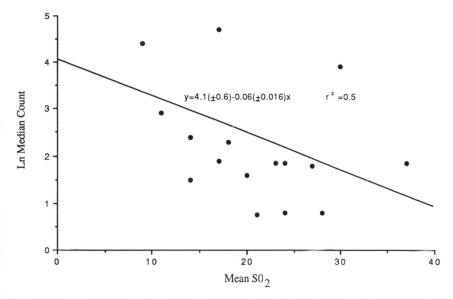

Figure 8. The regression line and scatter plot of the relationship between the natural log. of the medians of samples of 45 - 81 leaf yeast colony counts arising from discs cut from leaves of *F. excelsior* trees within 500 m of fixed specific ambient sulphur dioxide monitors and mean sulphur dioxide concentration over the preceding five days. Locations in Hamburg, Germany. Date 8/10/88.

Laboratory exposures

As reported above, a few field studies burning elemental sulphur have demonstrated decreased counts (Curran 1984; Dowding 1986), as have controlled fumigations with SO_2 (Magan and McLeod 1988) and a laboratory study which showed SO_2 to inhibit growth in solution (Curran 1984). Most of the exposure studies on yeasts in the laboratory have dealt with the effect of SO_2 as a means of food preservation (Balatsouras and Polymenacos 1963; Warth 1985; Stratford *et al.* 1987). Of these only one study refers to *S. roseus* (Balatsouras and Polymenacos 1963) where it was observed to be one of the least tolerant species. There is, therefore, very little controlled exposure of the leaf yeast to support field data. A method has subsequently been devised in the laboratory whereby the growth and reproduction of the leaf yeast *S. roseus* can be measured on a simple, simulated leaf surface. Leaf yeasts can be exposed in this system to inorganic pollutants in solution (SO_4, SO_3, NO_3) either singularly or in mixtures.

Methods
Small aliquots of suspension of *S. roseus* vegetative cells in 0.5% malt were pipetted onto sterile cellophane squares (1.7cm x 1.7cm) which were placed on a sterile surface consisting of 4 layers of 7 cm Whatman No. 1 filter paper soaked

in 4 mls of 0.5% malt in a glass petri-dish. Figure 9 shows the section and plan view of the arrangement of cellophane and filter paper in a model exposure chamber. Inoculation was then carried out on a clean air bench to minimize the risk of contamination. Each petri-dish was replaced and sealed with parafilm so that the risk of contamination was further reduced and then incubated at 21°C. Cellophane squares were destructively sampled every 6 hours up to 36 hours and the growth of the leaf yeast determined by measuring colony diameters under a light microscope at x400 magnification.

In order to examine the effects of acidity, phosphate buffers (0.1 M) were then tested at pH values from 2 to 5, together with different types of filter paper and with different methods of preparation of cellophane, to find out which materials and treatments were most suitable at holding the pH steady throughout the entirety of the experiment. The results indicated that use of buffers in the nutrient media and the soaking of the cellophane squares in the buffer provided stable pH values throughout the entirety of the experiment. The exception was pH 4 which may have a reduced buffering capacity as it is midway between two dissociation points (pK_a 1.96 and pK_b 6.7). As a consequence a pH of just below 4 was made up in solution which resulted in a final pH of approximately 4 after autoclaving.

The experiment was then run with the phosphate buffers plus 0.5% malt at pH 2, 3, 4, and 5. Results are presented as log (fm) against time (hours).

Results
(i) 0.5% malt (Fig. 10)
The results of the four runs were in good agreement showing an initial lag-phase up to 12 hours followed by an increase in the colony diameter up to 36 hours.

(ii) Phosphate buffers (0.1 M) (Fig. 11)
The results show that the leaf yeast failed to grow at pH 2 but at pH values of 3, 4 and 5 growth was not significantly different to that of the control (0.5% malt).

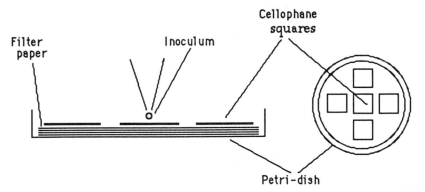

Figure 9. Section and plan view of the arrangement of cellophane and filter paper in model exposure chamber for the exposure of *S. roseus* to pollutants in solution.

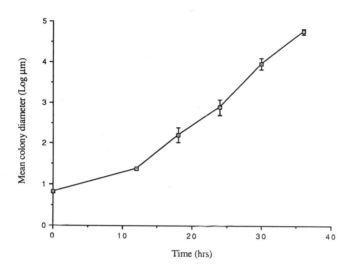

Figure 10. Graph showing mean colony diameter of S. roseus grown on 0.5% malt (mean of 4 samples).

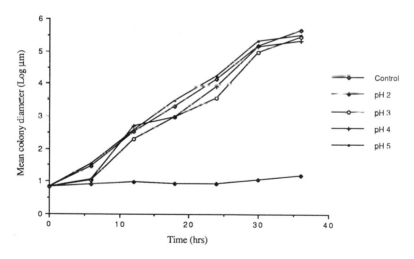

Figure 11. Graph showing mean colony diameter of *S. roseus* grown in phosphate buffer (0.1 M) at a range of pH values.

Acidity does not, therefore, seem to be an inhibiting factor of the leaf yeast growth from these studies except at very low pH values, which would also damage the leaf surface.

Future laboratory work

Experimental work is now progressing to examine the effects of the anions sulphite, sulphate and nitrate, singularly and in combination. Using the method as above, the anions are to be added to the phosphate buffer solution in varying concentrations and at different pH values (3, 4 and 5). A range of pH values are to be continued to be employed even though the effects of acidity has been examined as, for example, the antimicrobial properties of SO_2 are well known to be enhanced at low pH values due to the change in dissociation (Vas and Ingram 1949; Schmiz 1980). The study would, therefore, determine which pollutant or pollutant combination has the most deleterious effect on the leaf yeast.

Preliminary experiments, however, are being carried out with sulphite as this is an unstable anion and oxidizes quite readily to sulphate, particularly rapidly at higher pH values and with low concentrations (Garsed 1981). Tests are being preformed to determine the amount of sulphite that remains in the solution on the filter paper over the duration of the experiment. Results have indicated that sulphite concentration is reduced considerably in contact with filter paper. Reductions also occurred due to autoclaving in both solution and on filter paper. Lastly, as the pH decreased a greater amount of sulphite was retrieved indicating, in agreement with other studies (Garsed 1981), that sulphite is more readily oxidized at higher pH values.

Polarographic techniques are now being investigated to provide a more sensitive and accurate determination of sulphite concentrations, especially at lower concentrations. The technique will then be incorporated into the exposure experiments to determine sulphite concentration at each time interval.

In addition, examination of the effects on sporulation of the leaf yeast is to be incorporated into the experiment. At each of the 6 hour time intervals, the cellophane squares that have been used for colony diameter measurements will be attached to the lid of a petri-dish (colonized surface facing downwards) containing 1.0% malt agar. As in the spore-fall method, the petri-dishes will then be incubated at 21°C for 24 hours and then inverted and incubated for a further 48 hours. After 48 hours, colonies will be counted as an indication of the degree of sporulation which can be compared with growth of the yeast under the same conditions.

Possible modes of action of sulphur dioxide

The oxidation product of sulphur dioxide, sulphuric acid, may be the most important synergist of sulphur dioxide in polluted situations. The acidity of the solution of sulphur dioxide determines the proportions of the ions of sulphurous acid (Puckett *et al.* 1973). It is believed that the most toxic compound of sulphur dioxide with water is undissociated sulphurous acid (H_2SO_3) (Saunders 1966), which forms an increasing proportion of the dissolved sulphur dioxide as the pH falls below pH 4. Garsed (1981) stated that the resistance to solution of sulphur dioxide into water was minimal above pH 4 (from which point sulphurous acid ionizes). As more sulphur dioxide is taken up, and particularly if its oxidization rate in the solution is accelerated by an initially high pH (Baddeley *et al.* 1971), catalytic cations (Dlugi *et al.* 1982) or soot particles and/or higher than normal ozone concentrations (Martin and Barber 1985), the resistance to sulphur dioxide entering the solution will increase. Theoretically it will become infinite when the equilibrium constant between gaseous and dissolved undissociated sulphur dioxide has been attained. This is unlikely in nature as water films are never pure water. Studies that have been done on the chemistry of water on surfaces and in droplets have revealed a complicated chemistry in which neutralization of sulphuric acid by bases is as important in towns (Waller 1963) as in less polluted areas. Brimblecombe (1978) observed that dew was able to dissolve more sulphur dioxide than did distilled water because of dissolved cations in water films on leaf surfaces. Leaf surfaces and particles on them are reactive and the films and droplets are evanescent, so that the concentration of any soluble gas depends as much on the relative humidity of the air as on the concentration of the gas in the air surrounding the droplet/surface water film. Bewley and Campbell (1978) observed that microorganisms were more abundant on large amorphous particles on leaves polluted with heavy metals than on the leaf surface itself. Their results showed that most of the heavy metal contamination consisted of particles less than 1 µm diameter, and concluded that the large particles must have been providing additional nutrients. It is much more likely, however, that the large particles were basic and were neutralizing the acid and/or precipitating the metal ions in solution because of the low pH associated with emissions from the Avonmouth smelter. Commins (1963) found appreciable amounts of insoluble bases in London air. Martin and Barber (1978) reported that calcium sulphate forms a substantial part of deposited sulphate in the U.K. They speculated that this might have been derived directly from soil, but it is more likely that it derived from the reaction in the air of sulphurous/sulphuric acids in droplets with particles of lime (mixtures of calcium carbonate and calcium hydroxide) which became airborne as a result of both agricultural operations and building construction. Clay dust would also be neutralizing, but the cations would be a mixture of aluminium, iron, potassium and ammonium principally.

In rural areas, ammonia gas derived from biodegradation of high nitrogen wastes such as slurry from intensive cattle and pig farms would also neutralize some of the sulphate. Martin and Barber (1985) found that most of the ammonium in rural UK was in the small (<2.5 µm) particle fraction but that the amount of ammonium ions present in air were not sufficient to neutralize all the sulphate ions present.

Leaf surfaces themselves can affect the chemical properties of the water on them. Rowlatt *et al.* (1978) found that the pH of acidic rain was increased by two units after passage over senescent tree leaves in autumn. In polluted situations increased exudation of buffering capacity onto the senescing leaf surface may well permit populations of pollution-sensitive phylloplane organisms, such as *Sporobolomyces* spp., to increase for a short time. Huttunen and Soikkeli (1983) using scanning electron microscopy found that in polluted atmospheres minute roughnesses on the surface of leaf cuticles of conifers, *Rosa rugosa* and *Vaccinium* spp. were lost and that small epidermal folds in the surface of the leaves of *Aesculus hippocastaneum* had disappeared. It is possible that such changes in microhabitat make the urban leaf a less suitable place for phylloplane organisms, quite apart from the sensitivity of the phylloplane community to any particular pollutant.

The absorbtion of sulphur dioxide by leaf surfaces is greatly affected by the wetness of the leaf surface as well as by the opening of the stomata. Fowler and Unsworth (1979) reported that if leaves were covered in dew cuticlar resistance to uptake of sulphur dioxide was reduced to one-tenth of that of dry leaves. Fowler (1981) found that the canopy resistance of a senescent wheat crop to sulphur dioxide fell to zero when the leaves were wet. He estimated that in a wheat crop absorbtion of sulphur dioxide by the various surfaces was as follows: to wet cuticle 5 mm s^{-1}, through open stomata 5 mm s^{-1}, through dry live surface 2 mm s^{-1} and into dry dead leaves 3 mm s^{-1}. Taylor and Tingey (1983) estimated that the rate of sulphur dioxide uptake by the surface of the mesophyll cells was 3 μM m^{-2} h^{-1}. If the figures of Taylor and Tingey (1983) are combined with those of Fowler (1981) and an external/internal leaf surface ratio of 1/16 is accepted, then an uptake rate of 0.05 mM SO$_2$ m^{-2} h^{-1} into the surface water film appears to be possible. If the water film evaporates, the dissolved sulphur dioxide in it, in whatever form, will be concentrated. As so many possible reactions might occur to the dissolved sulphur dioxide it is probably not profitable to speculate on the concentrations of sulphurous acid that might be attained in the absence of impurities and of oxidation. However, in order-of-magnitude terms and in the absence of further reactions, a surface water film 20 μm thick on average will come to equilibrium with a sulphur dioxide concentration in air of 100 μg m^{-3} and a solution concentration of *c*.35 mg l^{-1} (0.5 mM) sulphurous acid in one hour. Saunders (1966) found by a combination of laboratory and field experimentation that these were threshold concentrations for toxicity to the infection of rose leaves by the spores of *Diplocarpon rosae*. He speculated that yeasts (as a class, rather than phyllosphere yeasts in particular) were more sensitive than *D.rosae* on the basis of work by Ingram (1948) who had found that fruit-spoilage yeasts were inhibited by sulphur dioxide concentrations of 10 mg l^{-1} in pure water (equivalent to 30 μg m^{-3} in air). By contrast, *Botrytis cinerea* (Couey and Uota 1961) and *Alternaria* spp. (Couey 1965) conidia were inhibited from germination when exposed to sulphur dioxide in air at concentrations above 2800 μg m^{-3}, nearly two orders of magnitude greater than the concentration toxic for yeasts. Both *B. cinerea* and *Alternaria* spp. were more sensitive to sulphur dioxide at 100% relative humidity than at lower humidities.

Evaluation of the spore-fall method

The use of the method as an estimator of past concentration of sulphur dioxide in the air at a particular locality is made difficult by several factors:

1. The seasonality of colony density and of sporulation potential, and the sensitivity of natural populations to relatively short-term (several days to several

weeks) changes in the duration of leaf wetness and in temperature. This is particularly true of rainfall frequency and amount. Even if host and leaf age are standardized the background unpolluted count level will be set by the stage in the season and by the wetness of the preceding days. It is possible that the use of a biological simulation model would allow fairly good estimates of background count levels once a body of count and weather data had been built up for an area. The values of rate parameters used in such as model could not be transferred without possible large error for use in other areas. It would not be worth the effort if the following problems can not be resolved at the same time.

2. The response of *Sporobolomyces* spp. to different concentrations of sulphur dioxide is dose dependent (i.e. there is a time function as well as a concentration function) and is non-linear at least in the vegetative phase on agar and in liquid culture and possibly in the sporulating phase. The augmentation of cell duplication and possibly of sporulation at low concentrations of sulphur dioxide is parallelled by Saunder's (1966) observations on *D. rosae* and makes the estimation of an average concentration of sulphur dioxide in the preceding time interval very difficult if the concentrations are subject to variation, as is the rule in the outside environment. Without further experiment it would be impossible to say if the augmentation effect occurs in the cell growth and pre-sporulation phases and at what level of sulphur dioxide it will occur. Direct observation of changes in spore-fall counts after the exposure of plants to controlled combinations of different sulphur dioxide concentrations, humidities, temperature and leaf wetness durations would give an empirical answer to this problem.

3. The extent of synergism and of antagonism from other airborne materials and from the substratum itself has not been investigated experimentally. If *Sporobolomyces* spp. is to be used for anything more than an indicator of general pollution levels (Bewley and Campbell 1980), these relationships will have to be addressed, possibly at the same time as empirical trials .

References

Baddeley, M.S., Ferry, B.W. and Finegan, E.J. (1971) A new method of measuring lichen respiration: response of selected species to temperature, pH and sulphur dioxide. *Lichenologist*, **5**, 18-25.

Bailey, M.F. and Dowding, P. (1985) *Final Report: Long-distance air pollution effects on rain chemistry and on leaf yeasts in the Dublin area*. Project no. B-83-69. Report to DG XII, European Commission, Brussels.

Balatsouras, G.D. and Polymenacos, N.G. (1963) Chemical preservatives as inhibitors of yeast growth. *Journal of Food Science*, **28**, 267-275.

Barnes, F. (1979) *Laboratory exposures of* Sporobolomyces roseus *and* S. salmonicolor *to sulphur dioxide, and a study of winter populations of* Sporobolomyces *spp. on ivy leaves*. B.A.Hons (Mod.) Thesis, University of Dublin.

Bashi, E. and Fokkema, N.J. (1976) Scanning electron microscopy of *Sporobo-lomyces roseus* on wheat leaves. *Transactions of the British Mycological Society,* **67,** 500-505.

Bashi, E. and Fokkema, N.J. (1977) Environmental factors limiting the growth of *Sporobolomyces roseus,* an antagonist of *Cochliobolus sativus,* on wheat leaves. *Transactions of the British Mycological Society,* **68,** 17-25.

Bewley, R.J.F. and Campbell, R. (1978) Scanning electron microscopy of oak leaves contaminated with heavy metals. *Transactions of the British Mycological Society,* **71,** 508-511.

Bewley, R.J.F. and Campbell, R. (1980) Influence of zinc, lead and cadmium on the microflora of hawthorn leaves. *Microbial Ecology,* **6,** 227-240.

Brimblecombe, P.R. (1978) Dew as a sink for sulphur dioxide. *Tellus,* **30,** 151-158.

Burrage, S.W. (1971) The microclimate at the leaf surface. In T.F. Preece and C.H. Dickinson (eds.) *Ecology of Leaf Surface Organisms,* 91-101. Academic Press, London.

Commins, B.T. (1963) The determination of particulate acid in town air. *Analyst,* **88,** 364-367.

Couey, H.M. (1965) Inhibition of germination of *Alternaria* spores by sulfur dioxide under various moisture conditions. *Phytopathology,* **55,** 525-527.

Couey, H.M. and Uota, M. (1961) Effect of concentration, exposure time, temperature and relative humidity on the toxicity of sulfur dioxide to the spores of *Botrytis cinerea. Phytopathology,* **51,** 815-819.

Curran, E.B. (1984) *The effects of sulphur dioxide on the reproduction and growth of* Sporobolomyces roseus. B.A.Hon. (Mod.) Thesis, University of Dublin.

Dickinson, C.H. (1971) Cultural studies of leaf saprophytes. In T.F. Preece and C.H. Dickinson (eds.) *Ecology of Leaf Surface Organisms,* 129-137. Academic Press, London.

Dlugi, R., Jordan, S. and Lindemann, E. (1982) Influence of particle properties on heterogeneous SO_2 reactions. In B. Versino and H. Ott (eds.) *Physico-chemical Behaviour of Atmospheric Pollutants,* 308-318. Dordrecht: Reidel.

Dowding, P. (1986) Leaf yeasts as indicators of air pollution. In N.J. Fokkema and J. Van der Heuvel (eds.) *Microbiology of the Phyllosphere,* 121-136. Cambridge University Press, Cambridge.

Dowding, P. (1987) The use of leaf yeasts as indicators of air pollution. In D.H.S. Richardson (ed.) *Biological Indicators of Pollution,* 137-155. Royal Irish Academy, Dublin.

Dowding, P. and Carvill, P.H. (1980) A reduction in counts of *Sporobolomyces roseus* Kluyver on ash (*Fraxinus excelsior* L.) leaves in Dublin city. *Irish Journal of Environmental Science,* **1,** 65-68.

Dowding, P. and Richardson, D.H.S. (1988) *Interim report No 2 on leaf yeasts as air quality indicators in the EEC,* Project No. B-71-58 (080886-004373). Report to DG XI European Commission, Brussels.

Dowding, P. and Richardson, D.H.S. (1989) *Final report on leaf yeasts as air quality indicators in the EEC,* Project No. B-71-58 (080886-004373). Report to DG XI European Commission, Brussels.

Dowding, P. and Richardson, D.H.S. (1990) Leaf yeasts as indicators of air quality in Europe. *Environmental Pollution*, **66**, 223-235.

Flannigan, B. and Campbell, I. (1977) Pre-harvest mould and yeast floras on the flag-leaf, bract and caryopsis of wheat. *Transactions of the British Mycological Society*, **69**, 485-494.

Fowler, D. (1981) Turbulent transfer of sulphur dioxide to cereals: a case study. In J. Grace, E.B. Ford and P.G. Jarvis (eds.) *Plants and Their Atmospheric Environment*, 139-146. Blackwell, Oxford.

Fowler, D. and Unsworth, M.H. (1979) Turbulent transfer of sulphur dioxide to a wheat crop. *Quarterly Journal of the Royal Meteorological Society*, **105**, 767-784.

Garland, J.A. and Branson, J.R. (1977) The deposition of sulfur dioxide to pine forest assessed by radioactive tracer method. *Tellus*, **29**, 445-454.

Garsed, S.G. (1981) The use of sulphite solutions for studying the effects of sulphur dioxide on higher plants. *Environmental Pollution (Series A)*, **24**, 303-311.

Gates, D.M. and Papian, L.E. (1971) *Atlas of Energy Budgets of Plant Leaves*. Academic Press, London.

Huttunen, S. and Soikkeli, M. (1983) The effects of various gaseous pollutants on plant cell ultrastructure. In M.J. Koziol and F.R. Whatley (eds.) *Gaseous Air Pollutants and Plant Metabolism*, 117-127. Butterworths, London.

Ingram, M. (1948) The germicidal effects of free and combined sulphur dioxide. *Journal of the Society of Chemical Industry*, **67**, 18-21.

Johnson, B. (1983) *An investigation into some of the external factors affecting phylloplane populations of* Sporobolomyces roseus, *a pigmented leaf yeast.* B.A.Hons. (Mod) Thesis, University of Dublin.

Last, F. (1955) Seasonal incidence of *Sporobolomyces* on cereal leaves. *Transactions of the British Mycological Society*, **38**, 221-239.

Magan, N. and McLeod, A.R. (1988) *In vitro* growth and germination of phylloplane fungi in atmospheric sulphur dioxide. *Transactions of the British Mycological Society*, **90**, 571-575.

Martin, A. and Barber, F.R. (1978) Some observations of acidity and sulphur in rain water from rural sites in Central England and Wales. *Atmospheric Environment*, **12**, 1481-1487.

Martin, A. and Barber, F.R. (1985) Particulate sulphate and ozone in rural air: preliminary results from three sites in central England. *Atmospheric Environment*, **19**, 1091-1102.

Mishra, R.R. and Dickinson, C.H. (1984) Experimental studies of phylloplane and litter fungi on *Ilex aquifolium*. *Transactions of the British Mycological Society*, **82**, 595-604.

Ni Lamhna, E., Richardson, D.H.S., Dowding, P. and Wells, J.M. (1983) *An air quality survey of Cork City and Great Island carried out by school children*. An Foras Forbartha, Dublin. 29pp.

Ni Lamhna, E., Richardson, D.H.S., Dowding, P. and O'Sullivan, A. (1984) *An air quality survey of the Shannon estuary carried out by school children*. An Foras Forbartha, Dublin. 27pp.

Ni Lamhna, E., Richardson, D.H.S., Dowding, P. and Molloy, K. (1985) *An air quality survey of the South Coast (Waterford and East Cork) carried out by school children*. An Foras Forbartha, Dublin. 33pp.

Ni Lamhna, E., Richardson, D.H.S., Dowding, P. and Kiang, S. (1986)*An air quality survey of the South East (Wexford and Wicklow) carried out by school children*. An Foras Forbartha, Dublin. 30pp.

Ni Lamhna, E., Richardson, D.H.S., Dowding, P., O'Dowd, N. and Downey, D. (1987) *An air quality survey of the East Coast of Ireland, north of Dublin, carried out by school children*. An Foras Forbartha, Dublin. 31pp.

Ni Lamhna, E., Richardson, D.H.S., Dowding, P. and Ni Grainne, E. (1988) *An air quality survey of the Greater Dublin area carried out by second level students*. An Foras Forbartha, Dublin. 28pp.

Park, D. (1982) Phylloplane fungi: tolerance of hyphal tips to drying. *Transactions of the British Mycological Society*, **79**, 174-178.

Pennycook, S.R. and Newhook, S.J. (1974) Diel periodicity and circadian rythms of ballistospore discharge in the Sporobolomycetaceae. *Transactions of the British Mycological Society*, **63**, 237-248.

Puckett, K.J., Nieboer, E., Flora, W.P. and Richardson, D.H.S. (1973) Sulphur dioxide: its effect on ^{14}C fixation in lichens and suggested mechanisms of toxicity. *New Phytologist*, **72**, 141-154.

Richardson, D.H.S., Dowding, P. and Ni Lamhna, E. (1985) Monitoring air quality with leaf yeasts. *Journal of Biological Education*, **19**, 299-303.

Rowlatt, S., Crawford, D.B. and Unsworth, M.H. (1978) Sulphur cycle in wheat and other farm crops. In J.C. Brogan (ed.) *Sulphur in Forages*, 1-14. An Foras Taluntais, Dublin.

Saunders, P.J.W. (1966) The toxicity of sulphur dioxide to *Diplocarpon rosae* Wolf causing blackspot of roses. *Annals of Applied Biology*, **58**, 103-114.

Schmiz, K.L. (1980) The effect of sulphite on the yeast *Saccharomyces cerevisiae*. *Archives of Microbiology*, **125**, 89-95.

Stratford, M., Morgan, P. and Rose, A.H. (1987) Sulphur dioxide resistance in *Saccharomyces cerevisiae* and *S. ludwigii*. *Journal of General Microbiology*, **133**, 2173-2179.

Taylor, G.E. and Tingey, D.T. (1983) Sulphur dioxide flux into leaves of *Geranium carolinianum* L.. Evidence for a non-stomatal or residual resistance. *Plant Physiology*, **72**, 237-244.

Vas, K. and Ingram, M. (1949) Preservation of fruit juices with less sulphur dioxide. *Food Manufacture*, **25**, 414-416.

Vietske, H. and Dowding, P.(1988) Der Rosafarbene Sporenwerfer (*Sporobolomyces roseus*) als Anzeiger der Luftbelastung. *Unteticht Biologie*, **131**, 32-35.

Waller, R.E. (1963) Acid droplets in town air. *International Journal of Air and Water Pollution*, **7**, 773-778.

Warth, A.D. (1985) Resistance of yeast species to benzoic and sorbic acids and to sulphur dioxide. *Journal of Food Protection*, **48**, 564-569.

Webster, J. (1970) *Introduction to Fungi*. Cambridge University Press, Cambridge.

Evaluation of a Chronic Toxicity Test Using Growth of the Insect *Chironomus riparius* Meigen

E.J. Taylor, S.J. Maund and D. Pascoe

School of Pure and Applied Biology, University of Wales College of Cardiff. PO Box 915, Cardiff, CF1 3TL, U.K.

Key words: chronic toxicity, growth, *Chironomus riparius*.

Abstract

A reproducible, sensitive bioassay based upon growth of the aquatic larval stages of the insect *Chironomus riparius* Meigen is described. Growth of 2nd instar larvae during a 10 day period (1/3 of the life cycle at 20°C) was investigated to show sub-lethal effects caused by chronic exposure to test chemicals. The applicability of the bioassay was assessed using the chemicals 3,4,-dichloroaniline (DCA), atrazine, copper and lindane. The no observed effect concentrations (NOEC) and lowest observed effect concentrations (LOEC) were obtained for each chemical and a comparison of lethal and sub-lethal toxicant effect concentrations made. During the copper study, metal bioaccumulation was also examined.

Growth of chironomid larvae is proposed as a laboratory bioassay to help provide information for the accurate definition of water quality for ecosystem protection. The method can be modified for field use, with growth changes indicating the presence of environmental stressors.

Introduction

Growth and development rates indicate the fitness of organisms and populations. Growth of an individual represents an integration of physiological processes which depends on the balance between metabolic demands and supplies, which in turn are affected by environmental conditions e.g. temperature and diet (Sutcliffe *et al.* 1981; Gauss *et al.* 1985). Growth inhibition is known to be a sub-lethal response to chronic toxicant exposure (Kosalwat and Knight 1987b).

In this study, larval growth of the insect *Chironomus riparius* was investigated as a criterion for developing a bioassay of chronic pollutant toxicity. This benthic species was chosen for the reasons discussed by Pascoe *et al.*(1989), including the ease with which it is cultured (McCahon and Pascoe 1988) and the

Bioindicators and Environmental Management
ISBN 0-12-382590-3

relatively short life cycle (approximately 25 days at 20°C). The reported tolerance of *C. riparius* to toxicants has largely been based on results from acute lethal studies performed with the 4th larval instar (Williams *et al.* 1984). However, studies with *Agapetes fuscipes* Curtis (McCahon *et al.* 1989) and *C. riparius* (Williams *et al.* 1986) have shown that early life-cycle stages of insects are more sensitive than late instars. In this bioassay the growth of 2nd instar larvae over 10 days, representing a chronic exposure period (more than 1/3 of the life cycle), at 20°C was determined. The selection of this stage, rather than the 1st instar, represents a compromise between sensitivity and ease of handling. Accurate determination of toxicant effects on growth was possible because larvae kept under control conditions develop to 4th instar stage during the test period.

Four common aquatic pollutants were investigated :

3,4-Dichloroaniline (DCA) - a hydrolysis product of several widely used herbicides (Wegman and Korte 1981)

atrazine - a non-persistent herbicide used in crop protection (Macek *et al.* 1976b; de Noyelles *et al.* 1982)

copper - associated with many industrial operations, e.g. power generation and mining (Kosalwat and Knight 1987) and used as copper sulphate to control aquatic vegetation (Taylor 1978)

lindane - used to control arthropod pests on food crops, timber, farm animals and humans (Harper *et al.* 1977).

The measurement of other parameters such as biochemical changes (Graney and Giesy 1987) or toxicant bioaccumulation (Dodge and Theis 1979), can be carried out in association with the growth bioassay to provide additional information on exposure effects. In this investigation the body burdens of larvae exposed to copper were determined and the data are discussed with respect to observed growth changes and potential ecological impact.

Materials and methods

For the investigation of each chemical ten egg masses were obtained, within 12 hours of oviposition, from a laboratory culture maintained as described by Holloway (1983). The eggs were incubated as a single culture at 20°C for ten days after which time the larvae (approximately 2000) had developed to 2nd instar stage.

Fifty individuals were used at all test concentrations and controls, with each larva randomly allocated to a compartment (6 cm^3) of a repli dish (Woodworth and Pascoe 1982) containing 4 ml of solution. Larvae were provided with a cellulose - mulch substrate (McCahon and Pascoe 1988), which had been soaked for one minute in the relevant test solution. Toxicity tests were carried out under static conditions with daily renewal of solutions, and each larva was provided with 0.5 mg/day of finely ground synthetic fish food (Tetramin - Tetra Werke, W.Germany) administered as 0.5 ml of a 1g l^{-1} suspension in the test solutions. This provided a slight excess to the requirement for optimal growth (Holloway 1983).

Table 1. 95% Confidence intervals for the mean toxicant concentration for 10 day growth of *Chironomus riparius* exposed to DCA, atrazine, copper and lindane.

DCA (mg l⁻¹)	Atrazine (mg l⁻¹)	Copper (µg l⁻¹)	Lindane (µg l⁻¹)
Control	Control	Control	Control
0.16-0.18	0.09-0.11	8.3-9.3	0.016-0.020
0.89-1.05	0.14-0.18	16.4-17.4	0.084-0.096
1.83-2.09	0.38-0.44	24.6-26.0	0.180-0.220
	0.72-0.88	30.3-35.5	0.780-0.800
	1.80-2.20	46.0-53.8	7.470-8.170
	4.00-5.80		

The tests were conducted in an incubator with a photoperiod of 16 hours light at 20°C. At the end of the exposure, larvae were blotted dry and individual wet weights were measured. After the exposure to copper, larvae were kept for an additional 16 hours in the test solutions without food or substrate. This starvation period was necessary for the egestion of contaminated gut contents, so preventing erroneous metal bioaccumulation values (Smock 1983). The larvae were then wet weighed for growth analysis, dried at 102°C and digested using AristaR Nitric Acid. The resulting solutions were diluted to 2% acid and analysed for copper by Atomic Absorption Spectrophotometry (AAS). It should be noted that as a consequence of the starvation period final wet weights of the larvae were reduced by approximately 15% (evidenced by the control weight value) compared to those of the other experiments.

Temperature, pH, dissolved oxygen and conductivity were determined daily by portable meters, while total hardness was determined by AAS. The analytical techniques used to measure the test concentrations of the four toxicants were : High Performance Liquid Chromatography (DCA and atrazine); Gas Liquid Chromatography (lindane) and AAS (copper). The 95% confidence intervals of the means of the toxicant concentrations recorded throughout the tests are shown in Table 1.

Results

Water quality parameters were similar for all four tests with temperature remaining at 20 ± 1°C, dissolved oxygen always exceeding 80% air saturation value, mean hardness at 151.2 ± 9.4 mg l⁻¹ as $CaCO_3$ and conductivity at 322.7 ± 5.5 µS cm⁻¹. The pH ranged from 6.8 to 7.2 between the four tests with variability within each individual bioassay of 0.2 pH units.

Final wet weights (means and 95% confidence intervals) of the larvae after the ten day exposure to the four toxicants are shown in Figure 1. Within each experiment the variances of the weights at the toxicant concentrations were not significantly different (95% level, Cochran's C test). The final wet weight of the larvae was found to be significantly (ANOVA) affected by the four toxicants, with an increase in exposure concentration leading to a reduction in growth : $F_{3,375} = 41.47$, $P < 0.001$ for DCA; $F_{6,346} = 3.82$, $P < 0.001$ for atrazine; $F_{5,264} = 11.58$, $P < 0.001$ for copper and $F_{5,264} = 156.58$, $P < 0.001$ for lindane. Multiple unplanned comparisons using Duncan's test ($P = 0.05$) were used to compare the mean wet weights of larvae at each test concentration to that of the control for the four chemicals so that no observed effect concentrations (NOEC) and lowest observed effect concentrations (LOEC) could be determined. These values and the derived maximum acceptable toxicant concentration (MATC) for each of the compounds are presented in Table 2.

(a)

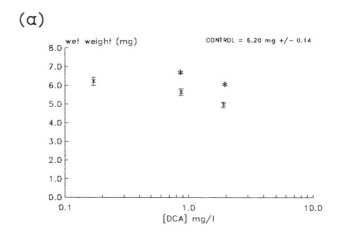

(c)

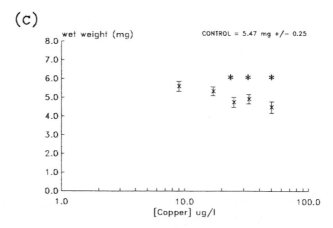

Figure 1. Mean wet weights with 95% confidence limits for *Chironomus riparius* after a 10 day exposure to (a) DCA, (b) atrazine, (c) copper and (d) lindane. *indicates a significant difference (P = 0.05) from the control.

In the copper bioassay, chironomid larvae are seen to have bioaccumulated the metal (Figure 2), the tissue levels increasing with the exposure concentrations. Although there is no direct supporting evidence from this study, previous work would suggest that the test period was adequate to allow bioconcentration to achieve a steady state (Kosalwat and Knight 1987a) so that bioconcentration factors (BCF) could be determined. These ranged from 150 to 390 in the 49.9 and 8.8 µg l⁻¹ test treatments respectively.

(b)

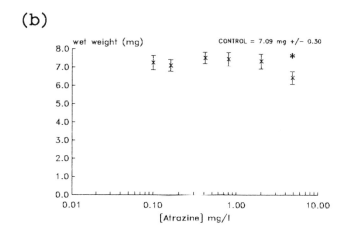

(d)

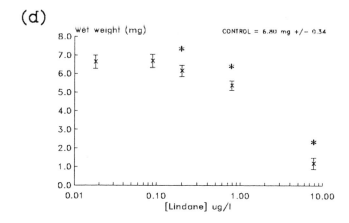

Figure 1. (continued).

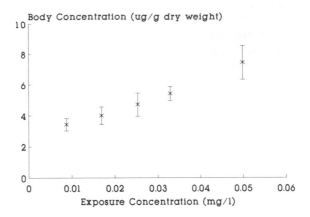

Figure 2. Mean body burden of copper with 95% confidence limits for *Chironomus riparius* exposed to various concentrations of copper for 10 days.

Table 2. Lowest (LOEC) and no observed effect concentration (NOEC) and the derived maximum acceptable toxicant concentration (MATC) for 10 day growth of *Chironomus riparius* exposed to DCA, atrazine, copper and lindane.

	DCA (mg l^{-1})	Atrazine (mg l^{-1})	Copper (μg l^{-1})	Lindane (μg l^{-1})
LOEC (growth)	0.97	4.9	25.3	0.20
MATC (growth)	0.43	3.1	20.7	0.13
NOEC (growth)	0.17	2.0	16.9	0.09

Discussion

The bioassay measures growth of *C. riparius* larvae over a 10 day period and has been successfully used to determine the NOEC and LOEC for four reference chemicals representing a range of toxicant types.

From lethal studies performed prior to the bioassay the 10 day LC50 for each chemical was found to be 4.2 mg l^{-1} for DCA, 18.9 mg l^{-1} for atrazine, 190 μg l^{-1} for copper and 13 μg l^{-1} for lindane. A ratio of each value to the corresponding MATC, a geometric mean of the NOEC and LOEC (Crossland and Hillaby 1985), was calculated for the chemicals. These ratios indicate the increase in sensitivity achieved when growth rather than mortality is investigated and are 9.7, 6.1, 9.2 and 100 for DCA, atrazine, copper and lindane respectively. The high ratio for lindane reflects this pesticides particular toxicity to insects.

The derivation of MATC's facilitates comparison with results obtained in other studies. The MATC calculated for DCA was 430 μg l^{-1} and this compares with 14 μg l^{-1} for *Daphnia magna* based on reproduction in a test conducted over 21 days (Crossland and Hillaby 1985), 90.5 and 27.2 μg l^{-1} for rainbow trout *Salmo gairdneri* Richardson based on relative growth rates of juveniles over 14 and 28 days respectively (Crossland 1985), and 13.7 μg l^{-1} for fathead minnow *Pimephales promelas* Rafinesque determined from a 32 day early life cycle test (Call *et al.* 1987).

The MATC of atrazine from this study was 3.1 mg l^{-1}, while Macek *et al.* (1976b) obtained the following values: 0.16 mg l^{-1} for *Chironomus tentans* based on changes in development rate and emergence during a 2 generation exposure (approximately 70 days); 0.12 mg l^{-1} for growth of brook trout fry *Salvelinus fontinalis* in a 44 week exposure and 0.25 mg l^{-1} for *Daphnia magna* production of progeny over 67 days. Although MATC's were not determined, de Noyelles *et al.* (1982) found that a concentration of 20 μg l^{-1} reduced phytoplankton within 2 days but that after a further period of 7 days growth had returned to the levels in the control. In the same study 500 μg l^{-1} of atrazine altered the pattern of phytoplankton succession.

The MATC obtained for copper, 20.7 μg l^{-1} compares with derived values from published data of: 113 μg l^{-1} for the chironomid *Paratanytarsus parthenogeneticus* based on reduced fecundity after life cycle exposure (Hatakeyama and Yasuno 1981); 22.2 and 13.9 μg l^{-1} for *Pimephales promelas*, from studies of reproduction and growth in water hardness of 200 and 30 mg l^{-1} CaCO$_3$ respectively (Mount 1968; Mount and Stephan 1969).

Sprague (1976) stated that measurement of tissue toxicant residues is more relevant if coupled to a determination of biological effects, which is possible within this bioassay. Final body burdens of copper indicated that the larvae had bioaccumulated the metal as was shown for *Chironomus decorus* by Kosalwat and Knight (1987a) who postulated that the thin chironomid integument facilitates copper penetration by active and/or passive transport. Other studies have shown the importance of surface adsorption in metal bioaccumulation by chironomids (Dodge and Theis 1979; Menzie 1980; Krantzberg and Stokes 1988; Smock 1983). The bioconcentration factors recorded in the present study (150-390) were lower than found by Kosalwat and Knight (1987a). However, their analysis of copper in chironomid larvae included gut contents which have been shown to contain over 50% of whole body burdens of metals (Smock 1983). The fact that copper reduces growth and is bioconcentrated by the larvae has consequences for the community structure. The consumers of *C. riparius*, both aquatic and terrestrial (Menzie 1980), may be affected by both the reduced chironomid biomass, limiting food availability, and the exposure to elevated concentrations of copper.

The MATC for lindane, 0.13 μg l^{-1}, compares with values found by Macek *et al.* (1976a) of: 3.3 μg l^{-1} for *C. tentans* based on development rate and emergence during a 2 generation exposure; 14.5 μg l^{-1} for *D. magna* reproduction over 64 days and 14.6 μg l^{-1} for *Pimephales promelas* growth in a 1 year investigation.

Comparison of the MATC's extrapolated from *C. riparius* growth studies with those for other species and other biological parameters provides a measure of the relative sensitivities of the tests but not the species. In many cases the reported MATC's were derived from particularly long exposure periods and the investigation of *C. riparius* for equivalent periods, with the inclusion of 1st instar larvae and reproductive parameters, would probably demonstrate effects at lower concentrations than those found in 10 days.

The results for the four reference substances indicate that the relative sensitivity of the 10 day bioassay to other tests is chemical dependent. For DCA and atrazine it is apparent that, as shown for other non-persistent herbicides (Buhl and Faerber 1989), the greatest risk to chironomids is acute exposure to toxic concentrations from field run-off shortly after application. Investigations of invertebrates for longer exposure periods, and also the responses of aquatic macrophytes and algae, are necessary to determine the chronic effect concentrations of such chemicals and their potential impact on aquatic communities.

. The results obtained in this study with copper and lindane indicate that growth reduction is a sensitive response criterion. The growth bioassay, which is relatively inexpensive, saves time and effort compared to longer tests. For some pollutants *C. riparius* is very sensitive to chronic exposure, and life cycle investigations should be performed to determine safe environmental concentrations.

Cairns and Dickson (1978) state that different classes of toxicants have different biological effects and that comprehensive hazard evaluation is not necessary for all chemicals. There is, however, an obvious need for the development of simple sub-lethal chronic tests with a number of species which represent potentially vulnerable groups. These methods would complement those currently used and provide assessments of toxicity intermediate to the acute and more complex chronic investigations. The 10 day growth bioassay is simple, cost effective and sensitive, providing information on the chronic effects of stressors which can be increased by parallel investigations and which may indicate the requirement for detailed life-cycle studies.

Acknowledgements

The work described in this report was partly funded by the Commission of the European Communities (Research Contract No. EV4V-0110-UK (BA) -Development and validation of methods for evaluating chronic toxicity to freshwater ecosystems).

References

Buhl, K.J. and Faerber, N.L. (1989) Acute toxicity of selected herbicides and surfactants to larvae of the midge *Chironomus riparius*. *Archives of*

Environmental Contamination and Toxicology, **18**, 530-536.

Cairns, J.Jr. and Dickson, K.L. (1978) Field and laboratory protocols for evaluating the effects of chemical substances on aquatic life. *Journal of Testing and Evaluation*, **6**, 81-90.

Call, D.J., Poirier, S.H., Knuth, M.L., Harting, S.L. and Lindberg, C.A. (1987) Toxicity of 3,4-dichloroaniline to fathead minnows, *Pimephales promelas*, in acute and early life-stage exposures. *Bulletin of Environmental Contamination and Toxicology*, **38**, 352-358.

Crossland, N.O. (1985) A method to evaluate effects of toxic chemicals on fish growth. *Chemosphere*, **14**, 1855-1870.

Crossland, N.O. and Hillaby, J.M. (1985) Fate and effects of 3,4-dichloroaniline in the laboratory and in outdoor ponds: II. Chronic toxicity to *Daphnia* spp. and other invertebrates. *Environmental Toxicology and Chemistry*, **4**, 489-499.

De Noyelles, F., Kettle, W.D. and Sinn, D.E. (1982) The responses of plankton communities in experimental ponds to atrazine, the most heavily used pesticide in the United States. *Ecology*, **63**, 1285-1293.

Dodge, E.E. and Theis, T.L. (1979) Effect of chemical speciation on the uptake of copper by *Chironomus tentans*. *Environmental Science and Technology*, **13**, 1287-1288.

Gauss, J.D., Woods, P.E., Winner,R.W. and Skillings, J.H. (1985) Acute toxicity of copper to three life stages of *Chironomus tentans* as affected by water hardness-alkalinity. *Environmental Pollution Series A*, **37**, 149-157.

Graney, R.L. and Giesy, J.P. (1987) The effect of short term exposure to pentachlorophenol and osmotic stress on the free amino acid pool of the freshwater amphipod *Gammarus pseudolimnaeus* Bousfield. *Archives of Environmental Contamination and Toxicology*, **16**, 167-176.

Harper, D.B., Smith, R.V. and Grotto, D.M. (1977) BHC residues of domestic origin: A significant factor in pollution of freshwater in Northern Ireland. *Environmental Pollution*, **12**, 223-233.

Hatakeyama, S. and Yasuno, M. (1981) A method for assessing chronic effects of toxic substances on the midge, *Paratanytarsus parthenogeneticus* - Effects of copper. *Archives of Environmental Contamination and Toxicology*, **10**, 705-713.

Holloway, M.T.P. (1983) *Factors controlling the productivity of a benthic detritivore* (*Chironomus riparius*). Unpublished Ph.D Thesis, UWIST, Cardiff.

Kosalwat, P. and Knight, A.W. (1987a) Acute toxicity of aqueous and substrate-bound copper to the midge, *Chironomus decorus*. *Archives of Environmental Contamination and Toxicology*, **16**, 275-282.

Kosalwat, P. and Knight, A.W. (1987b) Chronic toxicity of copper to a partial life cycle of the midge, *Chironomus decorus*. *Archives of Environmental Contamination and Toxicology*, **16**, 283-290.

Krantzberg, G. and Stokes, P.M. (1988) The importance of surface adsorption and pH in metal accumulation by chironomids. *Environmental Toxicology and Chemistry*, **7**, 653-670.

Macek, K.J., Buxton, K.S., Derr, S.K., Dean, J.W. and Sauter, S. (1976a) Chronic toxicity of lindane to selected aquatic invertebrates and fishes. *U.S. Environmental Protection Agency. Ecological Research Series EPA-600/3-76-046*.

Macek, K.J., Buxton, K.S., Sauter, S., Gnilka, S. and Dean, J.W. (1976b) Chronic toxicity of atrazine to selected aquatic invertebrates and fishes. *U.S. Environmental Protection Agency. Ecological Research Series EPA-600/3-76-047.*

McCahon, C.P. and Pascoe, D. (1988) Culture techniques for three freshwater macroinvertebrate species and their use in toxicity tests. *Chemosphere*, **17**, 2471-2480.

McCahon, C.P., Whiles, A.J. and Pascoe, D. (1989) The toxicity of cadmium to different larval instars of the trichopteran larvae *Agapetus fuscuscipes* Curtis and the importance of life cycle information to the design of toxicity tests. *Hydrobiologia*, **185**, 153-162.

Menzie, C.A. (1980) Potential significance of insects in their removal of contaminants from aquatic systems. *Water Air and Soil Pollution*, **13**, 473-480.

Mount, D.I. (1968) Chronic toxicity of copper to fathead minnows (*Pimephales promelaslas*, Rafinesque). *Water Research*, **2**, 215-223.

Mount, D.I. and Stephan, C.E. (1969) Chronic toxicity of copper to the fathead minnows (*Pimephales promelas*) in soft water. *Journal of the Fisheries Research Board of Canada*, **26**, 2449-2457.

Pascoe, D., Williams, K.A. and Green, D.W.J. (1989) Chronic toxicity of cadmium to *Chironomus riparius* Meigen - Effects upon larval development and adult emergence. *Hydrobiologia*, **175**, 109-115.

Smock, L.A. (1983) Relationships between metal concentrations and organism size in aquatic insects. *Freshwater Biology*, **13**, 313-321.

Sprague, J.B. (1976) Current status of sub-lethal tests of pollutants on aquatic organisms. *Journal of the Fisheries Research Board of Canada*, **33**, 1988-1992.

Sutcliffe, D.W., Carrick, T.R. and Willoughby, L.G. (1981) Effects of diet, body size, age and temperature on growth rates in the amphipod *Gammarus pulex*. *Freshwater Biology*, **11**, 183-214.

Taylor, D. (1978) *A summary of the data on the toxicity of various materials to aquatic life. Volume V Copper.* Imperial Chemical Industries Ltd. Brixham Laboratory, U.K.

Wegman, R.C.C. and De Korte, G.A.L. (1981) Aromatic amines in surface waters of the Netherlands. *Water Research*, **15**, 391-394.

Williams, K.A., Green, D.W.J. and Pascoe, D. (1984) Toxicity testing with freshwater macroinvertebrates: Methods and application in environmental management. In D. Pascoe and R.W. Edwards (eds.) *Freshwater Biological Monitoring. Advances in Water Pollution Control*, 81-93. Pergamon Press, London.

Williams, K.A., Green, D.W.J., Pascoe, D. and Gower, D.E. (1986) The acute toxicity of cadmium to different larval stages of *Chironomus riparius* (Diptera : Chironomidae) and its ecological significance for pollution regulation. *Oecologia*, **70**, 362-366.

Woodworth, J. and Pascoe, D. (1982) Cadmium toxicity to Rainbow Trout *Salmo gairdneri* Richardson: A study of eggs and alevins. *Journal of Fish Biology*, **21**, 47-57.

An Assessment of Field and Laboratory Methods for Evaluating the Toxicity of Ammonia to *Gammarus pulex* (L.) - Effects of Water Velocity

Paul C. Thomas, Craig Turner and David Pascoe

School of Pure and Applied Biology, University of Wales College of Cardiff, U.K.

Key words: ammonia, toxicity tests, episodic pollution, water velocity, substrate, Gammarus pulex, field tests, laboratory experiments.

Abstract

Ammonia is a common freshwater pollutant originating principally from industrial, sewage and farm effluents. The lethal toxicity of unionized ammonia (nominal concentration 6.5 mg l^{-1} at pH 9.0) to the freshwater amphipod *Gammarus pulex* (L.) was assessed over a 24 hour period using a variety of laboratory and field based toxicity tests. The effects of current velocity and the presence or absence of substrate in the test containers were examined. As expected, static toxicity tests showed wide fluctuations in pH and consequently in the amount of unionized ammonia present. The toxicity data recorded from these static tests thus proved unreliable. In the laboratory, using a continuous replacement dosing system, toxicity was similar whether or not substrate was present, while in an artificial laboratory stream and in the field, toxicity was less when substrate was present and under conditions of low current velocity.

Introduction

Ammonia is a common pollutant originating from sewage, industrial and farm wastes. Farm waste discharges have doubled over the last decade rising from 1 500 in 1979 to 3 000 in 1989 (N.R.A. 1989) and make up nearly one fifth of all pollution incidents reported in England and Wales. Slurry, a major contributor to farm waste episodes, is characterized by a high biological oxygen demand (B.O.D), and by elevated concentrations of ammonia, sulphide and particulate organic matter. Typical total ammonia concentrations are in the order of 1 200 mg l^{-1}. However, it has been established that the toxicity of ammonia to aquatic animals is related to the unionized form (Wurhmann and Woker 1948; Alabaster and Lloyd 1980; Fava *et al.* 1982; Williams *et al.* 1986), the concentration of which depends on both temperature and pH.

Bioindicators and Environmental Management
ISBN 0-12-382590-3

Despite current trends towards greater environmental relevance in toxicity testing (Pascoe 1988; Nimmo *et al.* 1989; McCahon *et al.* in press), the importance of water velocity as an additional factor has been given limited attention. Static, static-with-replacement and flow-through toxicity tests may not satisfactorily simulate field conditions and standard invertebrate toxicity testing chambers, used in the field, fail to reflect the natural habitat which may help to protect invertebrates from the full effects of a toxicant. Conditions which allow natural behavioural activity, such as invertebrate drift, and species interactions cannot be easily reproduced using traditional laboratory testing techniques and consequently such studies may not accurately reflect the true nature of pollution incidents. The present study attempted to determine the effect of water velocity and the influence of substrate, upon the toxicity of ammonia to *G. pulex* during an assessment of field, artificial stream and standard toxicity testing procedures.

Materials and methods

1. Animals used in the test
Gammarus pulex (L.) were collected from a limestone stream near Creigiau, South Glamorgan and maintained in aerated, dechlorinated Cardiff tap water at 12°C (total hardness 96 ± 4 mg l^{-1} as $CaCO_3$; pH 7.8 ± 0.2; conductivity 270 ± 5 µS cm^{-1}). Only males and non-brooding females between 6-10 mm were used for tests and these were acclimated to laboratory conditions for 24 hours before use. Animals used in the field were transported to the study site in aerated laboratory water and acclimated in the stream for 24 hours before dosing.

2. Toxicity testing methods
(a) Field dosing experiment - using a natural stream
A second order stream in the River Ely catchment, the Nant Dowlais (NGR ST 074840), flows through villages and grazing pasture before draining into the coal-polluted Nant Myddlyn. Water quality classification is good (class 1B; N.W.C., 1981) and a small tributary (1st to 2nd order) was used for the field experiment. Unionized ammonia concentrations were raised to a nominal 6 mg l^{-1} (Table 1) in one section of the stream by introduction of ammonium chloride solution (200 g l^{-1}) using a peristaltic pump (Fig 1a). In order to produce ammonia in the toxic unionized state, pH was increased to nine by dosing with sodium hydroxide (5M). Water hardness was 116 mg l^{-1} as $CaCO_3$ and conductivity 206 µS cm^{-1}. An upstream reference zone remained untreated and water quality parameters were monitored in both zones for 24 hours before, during and after dosing. The pH, temperature and dissolved oxygen were recorded using pHox 100 DPM multiparameter instruments. In addition, surface water samples were taken for ammonia analysis and acidified in the field (<pH2) to prevent nitrification. Ammonia concentrations were determined in the laboratory by a colorimetric method (Chaney and Marbach 1962).

(a) field experiment.

(b) artificial stream
    —high velocity.

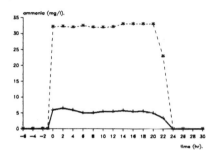

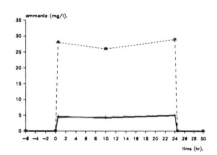

(c) artificial stream
    —low velocity.

(d) flow—through system.

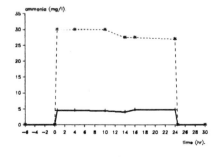

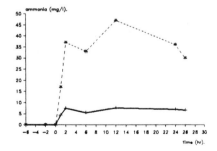

Figure 1. Total and unionized ammonia concentrations during field and laboratory experiments. (Total ammonia = --*--, unionized ammonia = --+--).
(a) Field experiment; (b) Laboratory artificial stream - high velocity; (c) Laboratory artificial stream - low velocity; (d) Laboratory flow-through system.

Table 1. Water quality parameters (mean and standard error) for the field and laboratory experiments. (* median and range reported due to wide variations).

Experiment	Total ammonia (mg l^{-1})	Unionized ammonia (mg l^{-1})	pH	Temperature (°C)
Field	30.6 ±0.75	6.47 ±0.17	9.00 ±0.01	10.3-14.3
Artificial Stream-high velocity	28.6 ±0.6	3.85 ±0.01	9.00 ±0.01	7.6
Artificial Stream-low velocity	29.0 ±0.58	4.06 ±0.16	9.02 ±0.02	7.5
Flow-through dosing system	38.2 ±3.26	5.95 ±0.29	9.00 ±0.03	10.0
Static * (i)	35.0 (nominal)	4.17-6.78 5.48	8.80-9.37 8.93	11.0-12.7
Static * (ii)	561.0 (nominal)	3.28-11.45 6.49	7.50-8.05 7.80	11.0

Toxicity testing was carried out using plastic containers with a nylon mesh (1 mm) section at either end to allow passage of water. Three sites were chosen for the test:- one fast flowing control site in the untreated reference zone, and two sites downstream of the dosing point, one with high (0.21 m s^{-1}) and one with low (0.13 m s^{-1}) water velocity. At each site four chambers were deployed, two provided with natural substrate from the stream and two without. Thirty G. pulex were counted into each of the 12 containers, together with a small quantity of shredded leaf from the stream which provided a source of food but not shelter. During the episode animals were checked regularly for mortality and all dead animals removed. Death was indicated by failure of an animal to respond to mechanical stimulation.

(b) Artificial Laboratory Stream
Experiments were carried out at high and low water velocities (0.15 and 0.06 m s^{-1}). Acclimated animals were placed into two adjacent 2.7m artificial streams (width 12cm x height 13cm). Four replicate sets of 30 animals were separated from each other by nylon mesh (1mm) placed at intervals along each channel. Water velocity was controlled by changing the slope of the channels and two sites along the stream were selected randomly to contain substrate (acid washed sand and five angular stones approximately 5 x 5 x 2 cm), while the remaining 2

sites were free from substrate. Animals were fed with finely chopped Horse Chestnut (*Aesculum hippocastanum* L.) leaves which had been allowed to 'condition' in aerated water for several weeks in order to establish populations of bacteria and fungi. Using a peristaltic pump to inject ammonium chloride, unionized ammonia was raised to a nominal concentration of 6 mg l^{-1} [actual concentration for the high velocity test 3.85 $\pm$ 0.01 mg l^{-1} (Fig. 1b), pH 9 $\pm$ 0.01, temperature 7.6°C, total hardness 95.2 $\pm$ 5.1 mg l^{-1} as $CaCO_3$; low velocity test ammonia (unionized) 4.06 $\pm$ 0.16 mg l^{-1} (Fig. 1c), pH 9.02 $\pm$ 0.02, temperature 7.5°C, total hardness 95.5 $\pm$ 3.2 mg l^{-1} as $CaCO_3$ (Table 1)]. As a control, animals were placed in the channels for 24 hours prior to dosing.

(c) Flow-through laboratory toxicity test
A continuous flow programmable dosing system (Greenham 1984) provided test tanks (volume 4 l) with an unionized ammonia concentration of 5.95 $\pm$ 0.29 mg l^{-1} (Fig. 1d) at a flow rate of 40 ml per minute giving a replacement of 99% every 8 hours. No significant water velocity was observed within the test chambers. Six tanks were used of which two contained stones, acid washed sand and test solution, a further two contained test solution alone, and two control tanks contained dechlorinated water. Thirty animals were placed in each tank. The pH of the test solution was maintained at 9.00 $\pm$ 0.03 using sodium hydroxide (NaOH), and temperature remained at 10°C throughout the experiments; total hardness was 96.4 $\pm$ 4.2 mg l^{-1} as $CaCO_3$.

(d) Static laboratory toxicity tests
(i) 30 animals were placed into each of eight 10 litre tanks. Four litres of toxicant were introduced into the six test tanks and dechlorinated tap water into the two control tanks. Two treatment tanks and one control contained stones and acid washed sand while all others were substrate free. Using NaOH the pH ranged from 8.80-9.37 throughout the 24 hour test period with a mean difference between tanks of up to 0.04 at any observation period and unionized ammonia ranged from 4.17-6.78. Total hardness was 96.2 $\pm$ 4.02 mg l^{-1} as $CaCO_3$ and temperature dropped from 12.7°C to 11°C during the test which continued for 24 hours with regular monitoring of mortality.

(ii) Conditions were as for Static Test (i) with the exception that the pH ranged from 7.50-8.05 with a maximum mean difference between tanks of 0.11, and a biological buffer (K_2HPO_4/KH_2PO_4) was employed in an attempt to control pH fluctuation. In view of the low pH it was necessary to raise the total ammonia concentration to 561 mg l^{-1} to ensure an adequate concentration of the unionized form. However, because of a pH effect the unionized ammonia varied from 3.28 to 11.45 mg l^{-1}. Temperature was maintained at 11°C.

Results

Analysis of mortality data
Mortality was < 5% in all the control experiments. Ammonia was found to be

toxic to *G. pulex* at the concentrations tested. However, mortalities in artificial stream experiments were insufficient for the calculation of median lethal response times (LT50s). No significant difference in mortality was found between replicates so mortality data were pooled and transformed to the normal equivalent deviate (N.E.D.) prior to analysis of covariance (ANCOVA; Sokal and Rohlf 1981). This allowed testing for statistically significant differences between mean mortalities when substrate was present or absent. All regressions of transformed mortality with time were significant (p<0.05, Figs. 2 to 4).

(a) high velocity.   (b) low velocity.

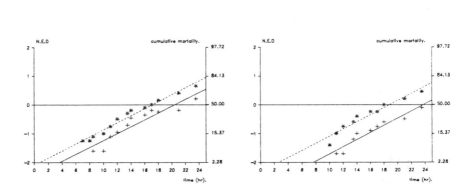

Figure 2. Regression lines of transformed cumulative percent mortality with time during field experiments. (Mortalities in containers without substrate = --*--, with substate = --+--). (a) High velocity; (b) Low velocity.

(a) high velocity.   (b) low velocity.

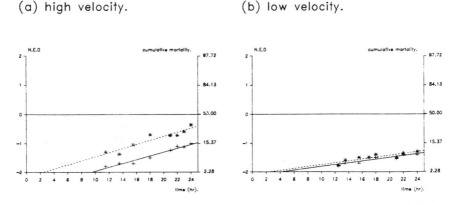

Figure 3. Regression lines of transformed cumulative percent mortality with time recorded in the artificial stream. (Mortalities in containers without substrate = --*--, with substrate --+---). (a) High velocity; (b) Low velocity.

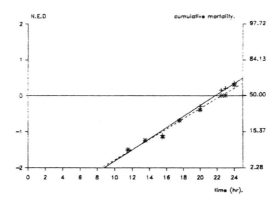

Figure 4. Regression lines of transformed cumulative percent mortality with time recorded in the flow-through dosing system. (Mortalities in containers without substrate = --*--, with substrate --+--).

(a) Field dosing experiment using a natural stream
A plot of the least significant differences (Sokal and Rohlf 1981) and mean mortalities (Fig. 5) shows that mortality was less in the low velocity section of the stream, and the presence of substrate significantly reduced the lethal response of *G. pulex* to ammonia, for both high and low water velocities.

(b) Artificial laboratory stream
Low mortality was observed in both high and low velocity streams (Figs. 3a and b), probably due to the lower temperature regime (7.5°C) which reduced the unionized ammonia concentration. Despite this, ANCOVA results showed mean mortality to be significantly lower in the high velocity stream when substrate was present (p<0.001). The presence of substrate in the low velocity stream had no significant effect (Fig 3b).

(c) Flow-through laboratory toxicity test
ANCOVA results showed no significant differences between mean mortalities in tanks with or without substrate (Fig. 4). Water velocity was undetectable.

(d) Static laboratory toxicity tests
These gave highly unsatisfactory results with large differences between replicates. Difficulties in maintaining a stable pH resulted in unacceptable changes in unionized ammonia concentration in test (i) and the use of a buffer (pH 8.00) in test (ii) necessitated a substantial increase in total ammonia (561 mg l^{-1}) to maintain the nominal unionized concentration of 6 mg l^{-1}. A small change of 0.1 pH unit was sufficient to cause large changes in unionized ammonia.

Static laboratory procedures were found to be inappropriate for assessing the toxicity of ammonia. However, water quality data have been included for comparison with other methodologies used (Table 1).

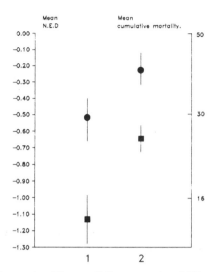

Figure 5. Plot of least significant differences (p<0.05, error bar) and mean mortalities recorded in the field experiment. (X-axis, 1 = low velocity; 2 = high velocity. N.E.D./Animal mortality with sustrate ■ , without substrate ●)

Discussion

The toxicity of ammonia to fish, particularly salmonoids has been well documented (Ball 1967; Thurston and Russo 1981; Arthur *et al*. 1987; Nimmo *et al*. 1989). Fewer studies, however, have examined the effect of this toxicant on invertebrates. Nevertheless, the Median Lethal Concentration (LC50) has been obtained for several aquatic invertebrates (Williams *et al*. 1986; Arthur *et al*. 1987; Hermanutz *et al*. 1987) and a range of sensitivities reported. Freshwater amphipods, including *Gammarus* species, have been widely used in aquatic toxicology for many years (Stephenson 1983; McCahon and Pascoe 1988). Williams *et al*. (1986) found *G. pulex* to be relatively insensitive to ammonia while Arthur *et al*. (1987) showed *Crangonyx* spp. to be moderately sensitive in comparison with the other invertebrates tested. It is likely that ammonia concentrations recorded in the present study would cause effects in other species, particularly fish.

Ammonia toxicity is not solely influenced by pH and temperature. Selesi and Vamas (1976) found that the lethal concentration was proportional to the oxygen content of the water, and Wuhrman and Woker (1948) noted that the toxicity of ammonia to the two fish species they used was correlated to the permeability of the gills.

The current study showed that the mortality of *G. pulex* due to ammonia was significantly (p<0.05) reduced at low water velocity and in the presence of substrate. There are several possible explanations for this observation:-

1)   Toxicity could be reduced because of lowered metabolic activity (and therefore less stress) in maintenance of position by the animal in the current.
2)   Velocity affects the interaction of toxicant with the boundary layer at the gill/body surface.
3)   Suppression of unionized ammonia concentration by direct interaction with the substrate could decrease toxicant availability.
4)   Provision of a semi-natural habitat (i.e. substrate) may have offset stress caused by the unnatural appearance of test chambers.

Animals tested in the absence of flow with and without substrate in the laboratory demonstrated no significant differences in mortality suggesting that factor (4) was unlikely to have played a major role. Factor (3) was discounted because ammonia samples from the artificial stream before and after the substrate section remained consistent.

The artificial stream experiment suggested that there may be a threshold water velocity below which no appreciable decrease in mortality would occur as there was no significant difference observed in tests carried out with and without substrate at low velocity. Further evidence of this was seen in the flow-through toxicity test, although these experiments provided no detail of the velocity required to increase mortality above a threshold level. All velocities used within the present study are comparable with those found in typical streams (Foster pers. comm.) and the results have obvious implications for *in situ* toxicity testing.

Direct comparison of field and laboratory tests cannot be made in this case because of pH and temperature differences. Adams (1983), however, found good agreement between laboratory and field experiments using *Daphnia magna* for acute and chronic tests in ponds. Hermanutz *et al.* (1987) found that, owing to dial and seasonal fluctuations in ammonia concentration, animal mortality in chronic tests was variable but acute test mortality was comparable.

Chapman (1983) specified that the importance of factors influencing toxicity in the laboratory depends upon the aims of the test. These could be i) to determine relative toxicity, ii) to predict actual effects, and iii) to estimate safe levels. The present study is a useful predictor for the effects of water velocities in the field and shows that these effects can be simulated in the laboratory.

It is recommended that for certain toxicity studies using lotic species, experiments should be designed to take account of water velocity because standard laboratory flow-through and static tests may fail to reflect animal behavioural patterns, which may only be expressed in the presence of water current and/or substrate. An appropriate laboratory flow through system is that of Green and Williams (1983) which incorporates test chambers with and without current velocity. Further work is clearly required to explain the physical and physiological mechanisms involved in the relationship between pollutant toxicity and water velocity.

Acknowledgements

We are grateful to WRc and NERC for financial support, to Stephen Maund for statistical guidance, to John Foster for stream velocity data and to Christine Gould for typing the manuscript.

References

Adams, W.J., Kimerle, R.A., Heidolph, B.B. and Michael, P.R. (1983) Field comparison of laboratory derived acute and chronic toxicity data. In W.E. Bishop, R.D. Cardwell and B.B. Heidolph (eds.) *Aquatic Toxicology and Hazard Assessment : Sixth Symposium*, 367-385. ASTM STP 802, American Society for Testing and Materials, Philadelphia.

Alabaster, J.S. and Lloyd, R. (1980) *Water Quality Criteria for Freshwater Fishes.* Butterworths, London, England.

Arthur, J.W., West, C.W., Allen, K.N. and Hedtke, S.F. (1987) Seasonal toxicity of ammonia to five fish and nine invertebrate species. *Bulletin of Environmental Contamination and Toxicology*, **38**, 324-331.

Ball, I.R. (1967) The relative susceptibilities of some species of fresh-water fish to poisons - I Ammonia. *Water Research*, **1**, 767-775.

Chaney, A.L. and Marbach, E.P. (1962) Modified reagents for the determination of urea and ammonia. *Clinical Chemistry*, **8**, 130-132.

Chapman, G.A. (1983) Do organisms in laboratory tests respond like organisms in nature? In W.E. Bishop, R.D. Cardwell and B.B. Heidolph (eds.) *Aquatic Toxicology and Hazard Assessment: Sixth Symposium*, 315-327. ASTM STP 802, American Society for Testing and Materials, Philadelphia.

Fava, J.A., Kapp, R.M. and Rue, W.J. Jr (1982) Evaluation of ammonia toxicity in site-specific environmental assessments. In P.J. Rand (ed.) *Land and Water Issues Related to Energy Development.*

Green, D.W.J. and Williams, K.A. (1983) A continuous flow toxicity testing apparatus for macroinvertebrates. *Laboratory Practice*, **32**, (11), 74-76.

Greenham, L.K. (1984) WRc Continuous flow dosing rig. PRS812-M/1, WRc Environment, PO Box 16, Marlow, Bucks.

Hermanutz, R.O., Hedtke, S.F., Arthur, J.W., Andrew, R.W., Allen, K.N. and Helgen, J.C. (1987) Ammonia effects on microinvertebrates and fish in outdoor experimental streams. *Environmental Pollution*, **47**, 249-283.

McCahon, C.P. and Pascoe, D. (1988) Use of *Gammarus pulex* (L.) in safety evaluation tests: Culture and selection of a sensitive life stage. *Ecotoxicology and Environmental Safety*, **15**, 245-252.

McCahon, C.P., Poulton, M., Thomas, P.C., Xu, Q., Pascoe, D. and Turner, C. (in press) Lethal and sub-lethal toxicity of field simulated farm waste episodes to several freshwater invertebrate species.

National Rivers Authority N.R.A. (1989) *Water pollution from farm wastes, England and Wales.* N.R.A. South West Region.

Nimmo, D.W.R., Link, D., Parrish, L.P., Rodriguez, G.J. and Wuerthele, W. (1989) Comparison of on-site and laboratory toxicity tests: derivation of site-specific criteria for unionized ammonia in a Colorado transitional stream. *Environmental Toxicology and Chemistry*, **8**, 1177-1189.

N.W.C., (1981) *River quality: the 1980 survey and future outlook.* London: National Water Council.

Pascoe, D. (1988) Episodic pollution incidents: Experimental studies in the field and laboratory. In M. Yasuno and Whitton B.A. (eds.) *Biological Monitoring of Environmental Pollution.* Proceedings of the fourth I.U.B.S. International symposium on biological monitoring of the state of the environment. 6-8 Nov, 1987, Tokyo, 145-156. Tokai University Press.

Selesi, D. and Vamas, R. (1976) The factors of lethality from ammonia in fish ponds. *Ichthyologia*, **8**, 115-121.

Sokal, R.R. and Rohlf, F.J. (1981) *Biometry*, 2nd edition. Freeman and Company, New York.

Stephenson, R.R. (1983) Effects of water hardness, water temperature, and size of the test organism on the susceptibility of the freshwater shrimp, *Gammarus pulex* (L.), to toxicants. *Bulletin of Environmental Contamination and Toxicology*, **31**, 459-466.

Thurston, R.V. and Russo, R.C. (1981) Ammonia toxicity to fishes. Effects of pH on the toxicity of the un-ionised ammonia species. *American Chemical Society*, **15**, 837-840.

Williams, K.A.W, Green, D.W.J. and Pascoe, D. (1986) Studies on the acute toxicity of pollutants to Freshwater macroinvertebrates 3. Ammonia. *Archiv für Hydrobiologie*, **106**, 61-70.

Wuhrmann, K. and Woker, H. (1948) Experimentelle Untersuchungen über die Ammoniak- und Blauäurevergiftung. *Schweizerische Zeitschrift fuer Hydrologie*, **11**, 210-244.

Effect of Ecological Factors on the Toxicity of $CuSO_4$ in Fish

J. Nemcsók[2], C. Albers[1], J. Benedeczky[3], K.H. Götz[1], K. Schricker[1], O. Kufcsak[2] and M. Juhasz[2]

[1]Department of Physiology, University of Regensburg, 8400 Regensburg, Germany.
[2]Department of Biochemistry and [3]Department of Zoology, Attila József University Szeged POB 533, H-6701 Szeged, Hungary.

Key words: fish, toxicity, $CuSO_4$, temperature, acid, hypoxia, physiological and biochemical processes.

Abstract

$CuSO_4$ increased the blood GOT, GPT, LDH activities and glucose level in three fish species in the following order - silver carp > carp > wels.

The effect of increased water temperature (from 4°C to 20°C) was greater accumulation of $CuSO_4$ in all studied organs of carp.

$CuSO_4$ had only a slight damaging effect on tissues of rainbow trout as indicated by measurements of biochemical parameters. The effects are significantly increased in the presence of sulphuric acid at pH 6.5.

Hypoxia increased the toxic effect of $CuSO_4$ in carp as it was reflected by the increased plasma GOT, GPT and LDH activities.

Introduction

Waste materials originating from intensive agricultural production exert harmful effects when they reach natural waters, since they may become concentrated in the organs of aquatic animals (Salánki et al. 1982; Nemcsók et al. 1987). The degree of accumulation is greater the higher the stage of the given organism in the food chain. Thus, fish are particularly sensitive to environmental contamination of the water and pollutants may significantly damage certain physiological and biochemical processes when they enter the organs of fish.

Pollutants are not only harmful to adult fish, but may also cause disturbances of development in embryonic stages. Several cases have been reported in which the toxic effect of pollutants may be decreased, or even increased by various water quality factors. These are the pH, temperature, hardness and dissolved oxygen content of the water (Zitko and Carson 1976; Pascoe et al. 1986).

Bioindicators and Environmental Management
ISBN 0-12-382590-3

The harmful effects, especially sub-lethal effects, retard the development of the survivors through influence on their normal metabolic processes.

In the light of the foregoing, studies were performed on the toxic effect of $CuSO_4$, used as an agricultural fungicide, whilst varying some ecological factors (water temperature, pH and dissolved O_2 content of the water). Our aim was to provide data on the effects of environmental factors on $CuSO_4$ toxicity with the purpose of predicting damage in the field.

Materials and methods

1. Sensitivity of fish species to $CuSO_4$

Common carp (*Cyprinus carpio* L.), silver carp (*Hypothalamichthys molitrix* V.) and wels (*Silurus glanis* L.) specimens of 350-400 g were obtained from the Fisheries Research Institute in Szarvas, Hungary, and held for a minimum of 7 days before experimentation in 100 litre aquaria (5 fish per aquarium) at a temperature of 20 ± 1°C. The length of exposure (2 h) to $CuSO_4$ pollution was chosen according to our previous investigations because most of the fish survived the treatment within this time. Two hours after the $CuSO_4$ administration, blood samples were taken from the hearts of both, the treated and control animal for GOT, GPT and LDH determination using Reanal (Hungary) kits.

2. $^{64}CuSO_4$ accumulation in carp depending on water temperature

Carp (*Cyprinus carpio* L.) weighing 350-400 g were obtained for experiments. After an appropriate adaptation period (3-7 days) each fish was placed into a 5 l aquarium with continuous aeration.

$^{64}CuSO_4$ specific activity: 2.7 mCi ml^{-1}. Copper content: 1.95 mg ml^{-1}. After $^{64}CuSO_4$ treatment, following an exposure time of 2 hours the incorporation of $CuSO_4$ was measured at 4°C and 20°C. Treatment of longer duration was not performed owing to the rapid decay of the $CuSO_4$ isotope used (half life: 12 hours).

The appropriate isotope concentration of the studied compounds was dosed into the water of the aquarium. After a given time the experimental fish were killed and the amount of $CuSO_4$ accumulated in the removed organs was determined on the basis of radioactivity measurements. About 1 g of the organs removed from the dead fish was dissolved in protosol solution /200 mg of organ/ 1 ml of protosol/, then made up to 5 ml with toluol-cocktail solution and the radioactivity measured by means of Packard liquid-scintillation apparatus. Each experiment was repeated 3-5 times.

3. Sensitivity of fish to low pH alone and combined with $CuSO_4$

Experiments were carried out on 17 rainbow trout (body weight 262 ± 8.1 (SEM) g) obtained from the Midland Trout Farm, Naulsworth, Gloucestershire, and kept in the closed circulation at the research unit for at least one week. Most procedures were identical to those described previously (Nemcsók and Hughes 1988). Fish were anaesthetized (MS222, 10 mg litre^{-1} before cannulation of the dorsal aorta using the method described by Hughes *et al.* (1983). Each fish was then kept in a bin containing 20-30 litres of aerated circulation water (P$_{O2}$ 150-155 mm Hg, 16°C) which was changed at least once each day. The pH and calcium hardness of the water was determined (214 mg litre $^{-1}$) and reductions in pH obtained by the addition of sulphuric acid. The mean pH (SD) of the different waters was as follows: before treatment, 8.324 (0.098); 24 h exposure to acid, 6.618 (0.094); 24 h recovery from these treatments, 8.229 (0.164). The lower pH values were adjusted every 4 h by the addition of H$_2$SO$_4$ during the 24 h exposure. Copper was added as copper sulphate to produce a final concentration of 0.2 ppm (μg litre^{-1}) and was checked by atomic absorption spectrophotometry at the end of the exposure period. Carbon dioxide was not measured, but the water was saturated with air and hence the partial pressure of oxygen was in the range 150-155 mm Hg.

At least three blood samples were taken in all cases:

(1) After 24 h recovery from cannulation

(2) Following 24 h exposure to the low pH or low pH and copper

(3) The bin was then thoroughly cleaned and washed and the fish allowed to recover in non-polluted water for 24 h before the 3rd blood sampling. All fish survived this treatment and in some cases samples were taken after longer recovery times.

Blood samples of 0.5 ml were centrifuged (10 min at 10 g) and the plasma stored at 4°C until the completion of the experiments, when the following biochemical parameters were determined using methods given previously (Nemcsók and Hughes 1988); GOT, GPT, LDH enzyme activities were made by Reanal kits.

The main experimental series was based on results using six fish exposed to acid alone, and six exposed to the combined treatment. Each of these twelve fish, and others, provided data for the control values.

4. Sensitivity of fish to lower O$_2$ concentration combined with CuSO$_4$

Male and female carp with weights of 1200-1500 g were used. After anaesthesia with MS222, a canula was introduced into the dorsal aorta. Immediately after this operation, the fish were put into a closed experimental box where the pH, temperature and partial oxygen pressure were regulated. After adaptation for 3 days, the fish were divided into two groups. In the first group, a 6 h period of hypoxia was applied (80 pHg oxygen) daily for 5 days. In the second group, hypoxia was applied as above and at the first day 20 ppb copper sulphate was injected via the canula. For comparative purposes, a control group not subjected to either treatment mode was also used. Blood samples were taken through a canula introduced into the dorsal aorta, and GOT, GPT, LDH activities were determined.

Results and discussion

1. Sensitivity of fish species to CuSO$_4$

The biochemical parameters in carp changed significantly after the CuSO^4 treatment (Figs. 1-4), compared to controls. The GOT activity increased by about 30% and the GPT activity was three times as high as in the control specimens. Blood glucose levels were about duplicated in the carp treated with CuSO$_4$.

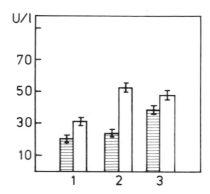

Figure 1. The effect of 10 ppm CuSO$_4$ on blood serum GOT activity of carp (1), silver carp (2), and wels (3). Water temperature $20 \pm 1°C$. The values represent averages for 8-12 fish specimens ($\pm$ S.D.). Hatched bars - control; clear bars - treated.

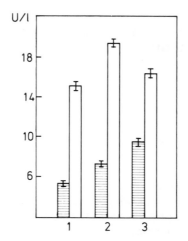

Figure 2. The effect of 10 ppm CuSO$_4$ on blood serum GPT activity of carp (1), silver carp (2), and wels (3). Water temperature $20 \pm 1°C$. Average for 8-12 fish specimens ($\pm$ S.D.). Hatched bars - control; clear bars - treated.

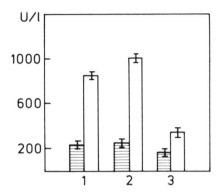

Figure 3. The effect of 10 ppm CuSO$_4$ on blood serum LDH activity of carp (1), silver carp (2), and wels (3). Water temperature $20 \pm 1°C$. Average for 8-12 fish specimens ($\pm$ S.D.). Hatched bars - control; clear bars - treated.

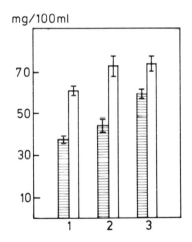

Figure 4. The effect of 10 ppm CuSO$_4$ on blood sugar level of carp (1), silver carp (2), and wels (3). Water temperature $20 \pm 1°C$. Average for 8-12 fish specimens ($\pm$ S.D.). Hatched bars - control; clear bars - treated.

The most significant changes were observed in silver carp. GOT activity doubled, GPT and the LDH activities were 2.5 and 4 times higher, and the blood glucose level was doubled.

CuSO$_4$ - similarly as in the carp and silver carp - enhanced serum GOT, GPT and LDH activities and blood glucose level were found in the wels. However,

$CuSO_4$ was less toxic to wels than to the other two species.

These results clearly show that the toxicity of $CuSO_4$ affected the three fish species differently. Bell (1968) studied the effect of hepatic poisons on GOT activity, using bromobenzene and carbon tetrachloride in high doses. He found that in the treated fish the GOT was significantly elevated. Even in fish with diseased kidneys he demonstrated a significant increase in plasma GOT activity over controls. Kristoffersson *et al.* (1974) reported that 5 ppm phenol increased the GOT and GPT activity in pike (*Esox lucius* L.). In our experiment, $CuSO_4$ enhanced GOT and GPT activities in all three investigated species, but the degree of changes was different. The highest GOT and GPT activities were measured in silver carp, reflecting a serious damage of tissues. The increase in GOT and GPT activities in carp was less significant. The slightly increased GOT and GPT activities of wels over the control refers to the fact that the tissue damage was not so significant. This slight difference between the treated and control animals might be due to the highest GOT and GPT activities in control wels. However, Onishi and Murayama (1970) demonstrated that hepatic GOT activity does not differ much between species. The elevated GOT and GPT activities were presumably due to damage of liver, but other organs may also have been damaged (kidney or/and gill). Reichenbach-Klinke (1972) reported gill damage after $CuSO_4$ treatment. Schreck and Lorz (1978) observed that kidneys of Cu-exposed fish had glomerular atrophy and epithelial necrosis of gills.

Blood glucose appeared to be a sensitive, reliable indicator of environmental stress in fish. On the basis of our results it is clear that $CuSO_4$, as is shown by the elevated blood glucose level, acted as a stressor on the fish. The order of magnitude of the stress effect is silver carp > carp > wels. The increased LDH activity showed the same order, presenting the metabolic changes in stressed fishes: the catabolism of glucose moved towards lactic acid, which is very dangerous and toxic to fish. The enhanced LDH activity in carp and silver carp might be due to the enhanced swimming activity in contrast to the $CuSO_4$-intoxicated wels which were resting on the bottom of the aquaria. So the enhanced LDH activity of carp and silver carp could be a consequence of increased swimming activity.

Our results demonstrated that wels tolerated the $CuSO_4$ pollution well while carp and silver carp were very sensitive to the mentioned environmental stressors.

The differences in the damaging effect of $CuSO_4$ between the three fish species might be due to the different rate of their operculum movement and also to the altered microsomal enzyme activities that might metabolize toxic metals.

2. $^{64}CuSO_4$ accumulation in carp depending on water temperature

With the exception of the brain, greater copper accumulation was measurable in every organ at 20°C (Fig. 5) compared to the values determined at 4°C (Fig. 6, Table 1). The order of the concentrations of copper taken up by the various organs was: skeletal muscle > liver > gills > intestine > kidney > heart > brain (Table 1).

Table 1. Effect of temperature (4-20°C) on *in vivo* accumulation of 10 ppm CuSO$_4$ into the various organs of Carp (*Cyprinus carpio*, L.). The values are the ratio of concentration at 20°C to that at 4°C expressed as a percentage.

Treatment time: 2 hours	concentration - 20°C ÷ concentration - 4°C x100
brain	75
gills	448
intestine	372
heart	250
kidney	360
liver	500
muscle	1050

The low CuSO$_4$ accumulation detected in the brain may be related to the presence of the blood-brain barrier. The high concentration of Cu in the gills and liver is related to the important role that the gills play in copper uptake, and the liver in its accumulation and detoxication (Reichenbach-Klinke 1972). The high copper-content of the skeletal muscle may be correlated with the greater activity of the fish at 20°C. During the course of this movement more copper, which has not yet been transported towards other copper storing organs during the treatment period, may enter the organ, because of the enhanced blood-supply and accelerated metabolic process of the skeletal muscle. Treatments of longer duration also confirmed this concept, since the considerable acetylcholinesterase

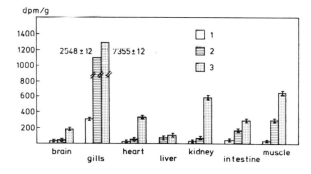

Figure 5. *In vivo* accumulation of 1.0 (1); 10.0 (2) and 100 (3) ppm CuSO$_4$ into various organs of Carp at 20°C following 2 hours treatment. Mean values of samples from 3-5 individuals are given, as dpm g^{-1} ($\pm$ S.E.M.). Specific activity: 2.7 mCi ml^{-1}. Copper content 1.95 mg ml^{-1}.

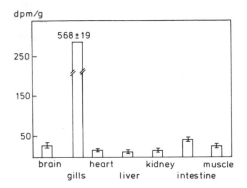

Figure 6. *In vivo* accumulation of 10 ppm CuSO$_4$ into various organs of Carp at 4°C following 2 hours treatment. Values are the average of the samples measured from 3-5 individuals, expressed in dpm g^{-1} ($\pm$ S.E.M.). Specific activity: 2.7 mCi ml^{-1}.

inhibition in carp, manifested at 5 ppm CuSO$_4$ treatment for two weeks, could only be measured in the first 24 hours. Following this, the acetylcholinesterase activity returned to normal level as before treatment (Fig. 7). The inhibition of acetylcholinesterase by CuSO$_4$ contamination has also been reported by Olson and Christensen (1980). According to other workers copper may also display its damaging effect elsewhere. On the basis of experiments performed on carp, damage to the gills, liver, kidney and nervous system has been recorded at a concentration of 1.5 ppm. At the same concentration, CuSO$_4$ caused adverse alterations in certain blood parameters of the carp: significant increase in

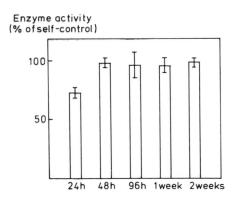

Figure 7. *In vivo* effect of 5 ppm CuCO$_4$ on AChE activity in serum of carp depending on exposure time. Values expressed in the percentage of the control are the averages of the samples measured from 3-5 individuals.

haematocrit value, haemoglobin content and protein, as well as in the concentrations of glucose. At the same time, a decrease was observed in the number of leucocytes (Svobodova 1982).

According to Reichenbach-Klinke (1972) the major target of copper in fish is the gill epithelium, while Schreck and Lorz (1978) consider that both the gills and kidney may become damaged by $CuSO_4$.

The electronmicroscopic studies of Rojik *et al.* (1983) have verified the foregoing conclusions. During the course of their studies, $CuSO_4$ damaged the gills, liver and kidney of carp, silver carp and silure, with damage mainly to the endoplasmic reticulum and mitochondria.

3. Sensitivity of fish to low pH alone and combined with $CuSO_4$

Table 2 contains a summary of mean values of biochemical parameters measured in blood plasma before and during exposure to acid alone and combined with 0.2 ppm copper sulphate. Mean values of measurements following recovery in normal water are also given. Means of percentages of the biochemical characteristics relative to self-controls (=100%) are plotted in Figure 8, together with corresponding data for the same concentration (0.2 ppm) of copper sulphate alone obtained in a previous study (Nemcsók & Hughes, 1988). Except for AChE, these show an increase following exposure, and the increase in percentage is greater for the combined copper and acid treatment. This is most marked in the case of ALAT. Analysis of variance showed significant differences between the columns in Table 2 and paired t tests between columns showed significance at the 5% level or better.

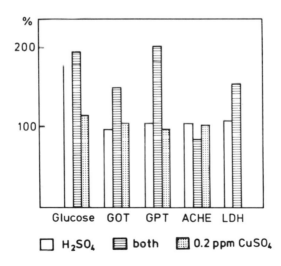

Figure 8. Mean levels of five biochemical measurements expressed as percentages of the control (=100%). Results of 24-h exposure to H_2SO_4 alone, and combined with 0.2 ppm $CuSO_4$, are given together with those for 0.2 ppm copper sulphate alone obtained in a previous study.

Table 2. Summary of results showing changes in biochemical characteristics of trout blood following 24 h exposure to H_2SO_4 pH 6.6 alone and together with 0.2 ppm $CuSO_4.5H_2O$.

24 h exposure to	$CuSO_4.5H_2O$ (0.2 ppm)(pH 8.4)	H_2SO_4 (pH 6.6)	H_2SO_4+ $CuSO_4.H_2O$ (0.2ppm)
Glucose	no change	+75%	+90%
GOT	no change	no change	+40%
GPT	no change	+5%	+100%
AChE	no change	no change	-10%
LDH	?	slight increase	+50%

One of the main findings from this series of experiments is that exposure to a combination of sulphuric acid and copper sulphate has a much greater effect than either treatment alone. This effect can be observed for most of the parameters determined, but is particularly clear for the biochemical measurements and especially GPT.

Acidification alone is these experiments did not cause any changes in the biochemical parameters during the 24 h treatment except for blood glucose level. The elevated blood glucose reflects the stress situation. If this adverse effect becomes long lasting, it can influence the condition of fish population very unfavourably, since the long-lasting stress effect may decrease the gammaglobulin level, which decreases the resistance of fish against infectious diseases (Wedemeyer 1970; Nemcsók et al. 1982).

In the present experiments, acidification significantly potentiated the toxicity of copper sulphate to fish, causing serious disturbance to physiological and biochemical processes, among others the inhibition of AChE activity. The latter is very harmful because it inhibits the normal nerve function and the various vital behaviour activities, which are essential in obtaining food and in defensive escape reactions (Baslow and Nigrelli 1961). The inhibition of AChE is especially dangerous to the heart, since the cholinergic system has a decisive role in the innervation of the heart in fish. Inhibition of acetylcholinesterase may lead to increased vagal tone which may cause severe disturbances in the metabolic processes dependent on the circulation. This is inevitable, since inhibition of the heart function has harmful influences on O_2 uptake and CO_2 releases; thus it may produce anoxia at the tissue level.

Other authors (Howarth and Sprague 1978) have observed that Cu is more toxic at pH 5.4 than at pH 7.3. However, Miller and Mackey (1980) found that Cu was less toxic at pH 4.7 to pH 4.34 than at pH 7.3. These authors suggest that although acid and metals both stimulate mucus production, the acid stimulus predominates, thereby chelating Cu and reducing its toxicity. It seems probable that, in the presence of acid, the particular ionic status of copper will be changed and, presumably, in the present experiments this renders it more toxic.

4. Sensitivity of fish to lower O$_2$ concentration combined with CuSO$_4$
Hypoxia increased the toxic effect of CuSO$_4$ in carp as it was reflected by increased plasma GOT, GPT and LDH activities. Hyperoxia, together with CuSO$_4$ treatment, did not cause any significant changes compared with the fish treated with CuSO$_4$ alone (Figs. 9, 10, 11). These data showed that lower O$_2$ concentration in water could potentiate the tissue damaging effect of CuSO$_4$ even at relatively low CuSO$_4$ concentrations.

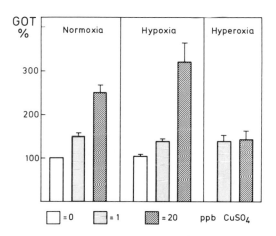

Figure 9. The effect of various CuSO$_4$ concentrations on the plasma GOT activity of Carp in the course of exposure time depending on the O$_2$ concentration in water.

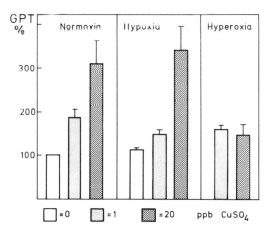

Figure 10. The effect of various CuSO$_4$ concentrations on the plasma GPT activity of Carp in the course of exposure time depending on the O$_2$ concentration in water.

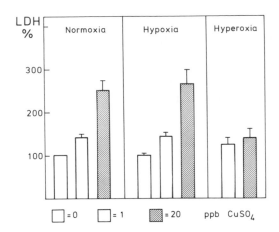

Figure 11. The effect of various $CuSO_4$ concentrations on the plasma LDH activity of Carp in the course of exposure time depending on the O_2 concentration in water.

Our results showed that ecological factors (water temperature, pH and dissolved O_2 content of the water) enhanced the $CuSO_4$ toxicity. These biochemical measurements could increase the possibility of predicting the damaging effects caused by ecological factors.

References

Baslow, M.H. and Nigrelli, R.F. (1961) Muscle acetylcholinesterase levels as an index of general activity in fishes. *Copeia*, **1**, 8-11.

Bell, G.R. (1968) Distribution of transaminases (aminotransferases) in the tissue of Pacific salmon (Oncorhynchus), with emphasis on the properties and diagnostic use of glutamic-oxalacetic transaminase. *Journal of the Fisheries Research Board of Canada*, **25**, 1247-1268.

Howarth, R.S. and Sprague, J.B. (1978) Copper lethality to rainbow trout in waters of various hardness and pH. *Water Research*, **12**, 455-462.

Hughes, G.M., Albers, C., Muster, D. and Götz, K.H. (1983) Respiration of the carp, *Cyprinus carpio* L. at 10 and 20°C and the effects of hypoxia. *Fish Biology*, **22**, 613-628.

Kristoffersson, R., Broberg, S., Oskari, A. and Pekkarinen, M. (1974) Effect of a sub-lethal concentration of phenol on some blood plasma enzyme activities in the pike (*Esox lucius* L.) in brackish water. *Annales Zoologic Fennici*, **11**, 220-223.

Miller, T.G. and Mackay, W.C. (1980) The effects of hardness, alkalinity and pH of test water to the toxicity of copper to rainbow trout (*Salmo gairdneri*). *Water Research*, **14**, 129-133.

Nemcsók, J., Olah, J. and Boross, L. (1982) Studies on stress effect caused by malachite green and formaline treatments of common carp (*Cyprinus carpio* L.). *Aquacultura Hungarica (Szarvas)*, **III**, 57-61.

Nemcsók, J., Orbán, L., Asztalos, B. and Víg, É. (1987) Accumulation of pesticides in the organs of carp (*Cyprinus carpio* L.) at 4°C and 20°C. *Bulletin of Environmental Contamination and Toxicology*, **39**, 370-380.

Nemcsók, J. and Hughes, G.M. (1988) The effect of copper sulphate on some biochemical parameters of rainbow trout. *Environmental Pollution*, **49**, 77-85.

Olson, D.L. and Christensen, G.M. (1980) Effect of water pollutants and other chemicals on fish acetylcholinesterase *in vitro*. *Environmental Research*, **21**, 327-335.

Onishi, T., Murayama, S. (1970) Studies on enzymes of cultivated Salmonoid fishes. II. Activities of protease, amylase, arginase GPT and GOT in various growth stage. *Bulletin of the Tokai Regional Fisheries Research Laboratory*, **63**, 123-132.

Pascoe, D., Evans, S. and Woodworth, J. (1986) Heavy metal toxicity to fish and the influence of water hardness. *Archives of Environmental Contamination and Toxicology*, **15**, 481.

Reichenbach-Klinke, H.H. (1972) Histologische und enzymatische Veranderungen nach Schadstoffeinwirkung beim Fisch. *Veröffent. Int. Kus. Binnenfischerei. (Hamburg)*, **53**, 113-120.

Rojik, I., Nemcsók, J. and Boross, L. (1983) Morphological and biochemical studies on liver, kidney and gill of fishes affected by pesticides. *Acta Biologica Academiae Scientiarum Hungaricae*, **34**, 81-92.

Salánki, I., V.-Balogh, K. and Berta, E. (1982) Heavy metals in animals of Lake Balaton. *Water Research*, **16**, 1147-1152.

Schreck, C.B. and Lorz, H.W. (1978) Stress response of coho salmo (*Onchorhynchus kisutsch*) elicited by cadmium and copper of stress. *Journal of Fisheries Research Board of Canada*, **35**, 1124-1129.

Svobodova, Z. (1982) Changes in some haematological parameters of the carp after intoxication with CuSO₄ x 5H₂0. *Bulletin VURH Vodnany*, **2**, 26-28.

Wedemeyer, G. (1970) The role of stress in the disease resistance of fishes. In S.F. Sniesko (ed.) *A Symposium on Diseases of Fishes and Shellfishes. American Fisheries Society Special Publications*, **5**, 30-35.

Zitko, V. and Carson, W.G. (1976) A mechanism of the effects of water hardness on the lethality of heavy metals to fish. *Chemosphere*, **5**, 299-303.

Effect of Hypoxia and Copper Sulphate on the Structure of Liver and Kidney of Carp

[1]I. Benedeczky, [2]J. Nemcsók, [3]C. Albers, [3]K.M. Götz

[1]Department of Zoology and [2]Biochemistry, József Attila University, Szeged, P.O.Box 659, Hungary.
[3]Department of Physiology, Regensburg University, Regensburg, Germany.

Key words: hypoxia, copper sulphate, carp, liver, kidney, cytopathology, ultrastructure.

Abstract

The effect of hypoxia alone and in combination with copper sulphate was investigated on carp tissues by electron microscopy. Several cytopathological alterations manifested in the liver cells after hypoxia: mitochondria and rough surfaced endoplasmic reticulum became swollen, large lipid droplets accumulated in the rEr cisternae, glycogen was absent. Ultrastructural alterations were mild in the tubular epithelial cells of kidney. The combined treatment caused the development of large vacuoles in the cytoplasm. Moreover, unusual horseshoe shaped nuclei appeared in the liver cells. Cytopathological alterations were more marked in the neighbouring pancreatic cells. The regular arrangement of the simple cuboidal epithelial cells was disintegrated in the kidney and large extracellular spaces developed among the epithelial cells. Noteworthy numbers of lymphocytes and granulocytes were observed in the liver and renal tubules after combined treatment.

Introduction

Essential and toxic heavy metals are taken up in excessive amount by animal tissues if the environment is polluted by these metals.

Heavy metal storage in different animal tissues is a well known fact, but the pathological effects of such metal ions on animals and humans are not always fully understood (Rojik *et al.* 1983; Benedeczky *et al.* 1986; Tort *et al.* 1987). Certain specific genetic defects may cause an excessive accumulation of copper in both human (Sternlieb 1980; 1982) and fish (Bunton *et al.* 1987) liver cells. Excessive copper is definitely harmful for both human and fish tissues and cells. Even in a low, sub-lethal concentration, copper sulphate is noxious to the liver

Bioindicators and Environmental Management
ISBN 0-12-382590-3

and kidney in carp (Benedeczky *et al.* 1986).

The toxic effects of heavy metals and other chemical pollutants may increase strongly if the environmental circumstances (temperature, pH, oxygen content, etc.) favour this. Phenol in water, for example, is much more toxic if the dissolved oxygen content of the water is low (Gary *et al.* 1981). Certain heavy metal ions (e.g. zinc) can themselves cause hypoxia (Tort *et al.* 1984).

This paper describes the cytopathological alterations caused by simultaneously applied hypoxia and copper sulphate treatment in the liver and kidney of the carp, *Cyprinus carpio* L.

Materials and methods

Male and female carp with weights of 1200-1500 g were used. After anaesthesia with MS222, a cannula was introduced into the dorsal aorta.

Three days after this operation, the fish were put into a closed experimental box where the pH, temperature and partial oxygen pressure were regulated. After adaptation for 3 days, the fish were divided into two groups. In the first group, a 6 h period of hypoxia was applied (80 pHg oxygen) daily for 5 days. In the second group, hypoxia was applied as above and at the first day 20 ppb copper sulphate was injected via the cannula. For comparative purposes, a control group not subjected to either treatment mode was also used.

On the fifth day after treatment, the animals were killed with a blow to the head, and small pieces (about 1 mm^3) were excised from the liver and kidney. Specimens were fixed in a cold fixative containing 4% paraformaldehyde and 2.5% glutaraldehyde for 24 h. Osmium tetroxyde fixation was performed for a further 2 h. After the dehydration procedures, samples were embedded in Durcupan ACM resin. The ultra-thin sections were contrasted with lead and examined by electron microscopy.

Results

Effects of hypoxia

After hypoxia, a conspicuous density difference was observed in the cytoplasm of the liver cells. Around the nuclei, a very dense, often irregular cytoplasmic area was found. Compact, rough surfaced endoplasmic reticulum tubules, dilated cisternae and swollen mitochondria filled this area (Fig. 1a). Large, moderately electron dense lipid droplets were often accumulated in great numbers at the peripheral part of these dense cytoplasmic areas. Lipid droplets were sometimes embedded into the lumen of the strongly dilated rEr cisternae (Figs. 1a and 1b). The "outer" surface of the lipid inclusions often occupied large fields in the electron lucent cytoplasmic areas, but a dense rim (mostly membrane bound ribosomes) always surrounded their surface.

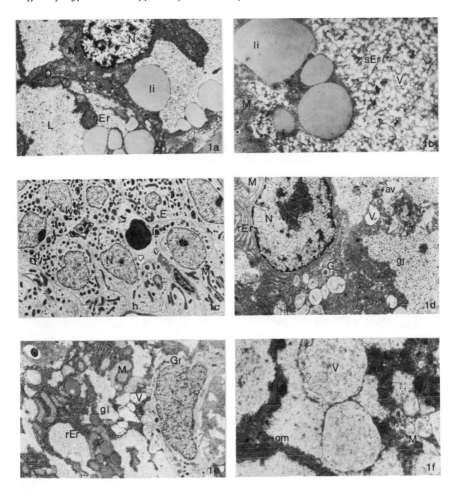

Figure 1. a. Liver cells of hypoxic carp. N=nucleus, D=dark cytoplasmic area, L=light cytoplasmic area, li=lipid, M=mitochondria, rEr=rough surfaced endoplasmic reticulum. x10.000. **b.** Detail of a liver cell of hypoxic carp. sEr=smooth surfaced endoplasmic reticulum, V=endoplasmic vacuolar system, li=lipid, M=mitochondria. x24.000. **c.** Epithelial cells of renal tubules of hypoxic carp. E=epithelial cells, ly=lysosomes, b=basal lamina, N=nucleus, M=mitochondria, d=cell detritus. x17.600. **d.** Liver cell, after hypoxia and copper sulphate treatment. N=nucleus, rEr=rough surfaced endoplasmic reticulum, V=vacuoles, M=mitochondria, G=Golgi apparatus, gl=glycogen, av=autophagic vacuole. x13.200. **e.** The same as Fig. 1d. . Gr=granulocyte, rEr=rough surfaced endoplasmic reticulum, M=mitochondria, gl=glycogen, V=vacuoles. x10.000. **f.** Details of liver cells after hypoxia and copper sulphate treatment. cm=cell mebrane, V=autophagic vacuoles, M=mitochondria. x16.600.

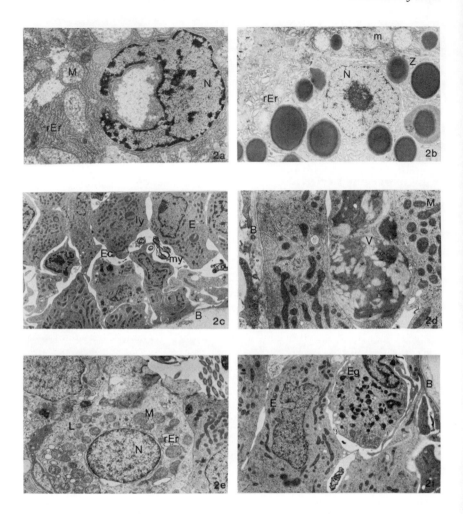

Figure 2 a. Horseshoe like nucleus (N) in the liver of carp after hypoxia and copper sulphate treatment. M=mitochondria, rEr=rough surfaced endoplasmic reticulum. x14.000. **b.** Pancreatic acinus cell after the same treatment. N=nucleus, Z=zymogen granules, rEr=rough surfaced endoplasmic reticulum., M=mitochondria. x10.000. **c.** Tubule cells of the carp kidney after hypoxia and copper sulphate treatment. E=epithelial cells, B=basal lamina, Ec=extracellular space, my=myelin figures, ly=lysosomes. x6.600. **d.** The same as Fig. 2c. Note large autophagic vacuoles (V) in the cytoplasm of tubule cells. B=basal lamina, M=mitochondria. x13.000. **e.** Light cells (L) among dark epithelial cells of renal tubules of carp. N=nucleus, M=mitochondria, rEr=rough surfaced endoplasmic reticulum. x10.000. **f.** Eosinophil granulocyte (Eg) among the tubule cells (E) of kidney of carp. The same treatment as above. B=basal lamina. x10.000.

A large volume of the liver cells involved electron lucent glycogen free cytoplasmic areas (Figs. 1a and 1b). A fine reticulum was seen evenly distributed in this area at low power magnification. High power magnification revealed a continuous, very complicated electron transparent vacuolar system (Fig. 1b). This corresponds to the glycogen storing compartment of the liver cells. Around this irregular vacuolar system, a smooth surfaced, moderately electron dense endoplasmic reticulum membrane compartment was identified. The latter also penetrated into the dark cytoplasmic area.

Relatively small amounts of dark heterochromatin and diffuse euchromatin were found in the karyplasm, and therefore the profile of the nucleus seemed to be spotted (Fig. 1a). Hypoxia did not seriously alter the fine structure of the tubular epithelial cells of the kidney: the structural preservation was good. A large number of lysosomes were bound to the apical pole of the cells. Occasionally, cell detritus and lymphoid cells were seen among the epithelial cells (Fig. 1c).

Effects of joint hypoxia and copper sulphate treatment

The fine structural alterations caused by hypoxia alone were also seen in the liver cells after hypoxia and copper sulphate treatment. Dilation of the rough surfaced endoplasmic reticulum and mitochondria was widespread (Figs. 1d and 1e). Some rEr cisternae were filled with a moderately dense homogeneous material (Fig. 1e). Not only the mitochondria were swollen, but the number of their cristae and the density of the mitochondrial matrix were strongly decreased (Figs. 1e and 2a). In other cells, shrinkage of the mitochondria was characteristic (Fig. 1f). The combined treatment caused the development of large vacuoles in the liver cells (Figs. 1d and 1e). These vacuoles varied in number , diameter and inner content in the different cells. Some of the vacuoles seemed to be empty (Fig. 1e), while others were filled with a fine granulated material (Fig. 1d). In some cases, both granular and filamentous material was found in the vacuoles (Figs. 1d and 1f). The Golgi apparatus appeared as a small, collapsed area (Fig. 1d). Most of the nuclei in its structure were unaltered, but very unusual, horseshoe shaped forms also occurred among them (Fig. 2a). The appearance of different granulocytes among the liver cells was common after the double treatment (Fig. 1e). Rarely, swollen, necrotic liver cells were also seen in the tissue.

Exocrine pancreatic cells were frequently found in the liver. The cytopathological alterations were more marked in these cells than in the liver cells. The rough surfaced endoplasmic reticulum was completely disintegrated. Only dilated cisternae and fragmented saccules were present in the glandular cells (Fig. 2b). The inner structure in the zymogen granules was inhomogeneous: the centre was dense, and the marginal part light. A large difference was observed in the diameter of the secretory granules. In contrast with the normal controls, the zymogen granules were evenly distributed in the cells, i.e. they were not accumulated at the apical pole of the glandular cells. All the mitochondria in the cytoplasm were destroyed.

Whereas the hypoxic kidney cells were relatively unaltered, the cytopathological alterations in the double treated kidney cells were frequent and extensive. The regular arrangement of the simple cuboidal epithelial cells was disintegrated and large, extracellular spaces developed in the convoluted tubules among the irregular cells. The regular foldings of the basal cell membranes in the tubular cells disappeared. Dense myelin figures were frequently seen in the large extracellular spaces (Fig. 2c). Structurally, the cell organelles of the tubular epithelial cells were usually well preserved, but large, dense spherical inclusions often occurred in the cytoplasm of the cells (Fig. 2d). Myelin figures were also found in the lumen of the tubules (Fig. 2e). The tubular epithelial cells sometimes included a few "light" cells (Fig. 2e). These cells seemed to be swollen, as were most of the cell organelles: the nuclear envelope, the rough surfaced endoplasmic reticulum and the mitochondria. Noteworthy numbers of lymphocytes, reticular cells and eosinophilic granulocytes were observed in the large extracellular spaces between the epithelial cells in the renal tubules (Fig. 2f).

Discussion

The biosphere contains increasing amounts of heavy metals from natural, industrial, agricultural and communal sources. Both plants and animals are able to take up these heavy metals and store them in their tissues (Johnels *et al.* 1967; Förstner and Prosi 1979; Salánki *et al.* 1982; Lodenius *et al.* 1983; V.-Balogh and Salánki 1987). Some of the stored heavy metals (cadmium, mercury, copper) may cause serious pathological alterations in both cells and tissues (Benoit *et al.* 1976; Beumer and Bacher 1982; Sternlieb, 1980; 1982). Copper sulphate is still used extensively as a fungicide in agriculture. Rains wash the copper sulphate into the waters, where it can act as a toxic agent on aquatic organisms, especially fish. Since hypoxia frequently arises in natural waters, we have studied the simultaneous effects of hypoxia and copper sulphate in different tissues of carp. The effects of hypoxia alone are well known in mammals (Popper 1982), but less so in fish. Mild hypoxia for 6 h daily during five days caused marked ultrastructural changes in the liver cells. Both the mitochondria and the rough surfaced endoplasmic reticulum were dilated. In the light cytoplasm, there was a conspicuous lack of glycogen particles and a high number of large lipid droplets. The mitochondrial and rEr alterations correlated with the decrease in the quantity of ATP. With the lack of ATP, the ion pump in the cell organelles is inhibited, and swelling develops not only in the cell organelles, but also in the whole cell (hydropic degeneration; Popper 1982). The large glycogen free light areas in the parenchyma cells are indicative of the utilization of the carbohydrate reserves during hypoxia. After this, lactate accumulates in the cells as a consequence of glycolysis. The elevated LDH activity (360%) also proves the increased lactate level (Asztalos *et al.* 1990). Morphologically it was surprising that lipid droplets always appeared in the lumen of the rEr cisternae or close to the rEr membrane. Vesicularization of the rEr and the accumulation of lipid droplets in the lumen of the cisternae were observed after orotic acid treatment

(Novikoff and Edelstein 1977), but these alterations can be induced also by alcohol, carbon tetrachloride and other chemical treatment (Novikoff 1982; Braunbeck *et al.* 1990). It is very probable that hypoxia (similarly to these drugs) through inhibition of the transport processes, induces a characteristic ultrastructural feature: "fat pharenozis". Such a lipid accumulation did not occur in the cells of the kidney during hypoxia. The reason may be that the protein synthetizing system is not so rich in the tubular epithelial cells as in the liver cells. In the kidney, the predominant myelin figures indicated the destruction of phospholipid metabolism.

Copper sulphate treatment in combination with hypoxia increased the degenerative alterations in the liver cells. Large vacuoles and autophagosomes appeared in the cytoplasm. This suggests that autophagocytosis (endocytosis) is induced in the liver by the combined treatment, whereas hypoxia alone does not lead to this phenomenon. However, in the absence of lysosomes, autophagy was incomplete and true autophagic vacuoles did not form. Not only lysosomes, but also peroxisomes were absent. The lack of these characteristic cell organelles from the liver cells may be connected with the damage to the rough surfaced endoplasmic reticulum, which was as serious as in the case of hypoxia. Since the swollen rEr and mitochondria are morphological signs of an inhibited protein synthesis, it seems clear that the formation of lysosomes and peroxisomes was also retarded. In contrast with Wilson's disease, we were not able to detect copper containing bodies in the liver cells in our material. With regard to the lack of lysosomes, this is not surprising. On the other hand, there is an active copper uptake system in the liver cells (McArdle *et al.* 1988). This uptake system might well be destroyed during hypoxia and copper treatment. Not only were the lysosomes absent, but the copper concentration was rather low and the exposure time was rather short. All these facts can explain the lack of copper containing dense bodies in the liver cells. Diffusely distributed copper ions might be present in the liver, however, since very irregular nuclei appeared in the cells, as observed earlier after copper sulphate treatment (Benedeczky *et al.* 1986). Similarly, as after phenol treatment (Benedeczky and Nemcsók 1990), very serious and extensive cell damage was observed in the exocrine pancreatic cells after the double treatment.

This corroborates our earlier finding (Benedeczky *et al.* 1986) that the exocrine pancreatic cells are very sensitive indicators of environmental cell damaging effects.

As regards the kidney, the hypoxia and copper sulphate treatment caused much more serious alterations in the tubular epithelium than in the liver cells. The large extracellular vacuoles, myelin figures and cell detritus all suggest a toxic necrotic cell injury caused by copper ion. Nevertheless, the presence of the large, dense spherical bodies in the epithelial cells may indicate an increased endocytotic activity, as a consequence of which these cells might incorporate the toxic copper ion from the serum. This supposition is based on the fact that most of the epithelial cells were quite intact after hypoxia, and the cytoplasm was filled with a large number of lysosomes. It appears very probable that, instead of the liver cells (where the lack of lysosomes was conspicuous), the tubular epithelial cells of the kidney absorbed and eliminated the toxic copper ion.

References

Asztalos, B., Nemcsók, J., Benedeczky, I., Gábriel, R., Szabó, A. and Refaie O.J. (1990) The effects of pesticides on some biochemical parameters of carp (*Cyprinus carpio* L.). *Archives of Environmental Contamination and Toxicology*, **19**, 275-282.

Benedeczky, I., Nemcsók, J. and Halasy, K. (1986) Electron microscopic analysis of the cytopathological effect of pesticides in the liver, kidney and gill tissues of carp. *Acta Biologica Szeged*, **32**, 69-91.

Benedeczky, I. and Nemcsók, J. (1990) Detection of phenol induced subcellular alteration by electron microscopy in the liver and pancreas of carp. *Environmental Monitoring and Assessment*, **14**, 385-394.

Benoit, D., Leonard, E.N., Christensen, G.M. and Fiandt, J.T. (1976) Toxic effects of cadmium on generations of brook trout (*Salvenius fontinalis*). *Transactions of the American Fisheries Society*, **105**, 550-560.

Beumer, J.P., Bacher, G.J. (1982) Species of *Anguilla* as indicators of mercury in the coastal rivers and lakes of Victoria, Australia. *Journal of Fish Biology*, **21**, 87-94.

Braunbeck, T., Storch, V. and Bresch, H. (1990) Species specific reaction of liver ultrastructure in zebrafish (*Brachydanio rerio*) and trout (*Salmo gairdneri*) after prolonged exposure to 4 chloroaniline. *Archives of Environmental Contamination and Toxicology*, **19**, 405-418.

Bunton, T.E., Baksi, S.M., George, S.G. and Frazier, J.M. (1987) Abnormal hepatic copper storage in a teleost fish (*Morone americana*). *Veterinary Pathology*, **24**, 515-524.

Förstner, U. and Prosi, F. (1979) Heavy metal pollution in freshwater ecosystem. In O. Ravera (ed.) *Biological Aspects of Freshwater Pollution*, 129-161. Pergamon Press, Oxford.

Gary, L.P., Gary, W.H. and Fiandt, J.T. (1981) Acute toxicity of phenol and substituted phenols to the fathead minnow. *Bulletin of Environmental Contamination and Toxicology*, **26**, 585-599.

Johnels, A.G., Westermark, T., Berg, W., Persson, P.I. and Sjöstrand, B. (1967) Pike (*Esox lucius* L.) and some other aquatic organisms in Sweden as indicators of mercury contamination in the environment. *Oikos*, **18**, 323-333.

Lodenius, M., Seppanen, A. and Herranen, M. (1983) Accumulation of mercury in fish and man from reservoirs in Northern Finland. *Water, Air and Soil Pollution*, **19**, 237-246.

McArdle, H.J., Gross, Sh.M. and Danks, D.M. (1988) Uptake of copper by mouse hepatocytes. *Journal of Cellular Physiology*, **136**, 373-378.

Novikoff, P.M. (1982) Intracellular organelles and lipoprotein metabolism in normal and fatty livers. In I. Arias, H. Popper, D. Schachter and D.A. Shafritz (eds.) *The Liver: Biology and Pathobiology*. Raven Press, New York.

Novikoff, P.M. and Edelstein, D. (1977) Reversal of orotic acid induced fatty liver in rats by clofibrate. *Laboratory Investigation*, 36, 215-231.

Popper, H. (1982) Hepatocellular degeneration and death. In I. Arias, H. Popper, D. Schachter and D. A. Shafritz (eds.) *The Liver: Biology and Pathobiology*. Raven Press, New York.

Rojik, I., Nemcsók, J., Boross, L. (1983) Morphological and biochemical studies on liver, kidney and gill of fishes affected by pesticides. *Acta Biologica Hungarica*, 34, 81-92.

Salánki, J., V.-Balogh, K. and Berta, E. (1982) Heavy metals in animals of Lake Balaton. *Water Research*, 16, 1147-1152.

Sternlieb, I. (1980) Copper and the liver. *Gastroenterology*, 78, 1615-1628.

Sternlieb, I. (1982) Pathobiology of metals. In I. Arias, H. Popper, D. Schachter and D.A. Shafritz (eds.) *The Liver: Biology and Pathobiology*. Raven Press, New York.

Tort, L., Flos, R. and Balasch, J. (1984) Dogfish liver and kidney tissue respiration after zinc treatment. *Comparative Biochemistry and Physiology*, 77C, 381-384.

Tort, L., Torres, P. and Flos, R. (1987) Effects on dogfish haematology and liver composition after acute copper exposure. *Comparative Biochemistry and Physiology*, 87C, 349-353.

V.-Balogh, K. and Salánki, J. (1987) Biological monitoring of heavy metal pollution in the region of Lake Balaton. *Acta Biologica Hungarica*, 38, 18-30.

Ion Channels of Nerve Membrane as Targets for Environmental Pollutants

Katalin S.-Rózsa and János Salánki

Balaton Limnological Research Institute of the Hungarian Academy of Sciences, Tihany, Hungary H-8237.

Key words: *Helix* neurons, ion currents, modulation, transmitter effects, heavy metals.

Abstract

The effect of sub-lethal doses of heavy metals ($HgCl_2$, $CdCl_2$, $PbCl_2$) on neurons of *Helix pomatia* L. was studied. In the experiments, the two microelectrode voltage clamp method was used for measuring ion channel activity, monitoring their alterations following acute heavy metal treatment.

The heavy metals modulate the ion channel activity of nerve cells and are able to alter their firing pattern. Heavy metals were shown to modify the slow, K-dependent Ca-current, the unspecific cation current and the delayed potassium current. The channel modulation altered the acetylcholine-activated current in a voltage dependent manner. Different neurons showed specific reactions to heavy metals. Both an increase and decrease of the transmitter induced permeability occurred in different neurons, depending on the receptors the neuron shared. In addition, heavy metals modulated the voltage dependent ion currents.

The ion channel modulation by heavy metals was shown to cause disfunction of such vital physiological processes as circulation, feeding, reproduction, etc. as a result of damage to nerve regulation at cellular level.

Introduction

In the past decade, studies of the cellular and molecular mechanisms of the effects of environmental pollutants, have come into the limelight. It became evident that membrane disturbances may be considered as one of the biological effects of pollutant-induced stress at cellular and molecular level. Heavy metals are known to be permanent components of the environmental pollutants causing a range of neurological disorders in humans (Prasad 1982). Acute or chronic

Bioindicators and Environmental Management
ISBN 0-12-382590-3

exposure to heavy metals were shown to elicit, among others, Minamata disease (Damstra 1977; Clarkson 1978), decreased mental awareness, memory loss, convulsion, seizure (Adler and Adler 1977) and paralysis (Avery and Cross 1974). Although the cellular physiological substrates of these neurobehavioural effects are not known, the majority of authors connect them to the modulation of transmitter release at the synapses (Atchinson and Narahashi 1984), appearing as a result of alteration to various types of calcium-mediated function (Cooper and Manalis 1983; Audesirk 1987).

As the molecular mechanisms leading to neurological abnormalities following exposure to heavy metals are difficult to study in the central nervous systems of vertebrates, due to heterogeneous cell populations and complexities, the model system was developed from the brain of gastropods. This permits studies of the effects on membranes of heavy metals and their interactions with neurotransmitters (S.-Rózsa and Salánki 1985; 1987). Because of the large size, easy accessibility and identifiability of the central neurons of gastropods, these are used as model systems for studying detailed mechanisms of ion flow through a variety of ion channels of the nerve membrane.

In these experiments particular central neurons of Helix pomatia were used for testing the effect of heavy metals. Both the Helix and Lymnaea neurons were shown to be suitable for studying the effects of heavy metals on the ion channels of nerve membrane (S.-Rózsa and Salánki 1985; 1990). It has been shown earlier that heavy metals modify the effects of neurotransmitters, such as acetylcholine, 5-hydroxytryptamine and dopamine on the snail neurons (S.-Rózsa and Salánki 1987; 1990).

The aim of the present investigation was to characterize further the ion channels affected by heavy metals in acute exposure and to verify the cellular targets of their effects.

Materials and methods

Experiments were carried out on the isolated suboesphageal ganglionic ring of Helix pomatia L. The connective tissue sheath of the ganglia were removed and the ganglia were pinned in a sylgard-lined dish perfused with physiological saline. At the start of the experiment, the ganglia were perfused with normal Helix saline, then the perfusion was switched to saline with heavy metals. The metal-containing solution was washed out again with normal saline. The solution in the bath were changed with a rapid perfusion at a rate of about 2 ml min^{-1}, which gives an almost instantaneous change of saline around the impaled neurons. The heavy metals were used at low concentrations causing significant but slight changes in firing pattern of cells (S.-Rózsa and Salánki 1987).

The transmitters were applied by pressure injection from micropipettes of 5 μm tip diameter to the surface of the neurons studied. The response of the neurons to the transmitters was studied in control conditions and following heavy metal treatment. Cell identification introduced earlier was used here (S.-Rózsa and Salánki 1990).

Both current clamp and voltage clamp mode of recording was used. The electrical activity of the neurons in current clamp mode was recorded with conventional electro-physiological method using KCl-filled glass microelectrodes, as reported earlier (S.-Rózsa and Salánki 1987). For measuring ion currents a two-microelectrode voltage clamp was used with a DAGAN-8500 amplifer (S.-Rózsa and Salánki 1990). The microelectrodes had resistances between 1-5 MOhm and were shielded with aluminium foil to avoid oscillation during the voltage step. The fluid level above the cells was only a few millimetres. The holding potential was routinely -50 mV, which is close to the resting membrane potential of these neurons. Voltage dependent currents were elicited by stepping from a holding potential to more negative or more positive levels. The family of currents were recorded and their I/V relationship was studied. The recording was made on a Gould-Brush recorder.

The heavy metals used were: $HgCl_2$, $CdCl_2$ and $PbCl_2$. Among neurotransmitters acetylcholine (Sigma) and 5-hydroxytryptamine (Sigma) were applied. The experiments were carried out at room temperature (20-22°C).

Results

As it was reported earlier, all the heavy metals studied alter ion-channel permeability of the *Helix* neurons alone, or in combination with neurotransmitters. Both the voltage dependent and transmitter induced currents were modified by heavy metals (S.-Rózsa and Salánki 1990).

1. Effect of acute exposure to heavy metals on the neuro-transmitter-induced responses

The individual *Helix* neurons have specific responses to various neurotransmitters. Both acetylcholine and 5-hydroxytryptamine were able to cause excitation, inhibition or biphasic membrane responses depending on the receptors possessed by the given neuron. The heavy metals can cause either synergistic or antagonistic modulation of these effects. Several kinds of heavy metal modulation was shown. Different heavy metals either had a range of modulatory influences on the same cells or influenced the transmitter effect in a similar way. For this reason the identified neurons were used to compare the results in these experiments.

On the neuron RPa2, located to the right parietal ganglion, both acetylcholine (Fig. 1) and 5HT (Fig. 2) cause an increase in firing frequency. The ACh effect was a short-term one but more intensive, while the 5HT-effect can be regarded as long-lasting, but with low-frequency firing (Figs. 1 and 2). As can be seen on Figure 1, $HgCl_2$ in acute treatment, eliminates the ACh-elicited high frequency firing. However, the increased membrane depolarization is maintained, even if $HgCl_2$ was present in the bath (Fig. 1C). Further treatment with $HgCl_2$ can increase the basic firing pattern of the RPa2 neuron and the effect of ACh can be eliminated entirely (Fig. 1D).

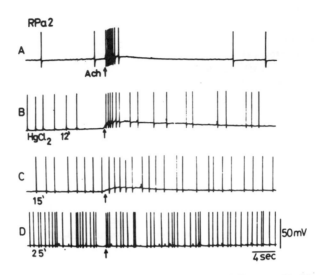

Figure 1. Modification of the ACh-evoked response following $HgCl_2$ treatment on the *Helix* neuron RPa2. A - control effect of ACh (10^{-4} M). B, C and D - decrease and elimination of the response of the neuron to ACh (10^{-4} M) at the 12, 15 and 25 min of $HgCl_2$ (10^{-6} M) treatment. Arrow: ACh application.

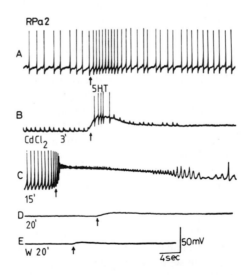

Figure 2. Modulation of the excitatory effect of 5HT following $CdCl_2$ treatment on the *Helix* neuron RPa2. A - control effect of 5HT (10^{-4} M). B, C and D - only the depolarizing effect was maintained to the 20th min of $CdCl_2$ treatment. E - wash out. The $CdCl_2$ effect was found to be irreversible. Arrow: 5HT application.

Acute treatment with $CdCl_2$ of neuron RPa2 caused a different effect from that of $HgCl_2$, on both the control firing pattern and the 5HT-induced increased firing (Fig. 2). The $CdCl_2$ treatment inhibited first the generation of the spontaneous action potentials, and subsequently the 5HT-induced responses, (Fig. 2c,d). Weak depolarization, however, remained throughout the $CdCl_2$ application, even following its wash out (Fig. 2e). The firing of the cell was not restored following a prolonged wash out of $CdCl_2$.

2. The heavy metals modulate ACh-induced currents

The nerve cell membrane contains a number of ion channels, with distinct ion selectivity contributing to the net current flow across the membrane (Hille 1984). Single components of ion currents, corresponding to the activity of individual channels, can be separated from each other by their ion selectivity, voltage dependence and sensitivity to various drugs.

In these experiments the family of currents evoked by ACh was measured at a range of holding potentials from +20 mV to -80mV. The voltage-dependent currents were also studied (S.-Rózsa and Salánki 1990).

The treatment of ganglia with $HgCl_2$ in the concentration range 10^{-6}-10^{-5} M modulated the ACh-evoked responses and currents in a specific way in each cell. The activity of neuron LPa3 was inhibited by ACh application, although a slight depolarization was present (Fig. 3), accompanied by an inward current at resting potential level. The current evoked by ACh (10^{-5} M) was increased by shifting the holding potential to more negative values. The peak current of ACh-elicited current roughly doubled when the holding potential was shifted from -40 mV to -80mV. In these cells, the ACh-elicited current reversed around +10 mV holding potential and became positive in sign (Fig. 3). Following $HgCl_2$ (10^{-5} M) treatment the inhibitory effect of ACh turned into an excitatory one (Fig. 3). In the presence of $HgCl_2$, the ACh-elicited current was nearly unaltered at a holding potential close to resting membrane potential or more positive values. By shifting the holding potential to a more negative value (-80 HP), the ACh-induced inward current was nearly doubled when compared to the control (Fig. 3). The reversal potential of the ACh-current remained unaltered in the presence of $HgCl_2$. Neuron RPa2, which usually reacts with an increase in firing frequency to ACh (Fig. 1), showed additionally an increase in the peak amplitude of the ACh-induced inward current at more negative holding potentials (-60-80 mV) during $HgCl_2$ treatment (Fig. 4).

The modulation of ACh-induced response following $CdCl_2$ treatment was similar to that elicited by $HgCl_2$. On neuron V2 the ACh caused a biphasic effect, where the short-term increase in firing frequency was followed by long-lasting inhibition (Fig. 5). Both phases of ACh-response were weakened in the presence of $CdCl_2$. Under the influence of $CdCl_2$, the ACh-elicited inward current did not change at the resting potential level, but decreased significantly at more negative holding potential (Fig. 6b). In contrast to this, on neuron RPah, the ACh-elicited inward current was potentiated following $CdCl_2$ treatment (Fig. 6).

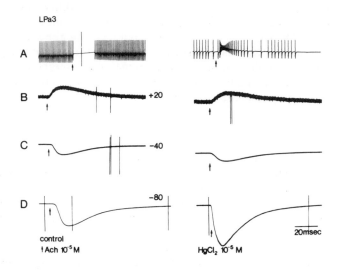

Figure 3. Effect of HgCl$_2$ treatment on the ACh elicited response as regard the firing (upper) and ion currents of the cell LPa3. Holding potentials are noted. Left panel: effect of ACh (10^{-5} M) in control saline. Right panel: effect of ACh (10^{-5} M) following HgCl$_2$ (10^{-5} M) perfusion. Here and on the following figures arrows show the moment of ACh-application.

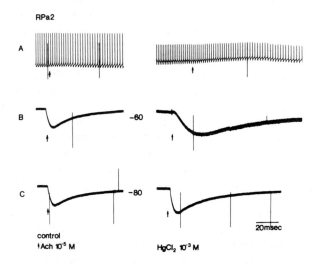

Figure 4. Modulation of AC-effect on the cell RPa2 following HgCl$_2$ treatment. A - current clamp recording. B and C - voltage clamp recording. HgCl$_2$ (10^{-3} M) increased the peak value of the ACh-induced current.

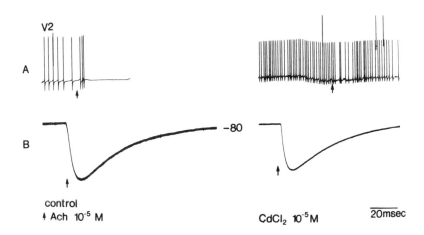

Figure 5. Modulation of the ACh-induced response on the cell V2 by CdCl$_2$ treatment. A - the effect of ACh (10^{-5} M) on the firing frequency of the cell in control and following CdCl$_2$ (10^{-2} M) treatment. B - changes of the inward current elicited by ACh before and after CdCl$_2$ treatment. The ACh elicited inward current decreased at -80 mV HP after CdCl$_2$ application.

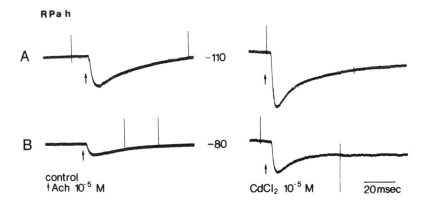

Figure 6. Modulation of the ACh-induced response by CdCl$_2$ on the cell RPah. A-B - ACh-elicited inward current in control and following CdCl$_2$ (10^{-5} M) treatment at two different holding potentials.

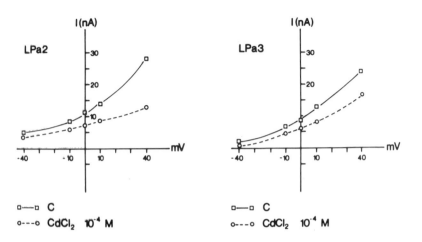

Figure 7. The I-V relationship of the voltage dependent current and its modification following CdCl$_2$ treatment on the neurons LPa2 and LPa3. C - control.

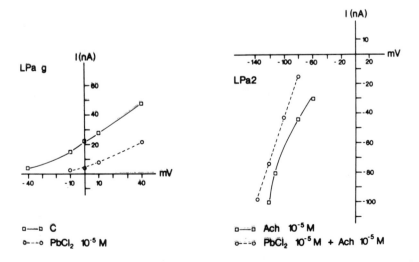

Figure 8. The effect of PbCl$_2$ treatment on the voltage dependent current on the neuron LPag and on the ACh-induced current on the neuron LPa2. C - control.

3. Modulation of voltage-dependent ion currents with heavy metals

Voltage activated currents were recorded from specified neurons between -40 and +40 mV holding potentials. The I-V relationship of these currents was found to be shifted either to the left or to the right depending on the heavy metal used or the membrane characteristics of the neuron.

The voltage dependent outward current was decreased on the neuron LPa2 and LPa3 following $CdCl_2$ (10^{-4} M) treatment (Fig. 7). The $CdCl_2$ treatment frequently caused the same modification of the voltage dependent outward currents in a number of the other specified cells.

The $PbCl_2$ treatment (10^{-5} M) also decreased the voltage dependent outward current on the cell LPag (Fig. 8). The ACh-induced inward current was also decreased in the presence of $PbCl_2$ on the cell LPa2 (Fig. 8).

Nevertheless, the voltage dependent currents can be potentiated by the other heavy metals, as it has been shown earlier (S.-Rózsa and Salánki, 1990).

Discussion

It is commonly accepted that heavy metals can be transported through lipid membranes using Ca or Na channels (Gutknecht 1981; Miyamoto 1983). Some heavy metals are bound to cytosolic components, while others enter to the synaptic terminals and interfere with transmitter release, leading to the irreversible loss of sensitivity of the synapses (Eriksen *et al.* 1990; Wang and Quastel 1990).

In acute treatment both the extracellular and intracellular sites of action are involved in the toxic responses to heavy metals. In the extracellular space, heavy metals can alter the ion channel function, as a result of interaction with surface charge groups, or by direct channel blockade (Hille 1984). Penetrating the cells, heavy metals produce widespread alterations in protein structure and function, including membrane phosphorylation (Biondi *et al.* 1989), leading to the modulation of ion-channel permeability of the membrane.

The results obtained in the present experiments support the idea that ion channels of nerve membranes can be targets of heavy metals. On the extracellular side they can compete at the agonist binding sites, they can allosterically modify receptor-channel coupling or block the channel itself. Most of the heavy metals were shown to react with sulfhydryl bonds (Müller *et al.* 1989).

Various types of ionic currents are activated by different heavy metals. A specific Ca-dependent K-current is activated only by Hg^{2+} and Pb^{2+}, while the Ca-dependent non-specific cation current is activated by Hg^{2+}, Zn^{2+} and Pb^{2+} alike in snail neurons (Müller *et al.* 1989). Furthermore, our earlier results showed the involvement of the outward potassium current into the heavy metal modulatory effect (S.-Rózsa and Salánki 1990). The decrease of potassium current during heavy metal treatment can contribute to the potentiation of the ACh-elicited currents demonstrated in our present experiments. Activation of various types of membrane conductance can be a result of heavy metal binding to different types of targets. The heavy metals can act directly on the channel proteins or closely attached proteins (Partridge and Swandulla 1987; 1988).

The alterations to membrane proteins by heavy metals are connected to the phosphorylation of ion channel constituents. It was shown that heavy metals are able to inhibit the activity of adenylate cyclase (Winder and Kitechen 1984), while their effects on phosphodiesterase are more contradictory, and can either be inhibitory or excitatory (Prasad 1982; Biondi *et al.* 1989). However, there are some indications that heavy metals modulate the transmitter effects through the second messenger system (Biondi *et al.* 1989).

In acute heavy metal treatment, compared with chronic administration, modulation of the synaptic transmission by alteration of the transmitter release has a less important role.

Modulation of transmitter effects (e.g. glutamate and GABA) with heavy metals has also been demonstrated in vertebrate brain (Prasad 1982; Westbrook and Mayer 1987). These results emphasize direct effect of heavy metals on the soma membrane of the central neurons, rather than their metabolic links, supporting our suggestion with regard to the target sites of acute heavy metal treatment. The disfunction of ion channels can cause damage to such vital physiological functions as circulation, feeding, reproduction and behaviour of the animal, resulting in changes to the whole population of the species.

References

Adler, M.W. and Adler, C.H. (1977) Toxicity of heavy metals and relationship to seizure threshold. *Clinical Pharmacology*, **22**, 774-779.

Atchison, W.D. and Narahashi, T. (1984) Mechanism of action of lead on neuromuscular junction. *Neurotoxicology*, **5**, 267-282.

Audesirk, G. (1987) Effect of *in vitro* and *in vivo* lead exposure on voltage-dependent calcium channels in central neurons of *Lymnaea stagnalis*. *Neurotoxicology*, **8**, 579-592.

Avery, D.D. and Cross, H.A. (1974) The effect of tetraethyl lead on behaviour in the rat. *Pharmacology Biochemistry and Behaviour*, **2**, 473-479.

Biondi, C., Fabbri, E., Ferretti, M.E., Sonetti, D. and Bolognani Fantin, A.M. (1989) Effects of lead exposure on cAMP and correlated enzymes in *Vivaparus ater* (Mollusca, Gastropoda) nervous system. *Comparative Biochemistry and Physiology*, **94C**, 327-333.

Clarkson, T.W. (1987) Metal toxicity in the central nervous system. *Enviromental Health Perspectives*, **75**, 59-64.

Cooper, G.P. and Manalis, R.S. (1983) Influence of heavy metals on synaptic transmission. *Neurotoxicology*, **4**, 69-84.

Damstra, T. (1977) Toxicological properties of lead. *Environmental Health Perspectives*, **19**, 297-307.

Eriksen, H.D.K., Andersen, T., Stenersen, J. and Andersen, R.A. (1990) Cytosolic binding of Cd, Cu, Zn, and Ni in four polychaete species. *Comparative Biochemistry and Physiology*, **95C**, 111-115.

Gutknecht, J. (1981) Inorganic mercury (Hg^{2+}) transport through lipid bilayer membranes. *Journal of Membrane Biology*, **61**, 61-66.

Hille, B. (1984) Ionic channels of excitable membranes. Sinauer Associates Inc. Sunderland, Massachusetts. 1-426.

Miyamoto, M.D. (1983) Hg^{2+} causes neurotoxicity at an intracellular site following entry through Na and Ca channels. *Brain Research*, **267**, 375-379.

Müller, T.H., Swandulla, D. and Lux, H.D. (1989) Activation of three types of membrane currents by various divalent cations of identified molluscan pacemaker neurons. *Journal of General Physiology*, **94**, 997-1014.

Partridge, L.G. and Swandulla, D. (1987) Single Ca-activated cation channels in bursting neurons of *Helix*. *Pflügers Archiv - European Journal of Physiology*, **410**, 627-631.

Partridge, L.D. and Swandulla, D. (1988) Calcium-activated non-specific cation channels. *Trends in Neurosciences*, **11**, 69-72.

Prasad, K.N. (1982) Tissue culture model to study the mechanism of the effect of heavy metals on nerve tissue. In K.N. Prasad and A. Vernadakis (eds.) *Mechanisms of Action of Neurotoxic Substances*, 67-94. Raven Press, New York.

S.-Rózsa, K. and Salánki, J. (1985) Effects of heavy metals on the chemosensitivity of neuronal somata of *Lymnaea stagnalis* L. *Symposia Biologica Hungarica*, **29**, 387-400.

S.-Rózsa, K. and Salánki, J. (1987) Excitable membranes - object for evaluating the effect of heavy metal pollution. *Acta Biologica Scientiarum Hungaricae*, **38**, 31-45.

S.-Rózsa, K. and Salánki, J. (1990) Heavy metals regulate physiological and behavioural events by modulating ion channels in neuronal membranes of molluscs. *Environmental Monitoring and Assessment*, **14**, 363-375.

Wang, Y.X. and Quastel, D.M.J. (1990) Multiple action of zinc on transmitter release at mouse end-plates. *Pflügers Archiv, - European Journal of Physiology*, **415**, 582-587.

Westbrook, G.L. and Mayer, M.L. (1987) Micromolar concentrations of Zn^{2+} antagonize NMDA and GABA responses of hippocampal neurons. *Nature*, **328**, 640-643.

Winder, C. and Kitchen, I. (1984) Lead neurotoxicity: a review of the biochemical, neurochemical and drug induced behavioural evidence. *Progress in Neurobiology*, **22**, 59-87.

Colonization of Artificial Substrata as a Multi-Species Bioassay of Marine Environmental Quality

Mark J. Costello[1] and Simon F. Thrush[2]

[1] Environmental Sciences Unit, Trinity College, Dublin 2, Ireland.
[2] Water Quality Centre, Department of Scientific and Industrial Research, Hamilton, New Zealand.

Key words: colonization, artificial substrata, field bioassay, sampling methods.

Abstract

The use of artificial substrata in the aquatic environment is reviewed. Artificial substrata have the advantages of allowing temporal and spatial replication, reproducibility, and experimental manipulation. Being identical in size and structure they can significantly reduce sample variation, be made of cheap materials, and be quicker to process than natural substrata. Community composition, diversity and structure on artificial and natural substrata can be similar. Colonization of artificial substrata can be considered a multi-species, *in situ*, choice-test. Although a standard technique in freshwater studies, the sensitivity and relevance (to larger scale patterns) of colonization of artificial substrata remains to be evaluated for marine pollution work.

Introduction

In the investigation of pollution, one should consider (a) the biological processes responding to the pollutant, (b) what techniques measure this response best, and (c) how one can relate the response to a cause and thereby assess the potential or actual pollutant impact (Underwood and Petersen 1988). Assessments of the potential and actual effects of pollution need to encompass the range of spatial and temporal scales apparent from laboratory to field studies (Fig. 1). In laboratory studies, environmental variables can be controlled and the results may be obtained within days (Fig. 1). Field surveys may appear more realistic measures of actual effects, but years of data could be sometimes needed to distinguish between anthropogenic and natural effects (Fig. 1). It may also be difficult to relate environmental changes to specific causes. Similarly, studies at a range of biological scales, from biochemical, to cellular, physiological, population, community and ecosystem levels are all necessary to develop a full understanding of the causes and mechanisms of ecotoxicity. Additionally, it is critical that studies at different scales can be related to each other.

Bioindicators and Environmental Management
ISBN 0-12-382590-3

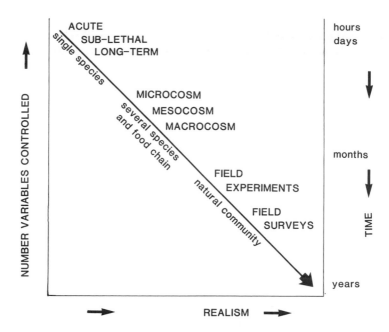

Figure 1. Diagram of relationships between the methods of pollution assessment, numbers of variables, and scales of study.

The gap between laboratory and field studies can be bridged by the use of laboratory maintained communities (e.g. mesocosms) and field experiments (Fig. 1) which control a few variables while allowing interactions between species (e.g. Hansen and Tagatz 1980; Nebeker *et al.* 1984; Oviatt *et al.* 1986; 1987; Frithsen *et al.* 1989; Pontasch and Cairns 1989; Rosemarin 1990). A recent workshop evaluated a variety of biochemical, cellular, physiological, and community analysis techniques, for marine organisms, but did not include studies using artificial substrata (Bayne *et al.* 1988).

One difficulty in sampling benthic marine communities is that they can consist of a mosaic of patches generated by historical disturbance effects, such as caused by predators and drift logs (Dayton 1971) decaying clumps of drift weed (Thrush 1986a). Sampling such habitats will therefore include patches in different stages of succession, thereby increasing variation between samples with concomitant increases in the numbers of samples necessary for statistical rigour. While these problems can be overcome by rigorously defined and intensive sampling programmes, it may be possible to obtain results more expediently by controlling for this variation. The simplest method of controlling for successional factors is to sample a community of known age. In this paper, we investigate whether artificial substrata could overcome these problems. While natural (e.g. floods, storms, volcanic activity) or artificial (e.g. dredging, trawling, explosions) catastrophes may occasionally provide such conditions, the exposure of artificial or defaunated natural substrata is a less destructive and more practicable alternative. The small scale of such sampling will also aid the replication and logistics of sample processing.

In this paper, we consider (a) whether colonization is a process sensitive to pollution, (b) if artificial substrata are a useful sampling technique, and (c) how one can analyse and interpret the results.

Colonization as a sensitive process

In some marine areas, succession in the marine benthos appears to be predictable (Pearson and Rosenberg 1978; Rhoads *et al.* 1977). Many immigrants may be larval and juvenile stages, and early life stages are often generally found to be more sensitive to environmental quality than older organisms (e.g. Hansen and Tagatz 1980; Weis and Weis 1989). The relative roles of active settlement and passive entrainment of larvae have been reviewed by Woodin (1986) and Butman (1987). However, particularly in small-scale (metres, days to weeks) studies adults may be the dominant colonizers of disturbed sediments (Thrush 1986b; Butman 1987; Frid 1989). Adults can still be very sensitive to sediment characteristics (Meadows 1964a; 1964b; Twomey 1977; Stoner 1980; Oaken *et al.* 1984), and some adult amphipods may be more sensitive than juveniles (DeWitt 1987). Later in succession, biotic effects, such as predator activity (Commito 1982; Ambrose 1984; Olafsson and Persson 1986), disturbance by burrowers (DeWitt and Levinton 1985; Jensen 1985), or density dependent effects (Ambrose 1986; Connell 1963), will tend to dominate community composition and structure.

In the early stages of succession there will be fewer species and individuals. Thus we hypothesise that interspecific effects on species occurrence and abundance are likely to be less important than abiotic factors (e.g. substratum, water quality due to low population densities). However, in selecting a successional stage for sampling for environmental quality it is also desirable to collect as great a variety of phyla and species as possible. The best time for sampling may be as soon as equilibrium (sensu MacArthur and Wilson 1967) is reached; thus diversity (species richness) will be maximized, but abundance and interspecific interactions minimized.

A variety of species is particularly important where the impact of a pollutant may be selectively toxic to certain phyla and species within phyla. For example, arthropods are more sensitive to some organophosphate pesticides than molluscs and fish (Ross and Horsman 1988). There is the further option of examining particular populations or individuals within the colonizing community in detail. Functional, behavioural and physiological processes, could be studied (e.g. growth, feeding, bioaccumulation).

Hansen and Tagatz (1980) allowed flow-through marine aquaria, some dosed with toxicants, to be colonized by organisms from the water supply. The developing communities contained seven phyla and averaged fifty species. These communities were at least as sensitive to toxicant concentration as single species acute and chronic tests. Both physical and chemical impacts of pollutants were detectable, and differing colonization patterns between species assisted prediction of probable impacts at sea. A colonization test is therefore a multi-species bioassay. Mesocosm tests are also multi-species bioassays but the biota is artificially introduced. Additionally, the colonization could be considered an *in situ* choice test.

Colonization would therefore appear to be a sensitive process, and hence valuable for marine environmental studies, because (a) abiotic factors will be more important than biotic in determining community structure, (b) early colonists are sensitive to substratum quality, and (c) a variety of taxa would be present.

Using artificial substrata

Artificial substrata are here defined as solid material placed in the environment and used for scientific purposes. As with natural substrata, artificial substrata may be destructively or non-destructively sampled, and may be two or three dimensional. For the purposes of this review, artificial substrata are further separated into "hard" and "soft" types.

Most "hard" marine artificial substrata have been two dimensional, be they experimental panels (Osman 1977; 1982; Withers and Thorpe 1977; Schoener *et al.* 1978; Russ 1980; Shin 1981; Harris and Irons 1982; Buss 1986), or fouling on pilings (Karlson 1978; Kay and Butler 1983) and offshore platforms (Goodman and Ralph 1981; Southgate and Myers 1985). Materials used in the experimental marine panels include asbestos (Fager 1971; Jackson 1977; Dean 1981), polyurethane (Goren 1979), perspex (Schmidt 1982), glass (Borowitzka *et al.* 1978), ceramic tiles (Sutherland 1974; 1981; Mook 1981), concrete (Keen and O'Neill 1980), and slate (Kensler and Crisp 1965). Three dimensional hard substrata have been less widely used in marine studies. Plastic mesh pads (Schoener 1974a; 1974b; Myers and Southgate 1980; Costello 1987) have been used in studies of marine communities, and plastic mesh is used in collecting commercial scallop spat (Rodhouse and Burnell 1979).

In freshwater habitats, artificial substrata have been successfully used as a means of sampling (e.g. Arthur and Horning 1969; Dickson *et al.* 1971; Benfield *et al.* 1974; Crossman and Cairns 1974; Hocutt *et al.* 1976; Voshell and Simmons 1977; Shaw and Minshall 1980; Briggs 1983; Reynolds and Hunter 1985), particularly in environments where other sampling techniques are ineffective. Other uses include studies of community dynamics (e.g. Khalaf and Tachet 1977; Sheldon 1977; Cover and Harrel 1978; Giller and Campbell 1989), influence of substratum on community structure (e.g. Cairns *et al.* 1976; Wise and Molles 1979), and equilibrium theory (e.g. Cairns *et al.* 1969; Dickson and Cairns 1972). The types of artificial substrata used in freshwater include microscope slides, asbestos panels, polyurethane foam, plastic netting or mesh (e.g. plastic scouring pads), bricks, and trays and baskets of gravel (Beak *et al.* 1973; Flannagan and Rosenberg 1982). In freshwater pollution monitoring "multiple plate" and baskets of rocks or plastic webbing are now a standard technique in the United States of America (Beak *et al.* 1973; Meier *et al.* 1979; American Public Health Association *et al.* 1985).

Table 1. Comparison of taxa occurring on artificial substrata (plastic mesh pads) in (a) Bimini Lagoon, Bahamas (Schoener 1974) and (b) Lough Hyne, south-west Ireland (Costello, unpubl.), and live sponges (c) *Halichondria panicea* (Pallas) and (d) *Hymeniacidon perleve* (Montagu), in Lough Hyne (Costello, unpubl.).

Taxon	Artificial Substrata		Live Sponges	
	(a)	(b)	(c)	(d)
ALGAE	+	+	+	+
FORAMINIFERA	+	+	+	+
PORIFERA	+	+	+	+
HYDROZOA	+	+	+	+
ANTHOZOA Actinaria		+	+	+
TURBELLARIA		+	+	+
NEMERTINEA		+	+	+
NEMATODA	+	+	+	+
POLYCHAETA	+	+	+	+
MOLLUSCA Bivalvia	+	+	+	+
Opisthobranchia	+	+	+	+
Polyplacophora	+	+	+	+
Prosobranchia	+	+	+	+
PYCNOGONIDA	+	+	+	+
ACARINA	+	+	+	+
CRUSTACEA Amphipoda	+	+	+	+
Cladocera		+	+	+
Copepoda	+	+	+	+
Cumacea	+	+	+	+
Decapoda	+	+	+	+
Isopoda	+	+	+	+
Ostracoda	+	+	+	+
Mysidacea		+	+	+
Leptostraca	+	+	+	+
Tanaidacea	+	+	+	+
INSECTA Chironomidae		+	+	+
SIPUNCULA	+			
BRYOZOA	+	+	+	+
ECHINODERMATA Asteroidea		+	+	+
Crinoidea		+	+	+
Echinoidea		+	+	+
Holothuroida	+			
Ophiuroidea	+	+	+	+
CHAETOGNATHA	+	+	+	+
TUNICATA	+	+	+	+
STOMATOPODA	+			

Soft artificial substrata consist of trays and plugs of sediment placed on or in estuarine and marine sediments (Brunswig *et al.* 1976; Santos and Simon 1980; Hockin and Ollason 1981; Flint *et al.* 1982;, Zajac and Whitlach 1982a; 1982b; Hockin 1982; Arntz and Rumohr 1982; Gallagher *et al.* 1983; Levings 1976; Whitlach and Zajac 1985; Savidge and Taghon 1988; Thrush and Roper 1988; Frid 1989; Smith and Brumsickle 1989), and trays of sediment in freshwater studies (e.g. Mason 1976; Shaw and Minshall 1980; Flannagan and Rosenberg 1982; Rosenberg and Resh 1982; Giller and Cambell 1989).

Artifical substrata cannot be considered a substitute for natural substrata, although in some circumstances they may mimic the physical structure of freshwater (e.g. Glime and Clemons 1972; Macan and Kitching 1972; Reynolds and Hunter 1985) and marine vegetation (Ghelardi 1960), and sessile marine organisms (e.g. Dean 1981; Dauer *et al.* 1982) for experimental purposes. For marine fauna a similar range of phyla, classes and orders of biota may occur on artificial and natural substrata in the same and different areas (Table 1). In Lough Hyne, south-west Ireland, plastic mesh substrata collected a similar range of amphipod species as occurred on live sponges (Table 2), and rank abundance curves were also very similar for the species on natural and artificial substrata (Fig. 2). There were 23 and 24 species in common between each species of sponge and the artificial substrata, and 25 species in common between sponges (excluding two obligate commensal species). Therefore, differences in community composition and structure between natural and artificial substrata can be no greater than differences between different natural substrata.

Different types of substrata, and the position in which they are installed in the environment can have important effects on their effectiveness. There are marked differences between the use of soft and hard substrata. The logistics of using trays and plugs, namely preparation, placement, retrieval and analysis of biota, can be more time consuming than for sampling natural sediments. The method of defaunation of the sediment may alter it physically (if sieved) or chemically (if frozen), reducing its comparability with the original sediment. Trays of azoic sediment laid in the sea or river bed can be affected by natural factors, be they physical (e.g. hydrodynamics), chemical, meteorological or biotic (Thrush and Roper 1988; Frid 1989; Pontasch and Cairns 1989). Using a combination of open and predator protected (covered) trays, Bonsdorff *et al.* (1990) found that natural predation could mask the effects of oil exposure on the colonizing fauna. Thrush and Roper (1988) did not find significantly less variation between sediment trays and ambient communities in relation to sediment contamination. If such artificial substrata are subjected to all the same sources of variation as natural substrata, and require more effort to use, then their use must be limited to experimental studies and exclude routine monitoring programmes.

In summary, artificial substrata of a wide variety of materials and forms have been used as an experimental technique in aquatic environments. The use of artificial substrata, particularly hard surfaces and plastic mesh pads, may have several advantages over standard sampling techniques in their collection, processing, statistical analysis, and amenability to experimental manipulation

Table 2. Comparison of species of amphipod Crustacea occurring on artificial substrata (plastic mesh pads) and natural substrata, the sponges (a) *Halichondria panicea* (Pallas) and (b) *Hymeniacidon perleve* (Montagu), in Lough Hyne, southwest Ireland.

	Artificial Substrata	Natural Substrata	
		(a)	(b)
ACANTHONOZOMATIDAE			
Iphimedia nexa Myers & McGrath	+	-	-
Iphimedia perplexa Myers & Costello	+	-	-
AMPHILOCHIDAE			
Gitana sarsi Boeck	+	+	+
AMPHITHOIDAE			
Ampithoe gammaroides (Bate)	+	+	+
Ampithoe helleri Karaman	-	+	+
Ampithoe ramondi Audouin	+	+	-
Ampithoe rubricata (Montagu)	+	+	+
Sunampithoe pelagica (Milne-Edwards)	-	+	-
AORIDAE			
Aora gracilis (Bate)	+	+	+
Aora spinicornis Afonso	+	+	+
Lembos websteri Bate	+	+	+
Microdeutopus anomalus (Rathke)	+	+	+
Microdeutopus versiculatus (Bate)	+	+	+
CALLIOPIIDAE			
Apherusa bispinosa (Bate)	I	-	+
Apherusa jurinei (Milne-Edwards)	+	-	-
COROPHIIDAE			
Corophium bonnellii (Milne-Edwards)	+	+	+
Corophium sextonae Crawford	+	+	+
DEXAMINIDAE			
Dexamine spinosa (Montagu)	+	+	+
Dexamine thea Boeck	+	+	+
GAMMARIDAE			
Gammarus locusta (Linnaeus)	+	-	-
ISAEIDAE			
Gammaropsis maculata (Johnston)	+	-	-
Microprotopus maculatus Norman	+	+	+
ISCHYROCERIDAE			
Ericthonius punctatus (Bate)	+	+	+
Jassa falcata (Montagu)	+	+	-
Parajassa pelagica (Leach)	+	-	-
Microjassa cumbrensis (Stebbing & Robertson)	+	-	-

Table 2. continued.

LEUCOTHOIDAE			
Leucothoe spinicarpa (Abildgaard)	-	+	+
LYSIANASSIDAE			
Lysianassa ceratina (Walker)	+	+	+
Orchomene nana (Kroyer)	+	-	-
Perrierella audouiniana (Bate)	-	+	+
MELITIDAE			
Abludomelita obtusata (Montagu)	+	+	+
Ceradocus semiserratus (Bate)	-	-	+
Cheirocratus sundevalli (Rathke)	+	+	+
Elasmopus rapax Costa	+	-	-
Gammarella fucicola (Leach)	+	+	+
OEDICEROTIDAE			
Perioculodes longimanus (Bate & Westwood)	-	+	+
PHOXOCEPHALIDAE			
Harpinia crenulata (Boeck)	-	+	+
Metaphoxus fultoni (Scott)	+	+	+
PODOCERIDAE			
Podocerus variegatus Leach	+	-	-
STENOTHOIDAE			
Stenothoe monoculoides (Montagu)	+	+	+
CAPRELLIDAE			
Caprella acanthifera Leach	+	+	+
Phthisca marina Slabber	+	+	+
Total number of samples	108	58	40
Total number of individuals	47104	20197	7780
Total number of species	35	30	29

(Table 3). A particularly valuable attribute of artificial substrata is their uniformity in structure and size, and hence the ability to experimentally replicate colonization on temporal and spatial scales (Table 3). However, although used as a standard technique in monitoring freshwater systems, artificial substrata remain to be fully evaluated as a sampling technique in marine pollution studies.

Analysis of colonization data

Species richness is a fundamental component of most measures of community structure and an important index of diversity (Magurran 1988). If species richness is to be used, it is necessary to know the time it takes until species equilibrium (sensu MacArthur and Wilson 1967) is reached. It is also important

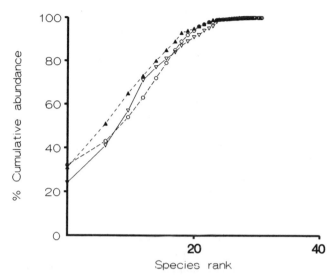

Figure 2. K-dominance curve of amphipod Crustacea on plastic mesh (o), and the sponges *Halichondria panicea* (▽) and *Hymeniacidon perleve* (▲) in L. Hyne, south-west Ireland.

Table 3. Advantages of artificial substrata in comparison to sampling natural substrata.

---

1. Replicability and reproducibility (standardized substratum type and sample size) in space and time
2. Low variation between samples (less replicates needed), because of (a) standard sample size and structure, and (b) biota in samples of same successional age
3. Rapid processing
4. Experimental manipulation (e.g. transfer, replacement, orientation)
5. No damage to natural substrata

---

to note potential sources of colonists, because colonization rate and species richness at equilibrium can be correlated with distance from the origin of colonists (e.g. Hockin and Ollason 1981; Gore 1982).

The simplest method of analysing colonization rate is to plot the number of species collected against time. If equilibrium is established, this colonization curve will reach an asymptote where the number of new species equals the number of species disappearing. Thus the composition of the community will continue to change (turnover) but the number of species will not. Due to variation between samples, there will be some variation in the measured number of species, the significance of which can be tested using a method described by Keough and Butler (1983).

Several studies have considered colonization rates in the establishment of the use of artificial substrata in freshwater monitoring programmes (e.g. Sheldon 1977; Voshell and Simmons 1977; Cover and Harrel 1978; Meier *et al.* 1979; Wise and Molles 1979). It is now recommended that the optimum time for retrieval of artificial substrata in North American rivers is six weeks after placement (American Public Health Association 1985). However, the limited information from marine studies suggests that two to four months are necessary for equilibrium to occur in temperate and sub-tropical areas (Osman 1982; Schoener 1982; Costello 1987).

The number of species at equilibrium can vary with season (Osman 1978; Hockin and Ollason 1981; Hockin 1982; Costello 1987). However, this effect on the colonization curve can be adjusted for by the calculation of turnover rate (after Brown and Kodrick-Brown 1977):

$$\text{Turnover rate} = \frac{\text{New species} + \text{Lost species}}{\text{Total number of species present in consecutive censuses}}$$

Hence seasonal effects on the number of species present are balanced by the same effects on colonization (Costello 1987). Species turnover is a largely unexplored process which may prove to be useful in understanding community dynamics. For instance, are there patterns in turnover rate for certain taxa, communities or habitats in the sea?

The relative abundance and biomass of species may also prove useful, allowing succession rate (Jassby and Goldman 1974; Lewis 1978), rank dominance and biomass (Warwick 1986; Warwick *et al.* 1987) and other parameters to be calculated. However, the use of such parameters is limited in seasonal marine environments where annual changes in abundance (and biomass) may not be distributed evenly between each species either within a year, between years, or between localities. Parameters based on species number (e.g. richness, turnover) may be less variable.

Although samples are from substrata of identical size, structure, and of the same successional age, caution is necessary when using the data to describe large scale patterns. However, it may be possible to calibrate for the effects of scaling by using artificial substrata of varying size. As for natural substrata, the sampling of artificial substrata should be scaled according to the objectives of the study (Underwood and Denley 1984; Livingston 1987).

Discussion

The process of colonization may be a sensitive and useful measure of environmental stress because,

(a) a variety of naturally co-existing phyla and species can be simultaneously assessed;

(b) immigrants are presented with the choice of whether or not to colonize;

(c) early life stages, whether immigrants or their progeny, will be an important component of the community;

(d) an early successional community may be more influenced by abiotic (e.g. substratum quality, pollution) than biotic (e.g. predation, competition) factors;

(e) it is an integrated response to natural processes occurring within the substratum and at the source of colonists.

Additionally, sampling at regular time intervals will enable temporal patterns in community structure to be analysed. Studies of island biogeographic and equilibrium theory indicate that predictable patterns do exist (e.g. Cairns *et al.* 1969; Dickson and Cairns 1972; Schoener 1974a; 1974b; Brown and Kodric-Brown 1977; Osman 1978; Hockin 1982; Costello 1987). Such patterns may elucidate factors controlling the rate and magnitude of change occurring in natural communities. Knowledge of the rate of colonization in different areas would provide a measure of the ability of an ecosystem to recover from disturbance. Quantitative data on the factors controlling the rate of change in natural communities is essential for prediction of the effects of anthropogenic disturbances.

Artificial substrata provide the most practicable means of measuring colonization. Furthermore, their use makes the destructive sampling of natural substrata and biota unnecessary, and they have the experimental advantages of temporal and spatial replication, reproducibility, and manipulation (Table 3). However, for in marine pollution studies the use of artificial substrata needs to be evaluated as has occurred in freshwater pollution monitoring. Thus colonization rate and time to equilibrium, and the factors influencing colonization (e.g. current, sedimentation, predation) need to be determined. Also, for monitoring programmes, the artificial substrata should be more cost effective to deploy, remove, analyse and interpret than are samples from natural substrata.

Sediment plugs are less susceptible to artifacts than are sediment trays. However, both methods are at least as time consuming as sampling natural sediments, are often subjected to the same sources of variation as natural sediments, and show less promise for routine monitoring at present. Plastic mesh substrata may not suffer from these problems to the same extent, and panels have the advantage that they can be non-destructively sampled (observed and photographed).

Acknowledgements

We would like to thank Julian Reynolds and Jim Wilson (Trinity College, Dublin) for their helpful criticism of this paper.

References

Ambrose, W.G. Jr. (1984) Increased emigration of the amphipod *Rhepoxynius abronius* (Barnard) and the polychaete *Nephtys caeca* (Fabricius) in the presence of invertebrate predators. *Journal of Experimental Marine Biology and Ecology,* **80,** 67-75.

Ambrose, W.G. Jr. (1986) Experimental analysis of density dependent emigration of the amphipod *Rhepoxynius abronius*. *Marine Behaviour and Physiology*, **12**, 209-216.

American Public Health Association, American Water Works Association, and Water Pollution Control Federation. (1985) *Standard Methods for the Examination of Water and Wastewater*. American Public Health Association, American Water Works Association, Water Pollution Control Federation.

Arntz, W.E. and Rumohr, H. (1982) An experimental study of macrobenthic colonization and succession, and the importance of seasonal variation in temperate latitudes. *Journal of Experimental Marine Biology and Ecology*, **64**, 17-45.

Arthur, J.W. and Horning, W.B.Jr. (1969) The use of artificial substrates in pollution surveys. *American Midland Naturalist*, **82**, 83-89.

Bayne, B.L., Clarke, K.R. and Gray, J.S. (eds.) (1988) Biological effects of pollutants: results of a practical workshop. *Marine Ecology - Progress Series*, **46**, 1-278.

Beak, T.W., Griffing, T.C. and Appleby, A.G. (1973) Use of artificial substrate samplers to assess water pollution. *Biological Methods for the Assessment of Water Quality*. ASTM STP 528, American Society for Testing and Materials, 227-241.

Benfield, E., Hendricks, A. and Cairns, J.Jr. (1974) Proficiencies of two artificial substrates in collecting stream macroinvertebrates. *Hydrobiologia*, **45**, 431-40.

Bonsdorff, E., Bakke, T. and Pedersen, A. (1990) Colonization of amphipods and polychaetes to sediments experimentally exposed to oil hydrocarbons. *Marine Pollution Bulletin*, **21**, 355-358.

Borowitzka, M.A., Larkum, A.W.D. and Borowitzka, L.J. (1978) A preliminary study of algal turf communities of a shallow coral reef lagoon using an artificial substratum. *Aquatic Botany*, **5**, 365-381.

Briggs, M.R. (1983) Algal sampling by use of artificial surfaces in Lough Neagh, Northern Ireland. *Irish Naturalists' Journal*, **21**, 151-155.

Brown, J.H. and Kodric-Brown, A. (1977) Turnover rates in insular biogeography: effect of immigration on extinction. *Ecology*, **58**, 445-449.

Brunswig, D., Arntz, W.E. and Rumohr, H. (1976) A tentative field experiment on population dynamics of macrobenthos in the Western Baltic. *Kieler Meeresforchungen*, **3**, 349-59.

Buss, L.W. (1986) Competition and community organization on hard surfaces in the sea. In J. Diamond and T.J. Case (eds.) *Community Ecology*. Harper and Row, New York.

Butman, C.A. (1987) Larval settlement of soft-sediment invertebrates: The spatial scales of pattern explained by active habitat selection and the emerging·role of hydrodynamical processes. *Oceanography and Marine Biology Annual Review*, **25**, 113-165.

Cairns, J.Jr., Dahlberg, M.L., Dickson, K.L., Smith, N. and Waller, W.T. (1969) The relationship of freshwater protozoan communities to the MacArthur-Wilson equilibrium model. *American Naturalist*, **103**, 439-454.

Cairns, J.Jr., Yongue, W.H.Jr. and Smith, N. (1976) The effects of substrate quality upon colonization by fresh-water protozoans. *Review of Biology,* **10,** 14-20.

Commito, J.A. (1982) Importance of predation by infaunal polychaetes in controlling the structure of a soft-bottom community in Maine, USA. *Marine Biology,* **68,** 77-82.

Connell, J.H. (1963) Territorial behaviour and dispersion in some marine invertebrates. *Research in Population Ecology,* **5,** 87-101.

Costello, M.J. (1987) *Studies on amphipod Crustacea in Lough Hyne, Ireland: colonization of artificial substrata, population dynamics, distribution and taxonomy.* Ph.D. thesis, National University of Ireland.

Cover, E.C. and Harrel, R.C. (1978) Sequences of colonization, diversity, biomass, and productivity of macroinvertebrates on artificial substrates in a freshwater canal. *Hydrobiologia,* **59,** 81-95.

Crossman, J.S. and Cairns, J.Jr. (1974) A comparative study between two different artificial substrate samplers and regular sampling techniques. *Hydrobiologia,* **44,** 517-522.

Dauer, D.M., Tourtelotte, G.H. and Ewing, R.M. (1982) Oyster shells and artificial worm tubes: the role of refuges in structuring benthic communities of the lower Chesapeake Bay. *Internatale Revue der Gesamten Hydrobiologie,* **67,** 661-677.

Dayton, P.K. (1971) Competition, disturbance and community organization: the provision and subsequent utilization of space in a rocky intertidal community. *Ecological Monographs,* **41,** 351-389.

Dean, T.A. (1981) Structural aspects of sessile invertebrates as organizing forces in an estuarine fouling community. *Journal of Experimental Marine Biology and Ecology,* **53,** 163-180.

DeWitt, T.H. (1987) Microhabitat selection and colonization rates of a benthic amphipod. *Marine Ecology - Progress Series,* **36,** 237-250.

DeWitt, T.H. and Levinton, J.S. (1985) Disturbance, emigration and refugia: How the mudsnail, *Ilyanassa obsoleta,* affects the distribution of an epifaunal amphipod, *Microdeutopus gryllotalpa. Journal of Experimental Marine Biology and Ecology,* **92,** 97-113.

Dickson, K.L. and Cairns, J.Jr. (1972) The relationship of freshwater macroinvertebrate communities collected by floating artificial substrates to the MacArthur-Wilson equilibrium model. *American Midland Naturalist,* **88,** 68-75.

Dickson, K.L., Cairns, J.Jr., and Arnold, J.C. (1971) An evaluation of a basket type artificial substrate for sampling macroinvertebrate organisms. *Transactions of the American Fisheries Society,* **100,** 553-559.

Fager, E.W. (1971) Pattern in the developement of a marine community. *Limnology and Oceanography,* **16,** 241-253.

Flannagan, J.F. and Rosenberg, D.M. (1982) Types of artificial substrates used for sampling freshwater benthic macroinvertebrates. In J.Jr. Cairns (ed.) *Artificial Substrates,* 237-266. Ann Arbor Scientific Publications Inc., Ann Arbor.

Flint, R.W., Duke, T.W. and Kalke, R.D. (1982) Benthos investigations: Sediment boxes or natural bottom? *Bulletin of Environmental Contamination and*

Toxicology, **28**, 257-265.

Frid, C.L.J. (1989) The role of recolonization processes in benthic communities, with special reference to the interpretation of predator-induced effects. *Journal of Marine Biology and Ecology*, **126**, 163-171.

Frithsen, J.B., Nacci, D., Oviatt, C., Strobel, C.J. and Walsh, R. (1989) Using single-species and whole ecosystem tests to characterise the toxicity of a sewage treatment plant effluent. In G.W. Suter II and M.A. Lewis (eds.) *Aquatic Toxicology and Environmental Fate*: 11th Vol., 231-250. ASTM STP 1007, American Society for Testing and Materials, Philadelphia.

Gallagher, E.D., Jumars. P.A. and Truebold, D.D. (1983) Facilitation of soft-bottom benthic succession by tube builders. *Ecology*, **64**, 1200-1216.

Ghelardi, R.J. (1960) *Structure and dynamics of the animal community found in* Macrocystis pyrifera *holdfasts.* Ph.D. thesis, University of California (La Jolla).

Giller, P.S. and Cambell, R.N.B. (1989) Colonization patterns of mayfly nymphs (Ephemeroptera) on implanted substrate trays of different size. *Hydrobiologia*, **178**, 59-71.

Glime, J.M. and Clemons, R.M. (1972) Species diversity of stream insects on *Fontinalis* spp. compared to diversity on artificial substrates. *Ecology*, **53**, 458-464.

Goodman, K.S. and Ralph, R. (1981) Animal fouling on the Forties platforms. In *Marine Fouling of Offshore Structures*, Society for Underwater Technology, London.

Gore, J.A. (1982) Benthic invertebrate colonization: source distance effects on community composition. *Hydrobiologia*, **94**, 183-193.

Goren, M. (1979) Succession of benthic community on artificial substratum at Elat (Red Sea). *Journal of Experimental Marine Biology and Ecology*, **38**, 19-40.

Hansen, D.J. and Tagatz, M.E. (1980) A laboratory test for assessing impacts of substances on developing communities of benthic estuarine organisms. In J.G. Eaton, P.R. Parrish and A.C. Hendricks (eds.) *Aquatic Toxicology*, 40-57. ASTM STP 707, American Society for Testing and Materials.

Harris, L.G. and Irons, K.P. (1982) Substrate angle and predation as in fouling community succession. In J. Jr. Cairns (ed.) *Artificial substrates*, 131-174. Ann Arbor Scientific Publications Inc., Ann Arbor.

Hockin, D.C. (1982) Experimental insular zoogeography: some tests of the equilibrium theory using meiobenthic harpacticoid copepods. *Journal of Biogeography*, **9**, 487-497.

Hockin, D.C. and Ollason, J.G. (1981) The colonization of artificially isolated volumes of intertidal estuarine sand by harpacticoid copepods. *Journal of Experimental Marine Biology and Ecology*, **53**, 9-29.

Hocutt, C.H., Dickson, K.L. and Masnik, M.T. (1976) Methodology developed for sampling macroinvertebrates by artificial substrates in the New River, Virginia. *Review of Biology*, **10**, 63-75.

Jackson, J.B.C. (1977) Habitat area, colonization, and developement of epibenthic community structure. In B.F. Keegan, P. O'Ceidigh and P.J.S. Boaden (eds.) *Biology of Benthic Organisms*. Pergamon Press, Oxford.

Jassby, A.D. and Goldman, C.B. (1974) A quantitative measure of succession rate and its application to the phytoplankton of lakes. *American Naturalist*, **108**, 688-693.

Jenson, K.T. (1985) The presence of the bivalve *Cerastoderma edule* effects on migration, survival and reproduction of the amphipod *Corophium volutator*. *Marine Ecology - Progress Series*, **25**, 269-277.

Karlson, R. (1978) Predation and space utilization patterns in a marine epifaunal community. *Journal of Experimental Marine Biology and Ecology*, **31**, 225-239.

Kay, A.M. and Butler, A.J. (1983) 'Stability' of the fouling communities on the pilings of two piers in South Australia. *Oecologia*, **56**, 70-78.

Keen, S.L. and Neill, W.E. (1980) Spatial relationships and some structuring processes in benthic intertidal animal communities. *Journal of Experimental Marine Biology and Ecology*, **45**, 139-155.

Kensler, C.B. and Crisp, D.J. (1965) The colonization of artificial crevices by marine invertebrates. *Journal of Animal Ecology*, **34**, 507-516.

Keough, M.J. and Butler, A.J. (1983) Temporal changes in species number in an assemblage of sessile marine invertebrates. *Journal of Biogeography*, **10**, 317-330.

Khalaf, G. and Tachet, H. (1977) La dynamique de colonization des substrats artificiels par les macroinvertebres d'un cours d'eau. *Annls Limnologie*, **13**, 169-190.

Levings, C.D. (1976) Basket traps for surveys of a gammarid amphipod, *Anisogammarus confervicolus* (Stimpson), at two British Columbia estuaries. *Journal of the Fisheries Research Board of Canada*, **33**, 2066-2069.

Lewis, W.M.Jr. (1978) Analysis of succession in a tropical phytoplankton community and a new measure of succession rate. *American Naturalist*, **112**, 401-414.

Livingston, R.J. (1987) Field sampling in estuaries: The relationship of scale to variability. *Estuaries*, **10**, 194-207.

MacArthur, R. and Wilson, E.O. (1967) *The Theory of Island Biogeography*. Princeton University Press, Princeton, New Jersey.

Macan, T.T. and Kitching, A. (1972) Some experiments with artificial substrata. *Verhandlungen der International Vereinigung für Limnologie. Limnologie*, **18**, 213-220.

Magurran, A.E. (1988) *Ecological Diversity and its Measurement*. London: Croom Helm.

Mason, J.C. (1976) Evaluating a substrate tray for sampling the invertebrate fauna of a small stream, with comment on general sampling problems. *Archive für Hydrobiology*, **78**, 51-70.

Meadows, P.S. (1964a) Experiments on substrate selection by *Corophium* species: films and bacteria on sand particles. *Journal of Experimental Biology*, **41**, 499-511.

Meadows, P.S. (1964b) Experiments on substrate selection by *Corophium volutator* (Pallas): depth selection and population density. *Journal of Experimental Biology*, **41**, 677-687.

Meier, P.G., Penrose, D.L. and Polak, L. (1979) The rate of colonization by macroinvertebrates on artificial substrate samplers. *Freshwater Biology,* **9,** 381-392.

Mook, D.H. (1981) Effects of disturbance and initial settlement on fouling community structure. *Ecology,* **62,** 522-526.

Myers, A.A. and Southgate, T. (1980) Artificial substrates as a means of monitoring rocky shore cryptofauna. *Journal of the Marine Biological Association of the United Kingdom,* **60,** 963-975.

Nebeker, A.V., Cairns, M.A., Gakstatter, J.H., Malueg, K.W., Schuytema, G.S. and Krawczyk, D.F. (1984) Biological methods for determining toxicity of contaminated freshwater sediments to invertebrates. *Environmental Toxicology and Chemistry,* **3,** 617-630.

Oaken, J.M., Oliver, J.S. and Flegal, A.R. (1984) Behavioral response of a phoxocephalid amphipod to organic enrichment and trace metals in sediment. *Marine Ecology - Progress Series,* **14,** 109-115.

Olafsson, E.B. and Persson, L-E. (1986) The interaction between *Nereis diversicolor* O.F.Müller and *Corophium volutator* Pallas as a structuring force in a shallow brackish sediment. *Journal of Experimental Marine Biology and Ecology,* **103,** 103-117.

Osman, R.W. (1977) The establishment and developement of a marine epifaunal community. *Ecological Monographs,* **47,** 37-63.

Osman, R.W. (1978) The influence of seasonality and stability on the species equilibrium. *Ecology,* **52,** 383-399.

Osman, R.W. (1982) Artificial substrates as ecological islands. In J.Jr. Cairns (ed.) *Artificial Substrates,* 71-115. Ann Arbor Scientific Publications Inc., Ann Arbor.

Oviatt, C.A., Keller, A.A., Sampou, P.A. and Beatty, L.L. (1986) Patterns of productivity during eutrophication: a mesocosm experiment. *Marine Ecology - Progress Series,* **28,** 69-80.

Oviatt, C.A., Quinn, J.G., Maughan, J.T., Ellis, J.T., Sullivan, B.K., Gearing, J.N., Gearing, P.J., Hunt, C.D., Sampou, P.A. and Latimer, J.S. (1987) Fate and effects of sewage sludge in the coastal marine environment: a mesocosm experiment. *Marine Ecology - Progress Series,* **41,** 187-203.

Pearson, T.H. and Rosenberg, R. (1978) Macrobenthic succession in relation to organic enrichment and pollution of the marine environment. *Oceanography and Marine Biology Annual Review,* **16,** 229-311.

Pontasch, K.W. and Cairns, J.Jr. (1989) Establishing and maintaining laboratory-based microcosms of riffle insect communities: their potential for multispecies toxicity tests. *Hydrobiologia,* **175,** 49-60.

Reynolds, J.D. and Hunter, C. (1985) Evaluation of the use of artificial substrates in sampling the invertebrate fauna of sewage fungus slimes in Irish rivers. *Verhandlungen der International Vereinigung für Limnologie. Limnologie,* **22,** 2239-2243.

Rhoads, D.C., Aller, R.C. and Goldhaber, M. (1977) The influence of colonizing benthos on physical properties and chemical diagenesis of the estuarine seafloor. In Coull, B.C. (ed.) *Ecology of Marine Benthos,* 113-138. University of South Carolina Press, South Carolina.

Rodhouse, P.G. and Burnell, G.M. (1979) *In situ* studies on the Scallop *Chlamys varia. Progress in Underwater Science,* **4,** 87-98.

Rosemarin, A. (1990) Land-based model ecosystems in the ecotoxicology of the Baltic Sea coast: pulp mill effluents as an example. In P.L. Chambers and C.M. Chambers (eds.) *Estuarine Ecotoxicology,* 11-18. Japaga, Ashford, Ireland.

Rosenberg, D.M. and Resh, V.H. (1982) The use of artificial substrates in the study of freshwater benthic macroinvertebrates. In J.Jr Cairns (ed.) *Artificial Substrates,* 175-236. Ann Arbor Scientific Publications Inc., Ann Arbor.

Ross, A. and Horsmann, P.V. (1988) *The use of Nuvan 500 EC in the salmon farming industry.* Marine Conservation Society, Ross-on-Wye, 24pp.

Russ, G.R. (1980) Effects of predation by fishes, competition, and structural complexity of the substratum on the establishment of a marine epifaunal community. *Journal of Experimental Marine Biology and Ecology,* **42,** 55-69.

Santos, S.L. and Simon, J.L. (1980) Marine soft-bottom community establishment following annual defaunation: larval or adult recruitment. *Marine Ecology - Progress Series,* **2,** 235-241.

Savidge, W.B. and Taghon, G.L. (1988) Passive and active components of colonization following two types of disturbance on an intertidal sandflat. *Journal of Experimental Marine Biology and Ecology,* **115,** 137-155.

Schmidt, G.H. (1982) Random and aggregative settlement in some sessile marine invertebrates. *Marine Ecology - Progress Series,* **9,** 97-100.

Schoener, A. (1974a) Colonization curves for planar marine islands. *Ecology,* **55,** 818-827.

Schoener, A. (1974b) Experimental zoogeography: colonization of marine mini-islands. *American Naturalist,* **108,** 715-738.

Schoener, A. (1982) Artificial substrates in marine environments. In J. Jr. Cairns (ed.) *Artificial Substrates,* 1-22. Ann Arbor Scientific Publications Inc., Ann Arbor.

Schoener, A., Long, E.R. and DePalma, J.R. (1978) Geographic variation in artificial island colonization curves. *Ecology,* **59,** 367-382.

Sheldon, A.L. (1977) Colonization curves: application to stream insects on semi-natural substrates. *Oikos,* **28,** 256-261.

Shaw, D.W. and Minshall, G.W. (1980) Colonization of an introduced substrate by stream macroinvertebrates. *Oikos,* **34,** 259-271.

Shin, P. (1981) The developement of sessile epifaunal communities in Kylesalia, Kilkieran Bay (west coast of Ireland). *Journal of Experimental Marine Biology and Ecology,* **54,** 97-111.

Smith, C.R. and Brumsickle, S.J. (1989) The effect of patch size and substrate isolation on colonization modes and rate in an intertidal sediment. *Limnology and Oceanography,* **34,** 1263-1277.

Southgate, T. and Myers, A.A. (1985) Mussel fouling on the Celtic Sea Kinsale Field Gas Platforms. *Estuarine and Coastal Shelf Science,* **20,** 651-659.

Stoner, A.W. (1980) Perception and choice of substratum by epifaunal amphipods associated with sea-grasses. *Marine Ecology - Progress Series,* **3,** 105-111.

Sutherland, J.P. (1974) Multiple stable points in natural communities. *American Naturalist,* **108,** 859-873.

Sutherland, J.P. (1981) The fouling community at Beaufort, North Carolina: a study in stability. *American Naturalist,* **118,** 499-519.

Thrush, S.F. (1986a) The sublittoral macrobenthic community structure of an Irish sea-lough: Effect of decomposing accumulations of seaweed. *Journal of Experimental Marine Biology and Ecology,* **96,** 199-212.

Thrush, S.F. (1986b) Spatial heterogeneity in subtidal gravel generated by the pit-digging activities of *Cancer pagurus. Marine Ecology - Progress Series,* **30,** 221-227.

Thrush, S.F. and Roper, D.S. (1988) Merits of macrofaunal colonization of intertidal mudflats for pollution monitoring: a preliminary study. *Journal of Experimental Marine Biology and Ecology,* **116,** 219-233.

Twomey, E. (1977) *Studies on the biology of* Corophium volutator *(Pallas).* M.Sc. thesis, National University of Ireland.

Voshell, J.R.Jr. and Simmons, G.M.Jr. (1977) An evaluation of artificial substrates for sampling macrobenthos in reservoirs. *Hydrobiologia,* **53,** 257-269.

Underwood, A.J. and Denley, E.J. (1984) Paradigms, explanations and generalizations in models for the structure of intertidal communities on rocky shores. In D.R. Strong Jr., D. Simberloff, L.G. Abele and A.B. Thistle (eds.) *Ecological Communities Conceptual Issues and the Evidence,* 151-180. Princeton University Press, Princeton.

Underwood, A.J. and Peterson, C.H. (1988) Towards an ecological framework for investigating pollution. *Marine Ecology - Progress Series,* **46,** 227-234.

Warwick, R.M. (1986) A new method for detecting pollution effects on marine macrobenthic communities. *Marine Biology,* **92,** 557-562.

Warick, R.M., Pearson, T.H. and Ruswahunyi (1987) Detection of pollutant effects in marine macrobenthos: further evaluation of the species abundance/biomass method. *Marine Biology,* **95,** 193-200.

Weis, J.S. and Weis, P. (1989) Effects of environmental pollutants on early fish development. *Review Aquatic Sciences,* **1,** 45-73.

Whitlatch, R.B. and Zajac, R.N. (1985) Biotic interactions among estuarine infaunal opportunistic species. *Marine Ecology - Progress Series,* **21,** 299-311.

Wise, D.H. and Molles, M.C. Jr. (1979) Colonization of artificial substrates by stream insects: influence of substrate size and diversity. *Hydrobiologia,* **65,** 69-74.

Withers, R.G. and Thorp, C.H. (1977) Studies on the shallow sublittoral epibenthos of Langstone Harbour, Hampshire, using settlement panels. In B.F. Keegan, P. O'Ceidigh and P.J.C. Boaden (eds.) *Biology of Benthic Organisms,* Pergamon Press, Oxford.

Woodin, S.A. (1986) Settlement of infauna: Larval choice? *Bulletin of Marine Science,* **39,** 401-407.

Zajac, R.N. and R.B. Whitlach, (1982a) Responses of estuarine infauna to disturbance. I. Spatial and temporal variation of initial recolonization. *Marine Ecology - Progress Series,* **10,** 1-14.

Zajac, R.N. and R.B. Whitlach, (1982b) Responses of estuarine infauna to disturbance. II. Spatial and temporal variation in succession. *Marine Ecology - Progress Series,* **10,** 15-27.

Evidence for Glutathione-S-Transferase Activity in *Mytilus edulis* as an Index of Chemical Pollution in Marine Estuaries

David Sheenan, Kay M. Crimmins and Gavin M. Burnell[1]

Departments of Biochemistry and [1]Zoology, University of Cork, Lee Maltings, Prospect Row, Cork, Ireland.

Key words: Glutathione-S-Transferase, xenobiotic, detoxification, mussel, induction, pollution, Cork Harbour, Bantry Bay.

Abstract

Glutathione-S-Transferase (GST) activity has been identified in the marine mussel *Mytilus edulis*. The activity is present in multiple forms which may be separated by ion exchange chromatography. The level of the enzyme appears to be unaffected by body weight but is dependent on growth habit and possibly on the pollution status of the habitat. We have found that the activity is significantly elevated in mussels taken from Cork Harbour as compared to those from Bantry Bay or those from Cork Harbour which have been subjected to depuration. GST activity in *M. edulis* may provide a useful additional index of water quality in estuarine environments.

Introduction

Among the factors likely to affect the pollution status of estuaries are agricultural residues, port activities, untreated domestic sewage and industrial effluent from chemical, pharmaceutical and other factories. Biochemical Oxygen Demand (BOD) measurements provide a useful index of total organic pollution but testing for low-level chemical pollutants such as Polychlorinated Biphenyls (PCB's), organochlorines and heavy metals is not as routine. Such chemicals may be concentrated in marine food-chains and have an environmental impact even at quite low concentrations (Moriarty 1990).

For this reason, the levels of xenobiotics have been measured in species such as *Mytilus edulis*, a marine mussel common in western European coastal waters (Jones *et al.* 1972). This species also has a commercial significance, being grown on the intertidal shore ("shore-grown") or on ropes ("rope-grown") in shellfish farms. Because of its sessile, filter-feeding lifestyle, this organism has been suggested as a bioindicator of pollution status (Bayne 1978). More recently,

Bioindicators and Environmental Management
ISBN 0-12-382590-3

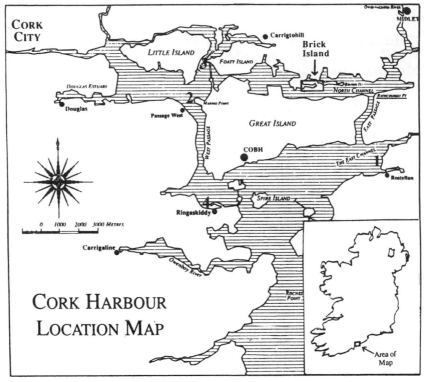

Figure 1. Environs of Cork Harbour. Sampling sites are donoted by 1 - 4.

enzyme levels in *M. edulis* have been suggested as subtle indicators of pollution (Goromosova *et al.* 1987). Livingstone (1988) has found 'Phase I' detoxification enzymes to be elevated in this and other species in response to xenobiotics, although this response is found to be somewhat variable (Livingstone 1990; Livingstone *et al.* 1989).

In more advanced organisms, the most quantitatively important 'Phase II' (i.e. conjugation) activity is that of the Glutathione-S-Transferases (GST's; E.C. 2.5.1.18), which are reviewed by Mannervik (1985) and Pickett and Lu (1989). This enzyme is widespread in animal, plant, fungal and bacterial sources. It is a dimeric, cytosolic protein which exists in multiple forms. The subunits are products of several genes and may be induced by a range of compounds some of which are also substrates (Ding *et al.* 1986).

We have studied the levels of this enzyme in *M. edulis* from a number of locations within Cork Harbour (Fig. 1). Using Bantry Bay as a comparative location, we have found the levels of *M. edulis* GST's in Cork Harbour to be significantly elevated. Our results are discussed in terms of the possible use of this enzyme as a further index of estuary pollution status with particular reference to xenobiotics, since these are the compounds which most often induce these detoxification proteins.

Experimental

Enzyme Assays: GST activity was measured spectrophotometrically using 1-chloro-2,4-dinitrobenzene (CDNB) and glutathione (GSH) as substrates (Habig *et al.* 1974) in a Kontron Uvikon 940 spectrophotometer. A unit (U) of activity is defined as that catalysing the formation of 1 nMole product/minute. Protein concentrations were estimated by the method of Hartree (1972). Specific activity is expressed as U mg^{-1} protein.

Chromatography: Ion exchange chromatography was performed in 10mM sodium phosphate, pH 6.7, on a column (11 x 4 cm) of CM-52 cellulose. The column was developed with a linear 0-75 mM KCl gradient.

Sampling: 5-10 individual mussels in winter condition (Bayne 1978) were harvested (October 1989 - February 1990). Soft tissue was homogenized in 0.35 M sucrose, 1mM EDTA, 10mM Na_2CO_3, 10 mM sodium phosphate, pH 7.2, centrifuged (20,000 X g for 30 mins) and desalted (sephadex G25) into 10 mM sodium phosphate, pH 7.2. Enzyme activity, protein concentration and specific activity were determined for each individual in a sample as described above.

Results

Mytilus edulis was found to possess GST activity. Ion exchange chromatography of mussel GST's is shown in Figure 2. This separated at least three distinct forms of the enzyme. This indicates that, in common with rat (Jakoby *et al.* 1976), human (Mannervik 1985) and other species, GST's exist in multiple isoenzyme forms in this marine mussel.

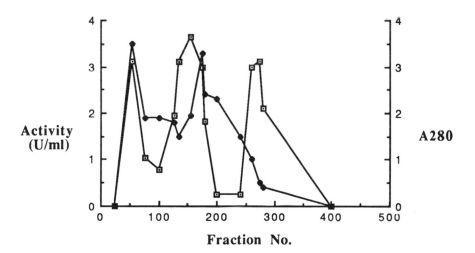

Figure 2. CM-cellulose chromatography of *M. edulis* GST's.

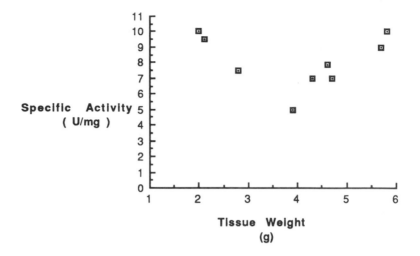

Figure 3. GST specific activity versus *M. edulis* soft tissue weight.

The possibility exists that this enzyme might be sensitive to a number of environmental variables. We therefore studied the effects of body-weight, growth habit and location on GST activity in these animals.

Figure 3 shows that, for shore-grown mussels taken from Cork Harbour in the weight range 2 - 5.8 g, GST specific activity varied over a range 5 - 10 U mg^{-1} in a non-correlated relationship (r=0.17). GST expression is therefore independent of this variable.

Our studies on the effect of growth habit concentrated on mussels taken from Bantry Bay (see Table 1). Rope-grown mussels displayed no detectable activity (n=10) while shore-grown mussels taken from nearby at the same time had 1.72 - 3.8 U mg^{-1} specific activity (n=9). There is a stark difference therefore, between the two growth habits in terms of GST expression.

The levels of GST activity were also determined in shore-grown mussels taken from a number of locations in Cork Harbour (Fig. 1). These results are also summarized in Table 1 together with data for depurated mussels from a commercial shellfish farm located in this estuary. The mean specific activity values for Bantry Bay and depurated mussels were the lowest measured and were not statistically different from each other. The values from the other locations in Cork Harbour, however, were all significantly elevated.

Discussion

GST's are the major component of 'Phase II' of detoxification (Mannervik 1985). The multiple isoenzymes of GST may be induced by a broad range of xenobiotics many of which are pesticides, herbicides, carcinogens and other potentially genotoxic components (Sheehan *et al.* 1984; McLellan and Hayes 1989).

Table 1. Levels of GST activity in mussels *Mytilus edulis* taken from Bantry Bay and Cork Harbour. GST specific activity is expressed as U mg^{-1}.

Location	n	Range (U mg^{-1})	x $\pm$ S.D. (U mg^{-1})
Bantry Bay			
Shore-grown	10	1.30-5.4	2.67 $\pm$ 0.35[*]
Shore-grown#	9	1.72-3.8	2.96 $\pm$ 0.79[*]
Rope-grown#	10	0	0
Cork Harbour			
Depurated	6	2.4-3.2	2.71 $\pm$ 0.82
1	5	4.4-6.2	4.92 $\pm$ 0.81[**]
2	8	4.4-9.7	6.34 $\pm$ 2.1[**]
3	8	4.0-14.6	7.93 $\pm$ 4.5[**]
4	8	5.2-12.2	8.04 $\pm$ 2.2[**]

[*]   Not significantly different to depurated sample
[**]   Significantly different to depurated sample
Harvested at same time for growth habit comparison

Such a sensitive response to chemical exposure is a common feature of detoxification proteins. However, little is known about the pathways of detoxification in lower life-forms such as molluscs. Livingstone *et al*. (1989) have studied the 'Phase I' enzymes of a number of species. These workers have found that these activities may be affected by exposure to xenobiotics as well as seasonal factors.

We have studied GST expression in *M. edulis* in order to assess its feasibility as an index of chemical pollution. *M. edulis* possesses GST activity which is expressed in several isoenzymes. The independence of the expression of this activity from body weight indicates that the enzyme is expressed at constant levels throughout the animal's life-cycle. However, we have identified two other variables which appear to affect the level of the enzyme.

More enzyme appears to be expressed in shore-grown than in rope-grown mussels. This may indicate that animals grown in contact with silt in estuarine environments are exposed to higher levels of xenobiotics than those grown on ropes. It is interesting that, while GST's display a very broad range of specifity to their second substrate, many of these molecules are hydrophobic in nature (Chasseaud 1979). Thus, it is possible that such molecules might be present in localized high concentrations in the silt of a marine estuary (Langstom *et al*. 1987).

The immediate environment of the mussel also affects its GST level. All the likely sources of pollution mentioned earlier are quantitatively more important in Cork Harbour - a major population and chemical industry centre - than in Bantry Bay (which may be regarded as relatively unpolluted). GST levels in *M. edulis* taken from four well-separated sites in Cork Harbour were elevated when compared to those of animals from Bantry Bay. This difference might be due to population heterogeneity. To test this, GST levels of commercially depurated *M. edulis* grown in Cork Harbour were also analysed. Our results indicate that depurated mussels taken from Cork Harbour and shore-grown mussels taken from Bantry Bay appear to express the same levels of GST, while animals shore-grown in Cork Harbour express elevated levels.

These findings are consistent with a 2 to 3.5-fold induction of GST's in mussels from Cork Harbour when compared to 'baseline' levels. Such induction of detoxification enzymes has been reported in mammals (Lechner *et al.* 1987), newts (Santagostino *et al.* 1989) and some plant species. Our study suggests, therefore, that GST levels in *M. edulis* may provide an additional criterion of water quality in estuaries.

References

Bayne, B.L. (1978) Mussel watching. *Nature* (London), **275**, 87-88.

Chasseaud, L.F. (1979) Compounds that conjugate with glutathione. *Advances in Cancer Research*, **29**, 200-255.

Ding, G.J.F., Ding, V.D.H., Rodkey, J.A., Bennett, C.D., Lu, A.Y.H. and Pickett, C.B. (1986) Rat liver Glutathione-S-Transferases. *Journal of Biological Chemistry*, **261**, 7952-7957.

Goromosova, S.A., Milovidiva, N.Y., Tamazhnyayu, U.A. and Shapiro, A.Z. (1987) Some ecological and biochemical indices of pollution tolerence in molluscs. *Hydrobiological Journal*, **23**, 65-69.

Habig, W.H., Pabst, M.J. and Jakoby, W.B. (1974) Glutathione-S-transferases: the first enzymatic step in mercapturic acid formation. *Journal of Biological Chemistry*, **249**, 7130-7139.

Hartree, E.F. (1972) Determination of protein: A modification of the Lowry method that gives a linear photometric response. *Analytical Biochemistry*, **48**, 422-427.

Jakoby, W.B., Habig, W.H., Keen, J.H., Ketley, J.N. and Pabst, M.J. (1976) In I.M. Arias and W.B. Jakoby (eds.) *Glutathione: Metabolism and Function*, 189-211. Raven Press, New York.

Jones, A.M., Jones, Y. and Steward W.P. (1972) Mercury in marine organisms of the Tay region. *Nature* (London), **238**, 164-165.

Langston, W.J., Burt, G.R. and Mingjiang, Z. (1987). Tin and organotin in water sediments and benthic organisms in Poole Harbour. *Marine Pollution Bulletin*, **18**, 634-639.

Lechner, M.C., Sinogas, C., Osorio-Almeida, M.L., Freire, M.T., Chaumet-Riffaud, P., Frain, M. and Sala-Trepat, J.M. (1987) Phenobarbital-mediated modulation of gene expression in rat liver. *European Journal of Biochemistry*, **163**, 231-238.

Livingstone, D.R. (1988) Responses of microsomal NADPH-cytochrome c reductase activity and cytochrome P-450 in digestive glands of *M. edulis* and *L. littorea* to environmental and experimental exposure to pollutants. *Marine*

Ecology - Progress Series, **46**, 37-43.

Livingstone, D.R. (1990) Cytochrome P-450 and oxidative metabolism in invertebrates. *Biochemical Society Transactions*, **18**, 15-19.

Livingstone, D.R., Kirchin, M.A. and Wiseman, A. (1989) Cytochrome P-450 and oxidative metabolism in molluscs. *Xenobiotica*, **19**, 1041-1062.

McLellan, L.I. and Hayes, J.D. (1989) Differential induction of class alpha glutathione-S-transferases in mouse liver by the anticarcinogenic antioxidant butylated hydroxyanisole. *Biochemical Journal*, **263**, 393-402.

Mannervik, B. (1985) The isoenzymes of glutathione transferase. *Advances in Enzymology*, **57**, 357-417.

Moriarty, F. (1990) Case studies. *Ecotoxicology: The Study of Pollutants in Ecosystems*, 2nd edition, 213-236. Academic Press, London.

Pickett, C.B. and Lu, A.Y.H. (1989) Glutathione-S-transferases: gene structure, regulation and biological function. *Annual Review of Biochemistry*, **58**, 743-764.

Santagostine, A., Colleoni, M., Arias, E., Zaffaroni, N.P. and Zavenella, T. (1989) Changes in catalase, glutathione peroxidase and glutathione-S-transferase activities in the liver of newts exposed to 2-methyl-4-chlorophenoxyacetic acid (MCPA). *Pharmacology and Toxicology*, **65**, 136-139.

Sheehan, D., Ryle, C.M. and Mantle, T.J. (1984) Selective induction of glutathione S-transferase D in rat testis by phenobarbital. *Biochemical Journal*, **219**, 687-688.

Electro-reception and Aquatic Biomonitoring

R.C. Peters, F. Bretschneider, W.J.G. Loos and I.S.A. Neuman

University of Utrecht, Laboratory of Comparative Physiology, Padualaan 8 NL-3584 CH Utrecht, The Netherlands.

Key words: electro-receptors, fish behaviour, pollutants, *Ictalurus nebulosus*.

Abstract

Ampullary electro-receptor organs of freshwater fish are cutaneous sensory organs, the receptor cells of which are directly exposed to the aquatic environment. Toxicants reach the electro-receptor cells within seconds. It has been demonstrated electrophysiologically that toxicants administered from the outside affect the sensory transduction in the receptor cells. This leads to loss of sensory information and as a consequence to behavioural disturbances.

To quantify the impact of aquatic pollution on sensory-motor performance a psychophysical experiment was designed in which electrosensitive catfish, *Ictalurus nebulosus*, were trained to locate a food dispenser in a controllable electric field. By decreasing the field strength after correct performances, sensitivity thresholds were determined. The threshold depends on the water quality. Traces of copper and cadmium increase the threshold considerably.

The experiment demonstrates that the ensemble of pollutants, temperature, and other factors influence sensory-motor performance. The sensitivity threshold gives indirectly a measure for the description of animal well-being.

Introduction

Survival of aquatic organisms is dependent on the quality of the habitat; water for aquatic animals. The habitat provides the means for feeding, sheltering, transportation, and reproduction. Variations in quality of the habitat are met by the adaptability of the organism. If the variations exceed critical limits, life becomes impossible. Survival limits for aquatic organisms are set by the properties of epithelia, the boundaries between the inner organism and the outer world. Organisms of all sizes, whether unicellular or multicellular, have to maintain electrochemical homeostasis and thus organismal integrity.

Electrochemical homeostasis is controlled by ionic channels and ion pumps.

Bioindicators and Environmental Management
ISBN 0-12-382590-3

These building blocks are located in specialized parts of membranes (unicellular organisms) or in specialized epithelia (multicellular organisms). Polluted habitats interfere with proper functioning of such channels and pumps, and thus make organismal survival impossible. Epithelial transport cells with apical microvilli are fast and sensitive monitors of disfunctioning membranes, because they are beset with channels and pumps. Electro-receptor cells can be considered as transport cells where the activation of the channels and the activity of the pumps is monitored by nerve fibres. The functioning of such electro-receptor cells is dependent on the quality of the aquatic environment, and is reflected in behaviour, via the nervous system.

The freshwater teleost *Ictalurus nebulosus* is sensitive to electrical fields (Parker and van Heusen 1917; Dijkgraaf 1968). Its electro-receptive abilities are used to localize prey or conspecifics (Peters and Meek 1973) and to navigate in aquatic electric fields (Peters and van Wijland 1974; Kalmijn *et al.* 1976). Catfish sense electric fields by means of cutaneous electro-receptor organs (Dijkgraaf 1968). The sensitivity of these organs is affected by exposure to mono- and divalent cations like cadmium, calcium, potassium and sodium (Roth 1971; Peters *et al.* 1989; Zwart 1988). Cadmium has been shown even to accumulate specifically in electro-receptor cells (Zwart 1988).

In order to investigate the possibility to quantify the effects of pollution on electro-receptor functioning a psychophysical experiment was designed in which catfish were trained to perform a forced two-choice electro-orientation task. This gave us the possibility to determine the lower limit of electrosensitivity of catfish in waters with varying degrees of pollution, and to infer how the behaviour is affected.

Materials and methods

Before the experiments catfish (15 to 20 cm, 80 to 100 g) are kept in 300 l aquaria containing Utrecht copper-free tap-water of ambient temperature. Main composition is in mmol l^{-1}: K$^+$ 0.02, Na$^+$ 0.57, Ca^{2+} 1.1, Mg^{2+} 0.16; pH 7.6; conductivity 23-34 mS/m (data: Waterleidingbedrijf Midden Nederland, 1988). The fish are subjected to a 12h light, 12h dark regime, and tested during the dark period when they are most active.

Testing takes place in a glass aquarium 30 cm x 90 cm filled with water up to 8 cm (Fig. 1). A strip of PVC, of approximately the same size as the fish, is fixed to one wall to provide shelter. Above this shelter a 9V bulb indicates when testing takes place. Plastic rods are mounted 18 cm to the left and right of the shelter across the floor of the aquarium to mark the position detection. The position of the fish is detected by infra-red detectors (Keyence PZ 42L, Osaka, Japan). The stimulus is delivered by a pair of silver strip electrodes at 10 cm behind the plastic rods. Sinusoidal electrical stimuli are generated by a function generator (Trio FG271, Tokyo, Japan) and attenuated by a 12 bit D/A A/D card (Rhothron, Homburg, FRG) switched as multiplier. The stimulus is split and fed into two isolation amplifiers and subsequently into two voltage-to-current

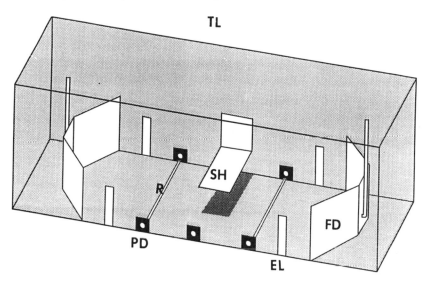

Figure 1. Diagram of the experimental tank. Overall dimensions 90, 30, 8 cm (l,w,h). TL top light; PD photodetector; R plastic rod; EL electrode; FD food dispenser.

converters connected to the silver strip electrodes. Because of the isolation amplifiers the electrode pairs behave as two independent floating dipoles. The field strength is measured at the plastic rods over the entire width of the aquarium with a cu-stom-built differential amplifier with silver chloride measuring electrodes. Peristaltic pump food dispensers are positioned at both ends of the tank. A modified Atari ST computer controls the experiment. Conductances are measured with a conductance meter (Dist 3, Hanna instruments, Woonsocket, USA).

Catfish can be trained to stay under the shelter while the top light is on. If the fish stays there for 2 seconds the light is dimmed and an electrical stimulus of 1 Hz, the optimum behavioural frequency (Peters *et al.* 1988), is given at either side of the aquarium. If the fish crosses the plastic rod at the correct side of the aquarium the stimulus is cut off and a piece of food delivered. Food consists of a mixture of agar-agar, gelatine and beef. After 20 seconds the top light is switched on again for the next trial. After an incorrect choice the top light comes on immediately. The stimulus amplitude for the next trial is always attenuated by 1dB following upon a correct choice; on an error the amplitude is increased by 3dB, in order to prevent demotivation. In this way the animal eventually reveals its lower detection limit. The pretraining of catfish does not require particular skills. If the fish gets the opportunity to accommodate over the weekend to the new environment after having been introduced into the test compartment the conditioning can begin. It turns out that a catfish orients itself involuntarily to the source of the electric dipole, and autoshapes itself. After a week's training real testing can begin.

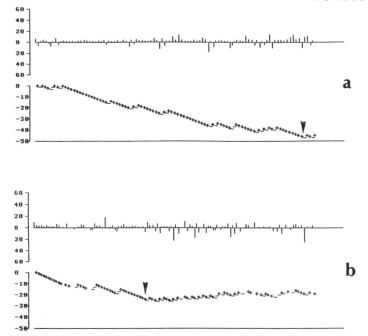

Figure 2. a) Reaction time and detection threshold of the electrosensitive catfish *Ictalurus nebulosus* in copper-free tap-water during an orientation task. Duration of the test is about 4 hours. The upper trace represents the reaction time in seconds before a correct response (up) and an incorrect response (down). The lower trace represents the stimulus strength. The Y-axis is calibrated in dB. O dB corresponds to 500 μV cm^{-1}. After correct choices the stimulus strength is decreased, after incorrect choices increased. This procedure results in the animal revealing its detection threshold. In this test medium the threshold is about 5 μV cm^{-1} (at the arrow). Specific conductivity 29 mS/m, temperature *c.* 17° C. Less than 1 ppb copper.
b) As a. The medium tested here is water from the Amsterdam-Rijn canal. The detection threshold is 50 μV cm^{-1} (at the arrow). Specific conductivity 66 mS/m; temperature *c.* 17 °C.

Toxicants are applied by replacing the standard water by test water from a separate buffer tank of 300 l. All test solutions are prepared in the buffer tank. Water samples for testing the concentration are taken from the test compartment at the end of a session. In the example presented here (Fig. 3), the cadmium concentration of water samples from the experimental tank was determined at the end of 48 hr exposure periods by means of atomic absorption spectrophotometry.

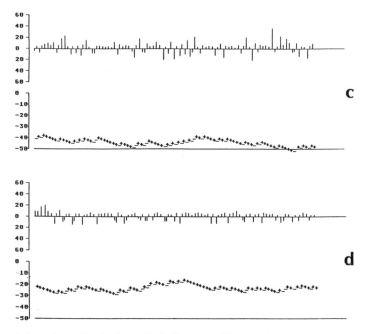

Figure 2 (continued). c) As a. Stability test. The medium is copper-free tap-water. In this session the stimulus strength remains at threshold level throughout the 4 hours lasting session. Threshold about 5 µV cm⁻¹. Specific conductivity 26 mS/m; temperature *c.* 17 °C. Less than 1 ppb copper.

d) As in a. Stability test. The medium tested is standard tap-water, which contains 16 ppb copper. The detection threshold is raised to about 50 µV cm⁻¹. Note that small traces of copper have the same effect as contaminated canal water (Fig. 2b).

Results

In this paper some examples are presented that show the kind of data that can be collected through this experiment. In Figure 2 and 3 results are given of sessions run under various conditions.

The first example shows reaction time and stimulus detectability during a test in tap-water from which copper ions had been removed. In the beginning the fish makes rapid choices, usually within 5 seconds, but near the threshold the number of mistakes increases as well as the reaction time. The experiment begins with stimulus strengths of 500 µV cm⁻¹; the threshold (arrow in Fig. 2a) is reached at about 5 µV cm⁻¹. This example is typical for an ordinary experiment, although better scores in the suprathreshold region were seen.

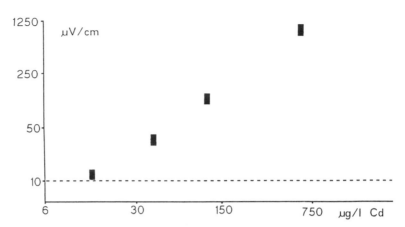

Figure 3. Behavioural electro-receptive thresholds of the catfish *Ictalurus nebulosus* after 48 hours of exposure to the indicated concentrations of cadmium.

The second example (Fig. 2b) shows the results of a threshold experiment with the same fish in water from the Amsterdam-Rijn canal. Again the stimulus strength is 500 μV cm^{-1} at the beginning of the experiment, but the threshold lies at about 50 μV cm^{-1}. The shift in threshold is partly due to the higher electrical conductivity, but also to the ionic composition of the water. The electro-receptive capacities of the catfish are impaired in this environment, and the contribution of the electric sense in locating prey is reduced.

The third example (Fig. 2c) shows the outcome of a session where the fish begins at its detection threshold, about 5 μV cm^{-1}, and stays at this level for almost 4 hours.

The fourth example shows the course of the threshold in regular tap-water which contains 16 ppb copper. The experiment begins at the threshold level, which has been reached in a previous session, and stays in the threshold zone for 4 hours.

The fifth example (Fig. 3) shows the outcomings of an experiment where the threshold is determined after 48 hours exposure to copper-free tap-water with various concentrations of cadmium chloride.

Discussion and conclusions

The electrophysiological experiments of Roth (1971) and Peters *et al.* (1989) now have a psychophysical follow up in the experiments described above. Although the experiment represents only the beginning of the development of a real biomonitor system for surface water, it is unmistakable that the quality of water can be characterized through the value of the detection threshold of the electric sense.

The results are promising. The test has been operational for more than one year without any major difficulty. In all 8 fish were tested. The test gives stable results over a period of more than 6 months. Various parameters influence the sensitivity. Among these are temperature, electrical conductivity, and the ionic composition of the water. The effect of the very toxicant can be determined after correction for temperature and conductivity. On the other hand it is an advantage that electro-receptors are simultaneously susceptible to more than one environmental factor. In real life the impact of a polluted environment on an organism depends on all factors. Temperature is a factor that modulates the effects of toxicants; and electrical conductivity represents the amount of dissolved ions, and thus for the load on the electrochemical balance.

In principle the system can operate without personal attendance. The only restriction in our set-up is the freshness of the food. If the set-up has to be operated for more than a couple of days without inspection, other feeding systems have to be installed.

In summary, electro-receptors present a promising and biologically relevant system for monitoring the quality of freshwater.

References

Dijkgraaf, S. (1968) Electroreception in the catfish, *Amiurus nebulosus*. *Experientia* (Basel), **24**, 187-188

Kalmijn, A.J., Kolba, C.A. and Kalmijn, V. (1976) Orientation of catfish (*Ictalurus nebulosus*) in strictly uniform electric fields: I. Sensitivity of response. *Biological Bulletin*, **151**, 415.

Parker, G.H. and van Heusen, A.P. (1917) The response of the catfish *Amiurus nobuloous* to metallic and non-metallic rods. *American Journal of Physiology*, **44**, 405-420.

Peters, R.C. and Meek, J. (1973) Catfish and electric fields. *Experientia* (Basel), **29**, 299-300.

Peters, R.C. and van Wijland, F. (1974) Electro-orientation in the passive electric catfish, *Ictalurus nebulosus*. *Journal of Comparative Physiology*, **92**, 273-280.

Peters, R.C., Evers, H.P. and Vos, J.J. (1988) Tuning mismatch between peripheral and central neurons reflects learning and adaptability. *Advances in Bio-Sciences*, **70**, 141-142.

Peters, R.C., Zwart, R., Loos, W.J.G. and Bretschneider, F. (1989) Transduction at electro-receptor cells manipulated by exposure of apical membranes to ionic channel blockers. *Comparative Biochemistry and Physiology*, **94C**, 663-669.

Roth, A. (1971) Zur Funktionsweise der Electrorezeptoren in der Haut von Welsen (*Ictalurus*): Der Eifluss der Ionen im Sßüßsswasser. *Zeitschrift fuer Vergleicheride Physiologie*, **75**, 303-323.

Zwart, R. (1988) Electrophysiological changes and histochemical demonstration of intracellular cadmium in catfish electro-receptors after exposure to cadmium. *Netherlands Journal of Zoology*, **38**, 215.

Distribution of the Otter *Lutra lutra* in Ireland, and Its Value as an Indicator of Habitat Quality

R.M. Lunnon and J.D. Reynolds

Department of Zoology, Trinity College, Dublin 2, Ireland.

Key words: *Lutra lutra*, distribution, river survey, habitat quality, water pollution bioindicator, Ireland.

Abstract

Otter populations in most parts of Western Europe are known to have experienced a sharp decline during the second half of this century. However, a survey carried out in 1980/81 indicated that Ireland still had a widely distributed population of otters. In this study a stratified sub-sample of the original sites surveyed in 1980/81 were searched for otter signs. The majority of these were in riverine habitats. Positive signs were found in 86% of the locations surveyed. A statistically significant relationship was found between the presence or absence of otters and two habitat quality factors, namely water pollution and bankside vegetation levels. It is suggested, therefore, that monitoring the distribution of otters over a period of time may be a useful way of indicating change in overall habitat quality and, in particular, water quality.

Introduction

The abundance of the otter (*Lutra lutra*) has been diminishing in Western Europe since the Industrial Revolution. In 1955, the organochlorine pesticide dieldrin was introduced for use in cereal seed dressings and sheep dips in Britain (Jefferies 1989). Dieldrin's properties of high lipid solubility and environmental persistence meant that it was passed on to birds and mammals from contaminated food (Lenton *et al.* 1980). Dieldrin has been identified as being the most probable cause of the crash in otter numbers which occurred in Britain in the late 1950's (Jefferies 1989; Chanin and Jefferies 1978).

Channin and Jefferies (1978) examined the records of otter hound packs in Britain and Ireland as an indication of how the otter population at large had been faring this century. The records showed that the success rate of hunts in Ireland remained high in the late 1950's and early 1960's, while those of their counterparts in Britain were falling dramatically.

Bioindicators and Environmental Management
ISBN 0-12-382590-3

The findings of Eades (1966; 1976) suggest a possible explanation for the continued success of the Irish hunts. After monitoring pesticides in the Irish environment in the 1960's and 1970's, he concluded that "the levels of organochlorine residues in most species of terrestrial and aquatic wildlife in Ireland are low" (Eades 1976, p. 347).

Several other factors are thought to have contributed towards the decline in otter abundance (cf. Mason and Macdonald 1986). These include increased disturbance from man, habitat destruction caused by drainage schemes, and worsening water pollution.

The relatively healthy state of the Irish otter population was confirmed in 1982 with the completion of a national survey, carried out by the Vincent Wildlife Trust. The survey was conducted using a method devised by the Joint Otter Group of the Nature Conservancy Council in Britain. This involved searches of both inland and coastal habitats for signs of otters, principally otter faeces or spraints. In the 1980/81 survey, a total of 2373 sites in alternate 50 km squares of the National Grid were examined for the presence of otter signs.

With resurveys planned for Britain approximately every seven years, it was felt in 1990 that a resurvey of Ireland would be timely. In order to obtain an indication of any significant change over the previous decade, a stratified subsample involving some 12% of the original sites was chosen for resurveying.

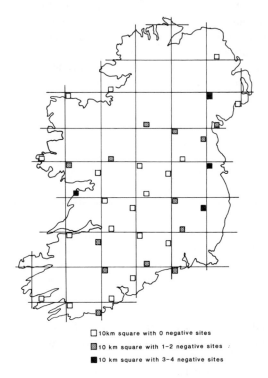

☐ 10km square with 0 negative sites
▨ 10 km square with 1–2 negative sites
■ 10 km square with 3–4 negative sites

Figure 1. Distribution of otter survey records for 1990.

Methods

Three out of the twenty-five 10 km squares (squares A, K, W in Chapman and Chapman 1982, p. 7) in each alternate 50 km square were chosen for resurveying (Figs. 1 and 2). Between 3 and 8 sites (mean = 5) were visited in each 10 km square.

The sites discussed here were surveyed from January to September 1990. When otter signs were found at a site, the search was terminated; if not, a minimum of 600 m of habitat was surveyed. A field sheet, recording habitat details was filled out at each site. Sites were either recorded as unpolluted, or were assigned to one or more of nine pollution types under the headings Domestic, Agricultural, or, Industrial. Vegetation cover was assessed as being "high" or "low" from records of types of vegetation present, tree species, and the surveyors' notes on cover.

The apparent main uses of a water body were noted at each site. Human disturbance levels were assessed by the surveyor at all sites surveyed. The presence of mink scats (faecea) at a site was also recorded. However, as the search for signs was terminated when otter spraint was found, the data on mink undoubtedly represents an under-recording of that species.

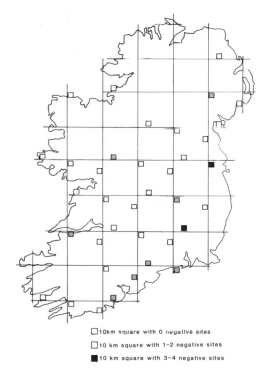

□ 10km square with 0 negative sites
□ 10 km square with 1–2 negative sites
■ 10 km square with 3–4 negative sites

Figure 2. Distribution of corresponding otter survey records for 1980/81.

Volunteers from the Irish Wildlife Federation and Ulster Wildlife Trust, and government employees (Wildlife Rangers of the Irish Office of Public Works) carried out about 60% of the field work. They either had previous experience of otter surveying, or attended a one day training session. All the surveyors were given an information pack containing details of the survey methodology and otter field signs. They were not informed of the 1980/81 status of the sites assigned to them. The survey work was coordinated and supervised by one of the authors (R.M.L.). Spraint samples from each positive site were checked, and sites initially recorded as negative were revisited by her where possible.

Results

We here present results for the 194 completed sites - some 70% of the total survey. Of these sites, 86% were riverine, 7% were coastal, and 7% were in other freshwater habitats. Thus, essentially a river survey was undertaken. Signs of otters were found in 166 (86 %) of the 194 sites visited. Figure 3 shows the pollution status of all sites visited in 1990, and the relative proportions of pollution types at sites where otters were present or absent respectively. 80% of the sites where otters were recorded were on unpolluted waterways, whereas in contrast, the water was polluted at almost half (46%) of the negative sites ($chi^2 =$ 8.8, p < 0.01).

Figure 4 shows water use in relation to otter occurrence. While more than half the sites had no apparent water use, angling was noted at 23% of sites where otters were recorded, and at only 11% of the sites where they were absent. Bankside shooting was recorded at 25% of the negative sites. Table 1 gives results of chi-square analysis of otter presence in 1980/81 and 1990 against pollution level, presence of mink, human disturbance levels, and an index of bankside vegetation cover. Both surveys found that neither disturbance levels nor presence of mink were significantly correlated with otter presence.

Table 1. Otter presence or absence in relation to habitat quality factors: Comparison of chi-square test results for 1980/81 and 1990.

	1990	1980/81
Pollution	X = 8.8	X = 215.9
	p < 0.01	p < 0.001
Mink	X = 1.3	X = 0.11
	p < 0.3 n.s.	p < 0.8 n.s.
Human	X = 0.05	X = 1.4
disturbance	p < 0.99 n.s.	p < 0.3 n.s.
Bankside	X = 6.2	--------
vegetation	p < 0.02	

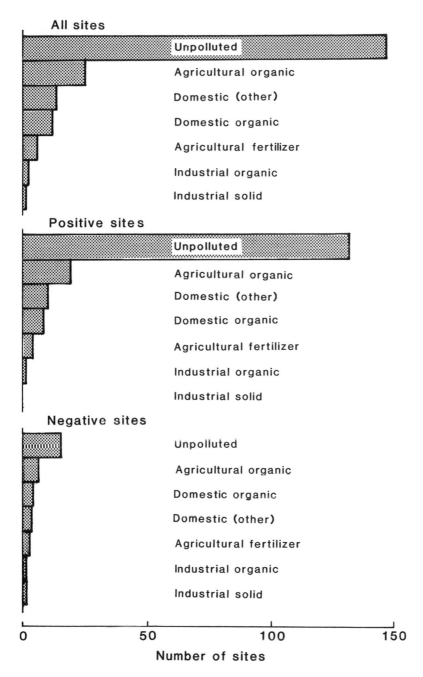

Figure 3. Pollution status of all sites searched for signs of otters, sites where otters were recorded, and sites where otters were not recorded.

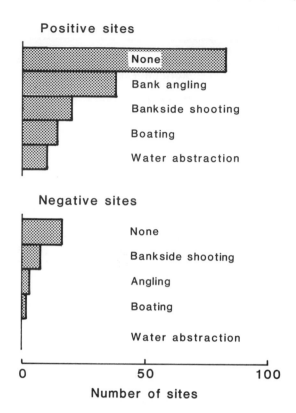

Figure 4. Water use at sites examined for presence of otters.

Discussion

The recording of signs at 86% of the sites visited in this survey indicates that otters are still widely distributed in Ireland. The percentage of positive sites compares favourably with the most recent surveys carried out in Great Britain. Otter signs were found in 65% of the sites surveyed in Scotland (Green and Green 1986), 33% of those visited in Wales (Andrews and Crawford 1986), and the most recent unpublished results for England indicate an improvement on the 6% positive of 1977-79 (Lenton *et al.* 1980). A number of the sites surveyed this year will be revisited, and based on the experience of the previous Irish survey , this is likely to increase the overall number of positive sites. Therefore, we feel that there is no evidence of any significant change in otter distribution since 1981.

There has been some debate about the value of spraint surveys. They have been criticized as a means of monitoring the density of otters, and their use of habitat (Kruuk *et al.* 1986). Kruuk and Conroy (1987) also argue that as spraints are a form of territorial marking, their rate of deposition is not constant and they should be used with caution for survey purposes. In the Irish context, a recent survey of a bogland lake system found that the level of sprainting activity was very low and was concentrated along otter paths, leaving long stretches of lake shoreline almost devoid of signs, (Kyne *et al.* 1990). This lead the authors to note (as have other workers) that in areas such as these, using a standard search length of 600 m, can result in an under-recording of the presence of otters .

However, Jefferies (1986), in his review of the value of spraint surveys, concludes that not-withstanding certain problems, they are the only practicable and repeatable method of monitoring changes in otter distribution on a national basis. Mason and Macdonald (1987) have also evaluated spraint surveys, and conclude that, using standard methodology, they give a reliable indication of the status of otters.

With respect to the effect of human disturbance on otter distribution, this survey and those of Lunnon (1989), who worked on suburban habitats, and Chapman and Chapman (1982) have found no significant statistical relationship between the two. However, there is evidence to suggest that otters do alter their sprainting behaviour seasonally in response to increased human activity during the summer (Chapman and Chapman 1982; MacDonald *et al.* 1978).

The American mink *Mustela vision* Schreber, is now known to be wide-spread throughout Ireland (Smal 1988), including the west of the country where its range is still expanding (Elizabeth and Martin Byrnes, pers. comm.). However, this survey and that of Chapman and Chapman (1982), indicate that there is no significant relationship between the presence of mink and otter distribution (see Table 1). This is in keeping with the conclusions of Erlinge (1969), who worked on the two species in Sweden and found that in their preferred habitats, the more specialist otters could outcompete mink. A recent study in the Irish midlands (Kyne *et al.* 1989) found that there was considerable dietary separation between the two species, indicating a minimum of competition for food.

Our 1990 survey shows a significant correlation between high density of bankside vegetation and otter presence (not tested in 1980/81). A statistically significant relationship has also been found between spraint density (not investigated here) and bankside vegetative cover in several studies in Greece and the United Kingdom (references in Mason and MacDonald 1987, p. 172-173).

Finally, two previous Irish surveys (Chapman and Chapman 1982; Lunnon 1989) agree with our findings in indicating that water pollution is an important factor affecting otter distribution in Ireland. Therefore, we suggest that the monitoring of changing distribution patterns of otter signs over a period of time may be a useful means of indicating change in overall habitat quality and, in particular, water quality.

Acknowledgements

We would like to express our gratitude to Avonmore Foods Plc. for their generous support, without which this project could not have been undertaken. Thanks are also due to the Wildlife Branch of the Office of Public Works, and in particular to Noreen O'Keeffe and Eamon Grennan for their financial and practical support. We are indebted to the following: Elizabeth and Martin Byrnes, John Culligan, Patrick Halpin, Michael O'Sullivan, Gerry Collery, all the other volunteers from the Irish Wildlife Federation, and to Tim Pearce and Sarah Feore of the Ulster Wildlife Trust for their enthusiastic work on the survey. Finally we would like to thank the Vincent Wildlife Trust and Peter Chapman for supplying us with the data from the 1980/81 survey.

References

Andrews, E. and Crawford, A.K. (1986) *Otter Survey of Wales 1984-85*. London: Vincent Wildlife Trust.

Channin, P.R.F. and Jefferies, D.J. (1978) The decline of the otter *Lutra lutra* L. in Britain : an analysis of hunting records and discussion of causes. *Biological Journal of the Linnean Society*, **10**, 305-328.

Chapman, P.J. and Chapman, L.L. (1982) *Otter Survey of Ireland 1980-81*. London : Vincent Wildlife Trust.

Eades, J.F. (1966) Pesticide residues in the Irish environment. *Nature,* **210,** 650-652.

Eades J.F. (1976) Organochlorine pesticide residues in the Irish environment. *Irish Journal of Agricultural Research*, **15,** 341-348.

Erlinge, S. (1969) Food habits of the otter *Lutra lutra* L. and the mink *Mustela vision* Schreber in a trout water in Southern Sweden. *Oikos, 20*, 1-7.

Green, J. and Green, R. (1986) *Otter Survey of Scotland 1984-85*. London: Vincent Wildlife Trust.

Jefferies, D.J. (1986) The value of otter *Lutra lutra* surveying using spraints: an analysis of its successes and problems in Britain. *Otters. Journal of the Otter Trust 1985*, **1(9),** 25-32.

Jefferies, D.J. (1989) The changing otter population of Britain 1700-1989. *Biological Journal of the Linnean Society*, **38,** 61-69.

Kruuk, H., Conroy, J.W.H., Glimmerveen, U. and Ouwerkerk, E.J. (1986) The use of spraints to survey populations of otters *Lutra lutra. Biological Conservation*, **35,** 187-194.

Kruuk, H, and Conroy, J.W.H. (1987) Surveying otter *Lutra lutra* populations: a discussion of problems with spraints. *Biological Conservation, 41*, 179-183.

Kyne, M.J., Kyne, M.J. and Fairley, J.S. (1990) A summer survey of otter sign on Roundstone Bog, South Connemara. *Irish Naturalists' Journal*, **23,** 273-276.

Kyne, M.J., Smal, C.M. and Fairley, J.S. (1989) The food of otters *Lutra lutra* in the Irish midlands and a comparison with that of mink *Mustela vision* in the same region. *Proceedings of the Royal Irish Academy*, **89 B,** 33-46.

Lenton, E.J., Channin, P.R.F. and Jefferies, D.J. (1980) *Otter Survey of England 1977-79*. Nature Conservancy Council, London.

Lunnon, R.M. (1989) *The Distribution and Food of Otters* (Lutra lutra*) in the Dublin*

Area. B.A. Mod. Thesis, Trinity College, Dublin.

Macdonald, S.M., Mason, C.F. and Coghill, I.S. (1978) The otter and its conservation in the River Teme catchment. *Journal of Applied Ecology,* **15,** 373-384.

Mason, C.F. and Macdonald, S.M. (1986) *Otters: Ecology and Conservation.* Cambridge University Press, Cambridge.

Mason, C.F. and Macdonald, S.M. (1987) The use of spraints for surveying otter *Lutra lutra* populations: an evaluation. *Biological Conservation,* **41,** 167-177.

Smal, C.M. (1988) The American mink *Mustela vision* in Ireland. *Mammal Review,* **18,** 201-208.

INDEX